Artistry and Ingenuity in Artificial Stone

Indiana's Concrete Bridges, 1900 - 1942

by
James L. Cooper
Professor Emeritus of History
DePauw University, Greencastle, Indiana

1997

ISBN 0-936631-13-9

Acknowledgements

An author gets his name printed in large letters on the title page and necessarily takes full responsibility for the ensuing text. Readers know, however, that the author does not act alone; he owes much to others.

I am grateful to the many who supplied copies of bridge plans or provided access to them. Margaret Barondess (Michigan Department of Transportation), Steve May (Hamilton County, Ohio, Engineer's Office), Steven Wegman (Morgan County Engineer), and Michael Wenning (American Consulting Engineers of Indianapolis) each furnished sets of plans. Maynard C. Schenk (Elkhart County Highway Engineer, now retired) alerted me to the invaluable cache of William S. Moore's drawings housed at Lang, Feeney and Associates in South Bend. Bernard Feeney and William D. Lang, in turn, graciously offered access to those plans. David Penturf (Putnam County Surveyor) let me comb his office files for old drawings.

Many shared data in their hands or under their custody. Eric DeLony (Historic American Engineering Record [HAER], Washington, D. C.) made possible the Iowa connection, forwarding copies of state highway articles and uncovering files on the state's attack in court on concrete bridge patents. Camille Fife (Westerly Group of Farmersburg) has shared HAER documentary reports on bridges in process of being removed. Helen Hanna (Monticello) loaned a copy of the dedicatory program for the Carrollton Bridge. James L. Kisling (Plus 4 Architects of Indianapolis) provided photographic prints of the reconstruction of the Country Club Bridge (Indianapolis). Daniel Luten's son and daughter (Daniel B. Luten, Jr. and Mary Luten Watson) on many occasions shared records of their father's work. Mary Ann Moore (Brownsburg) supplied all the materials for the boxed insert on a bridge-building family. Wayne Sanford (Crown Hill Cemetery, Indianapolis) offered access to records on the Maple Street Subway. David Simmons (Ohio Historical Society, Columbus) has responded to periodic requests for copies of notes, drafts, and final text of his articles on concrete bridges in Ohio. Stephen Weintraut (Butler, Fairman, and Seufert of Indianapolis) provided recent engineering assessments of the West Washington Street Bridge. Norris Whittaker (Woodsfield, Ohio) opened his home and shared the kitchen table for the reading of his father's Luten Bridge Company records.

Librarians have also played an important part. Those at the Danville, Fort Wayne, Plainfield, and Paoli Public Libraries were especially helpful. I also owe thanks to those in Special Collections and in the Compact Storage Department of the Purdue University Library (West Lafayette), to the Indiana State Archives and Library (Indianapolis), to the Indiana Historical Society (Indianapolis), and to the State Historical Society of Iowa Library and Archives (Des Moines).

Among the especially helpful people at the Indiana Department of Transportation in Indianapolis are Theo Bowman (Program Development), William Kelly (Graphics Engineering), Janie Marks (Records, Bridge Design), and Wayne Wiese (Technical Services). Important assistance came from the regional offices, notably from Randy Carie (Vincennes) and Pat Doyle (Crawfordsville).

To colleagues at DePauw University – Anthony Catanese (Professor of Economics), Ralph Gray (Professor of Economics), and John E. Morrill (Professor of Mathematics) – who read and critiqued the manuscript, so much is owed that I can speak of reducing only the deficit, not my overall debt. Paul Kissinger (Professor of Physics) reviewed boxed inserts on how beams and arches work and suggested some illustration.

Two from outside the DePauw University faculty played key editing roles, too. James Barker of AECON (Nashville) corrected structural explanations and helped me put them into layman's English. Sheila M. Cooper, my spouse who is also a professional historian, has done her best to eliminate flowery phrases, winding sentences, and *non sequiturs*.

Most of the photography in this volume is the author's. However, in printing almost all the selected negatives, Marsh Davis of Historic Landmarks Foundation of Indiana has focused and teased out greater clarity than the photographs probably would have had. Dian D. Phillips of Full Circle Graphics, Greencastle, did the cover design and page layout. Gary Grahn, Jr., of Pierson Printing, Indianapolis, once again took charge of the printing and binding.

This book is one product of a research project that has extended well over a decade and has been materially supported by a number of different groups and persons. Some of the early research was funded in part by grants from the National Park Service of the U. S. Department of the Interior, administered through the Division of Historic Preservation and Archaeology, Indiana Department of Natural Resources. Staff members in the Indiana Division of Historic Preservation and Archaeology have also provided helpful advice and encouragement over the years. Later research, more directly relevant to the following text, had financial support from a number of individuals and from the Shell Oil Company Foundation. On several occasions, DePauw University provided the author sabbatical leaves for research and writing, and throughout the whole period the University's accounting department has offered unstinting service.

Selection and interpretation of materials is finally and fully the responsibility of the author. The contents and opinions expressed herein do not necessarily reflect the views or policies of individuals, groups, or agencies in any way associated with the support, research, or editing of the ensuing text.

The best outcome from the labors that have gone into this and the first volume on the history of Indiana's bridges – *Iron Monuments to Distant Posterity* (1987) – would be vigorous efforts to preserve important parts of this heritage. Except for venerable timber-truss structures, Indiana has done altogether too little to retain its historic bridges. Over sixty percent of the structures listed in the inventory of *Iron Monuments* have been removed since its publication. The state's Division of Historic Preservation promises to take a more active hand in saving significant spans, and the Historic Landmarks Foundation of Indiana has invested time and resources in some projects for bridge preservation. If *Artistry and Ingenuity in Artificial Stone* encourages readers to support the work of public agencies and private groups committed to preservation and to spur those efforts onward, the author will feel compensated for his decade of labor.

Advice to the Reader

The interests and backgrounds of readers of the current study will likely vary as much as they did with readers of *Iron Monuments to Distant Posterity*. While engineers, for example, have been taught how reinforced-concrete spans function, many others have studied no physics nor taken courses in applied physical science. To accommodate this diversity, many simple explanations of how structural forms work have been placed in boxed inserts to be absorbed by some and skipped by others. The inserts on structure build on one another. Both the text and the boxes in later chapters presume the reader understands the materials provided in the earlier inserts. (Look for the symbol.)

While this volume has no formal index, it does contain several cross-referencing features. The inventory can be used as an index of relevant structures in Indiana, and the individual entries do note wherever a bridge is also discussed in the preceding chapters. The Table of Contents with its unusually large number of subdivisions should allow the reader to locate discussions of particular bridge forms and reinforcing systems as well as of major designers and their work. The endnotes for each chapter also contain a number of references to where related materials may be found in the text.

J. L. C.
Greencastle, Indiana

Contents

Part Two:
Consolidation and Standardization, 1917-1942

IX

Bridging Hoosier Byways with Private Design in Hard Times

Part Three:
An Inventory of Selected Concrete Spans

PROLOGUE

RENEWING CONNECTIONS

BRIDGES

by Amanda Southerland[1]

Big Four Arches.

The day before school started, my friend … and I decided to take a midnight drive. The air was cool and humid and smelled of freshly cut grass. We both knew why we had decided to go on the drive. We were saying goodbye to one another, so long to friendship that had lasted one humid Indiana summer. We drove past fields of corn taller than both of us and cruised down the gravel roads that led to the Four Arches, a beautifully designed concrete railroad bridge built in the early 1900s. The quarter moon cast a pale yellow glow against the concrete arches of the bridge, making them appear white and ghostlike. …

'Want to climb to the top?' he asked, motioning to a brush-covered trail that ran up beside the bridge. We hiked up slowly, checking our steps in the dim light. I could hear locusts and crickets chirping their night songs in the nearby trees and the tap of pebbles rolling down the steep hill beneath our feet. We pulled ourselves up onto the bridge and walked on every other railroad tie until we reached the spot where the creek ran directly below us. We tossed gravel into the dark and waited for the splash to know that we had hit the water. We looked up at the stars and inhaled deep breaths of stale summer air. We didn't speak. We knew what the other was thinking. This was the last time that we would be together as best friends. …

We heard a low rumbling sound in the distance like air being forced from a paper bag. We were silent as the rumbling grew louder and the iron tracks began vibrating in rhythm. … A blinding white light rounded the bend and began to move towards us. We jumped up and shouted as we heard the low whistle blow across the valley. … I raised up my arms and felt the rush of the wind almost knocking me backward, billowing out my white T-shirt like a ship's sail. I yelled into the wind, my voice lost in the awesome sound of machinery moving at 60 miles per hour. The wind died down as the last car cruised by, and we listened intently until the sound grew fainter and the white and red lights disappeared into the darkness.

We made our way down the hill clumsily. I scraped my knee on a rock but didn't utter a sound. We drove away from the bridge and didn't say a thing about the train. There was nothing really left for us to say.

The bridge was our connection that night.

Bridges are about making connections of different kinds. All surmount topographical obstacles. The rivers and ravines that provide the most common barriers are spanned often in mundane, sometimes in imaginative ways to carry physical burdens from one bank to another. A bridge's utility, though, need not be limited to transportation.

The designers of some spans, especially a number of cut stone or of reinforced concrete, have incorporated a second kind of connection within their structures. They have at times raised a flag, celebrated community accomplishment, commemorated war dead, or enshrined a citizen in sculpture on the spans over a stream. Bridges have served as monuments as well as carriers.

Not all the connections which bridges make, however, remain in the hands of their purchasers, designers, or builders. The associations which common citizens sometimes establish with bridges can overshadow their role as carriers. "In folklore and mythology," a well-known folklorist has concluded, "the bridge is a universal symbol of the passage way between this life and the world beyond."[2] Three concrete arch structures on the Big Four Railroad in central Indiana underline the power of bridges as symbols of passage from one life stage to another and from life to death, which spark considerable attention and activity. Like Amanda Southerland's female character, many a young person in Putnam County has made a ritual life-passage trip to the Big Four Arches.[3]

Tales of ghosts at two similar railroad structures in Hendricks County – Haunted Bridge and Twin Bridges – have grown into legend.[4] Visiting the bridges, where ghosts are reportedly often heard and seldom seen, "is a kind of test, one of the initiation rituals among the many [that] young males have to pass as they leave childhood and make their way toward adulthood." The chalk- and paint-covered inner walls of the arches speak to these rites of passage. In padlocking the entrances to the Haunted Bridge, railroad officials acknowledge the power of legend and liability as they try to reclaim these spans for their intended uses alone.[5]

Indiana's concrete bridges have, on the whole, evoked less notice than its timber ones. None have attracted large groups like the thousands of tourists who now annually visit and photograph the state's remaining covered bridges. The timbers of the latter are reminders of almost lost craftsmanship and inspire nostalgia for rural sim-

Far left, The chalk- and paint-covered inner walls of the arches speak to these rites of passage.

Probably to limit its liability, the Consolidated Railroad Corporation enclosed and padlocked the spandrel opening through which most visitors once entered the interior cells of Haunted Bridge.

plicity. Concrete spans also generally lack the heroic associations attached to larger and more monumental, metal suspension structures like the Brooklyn or Golden Gate Bridges. With little romance, less monumentality, and no nostalgia, most concrete bridges have been built as ordinary utilitarian structures. The planning, designing, and building of even these merely useful spans have nonetheless generated many stories worth recounting.

Behind and Beyond Bridge Lore

This study traces the design and building of concrete bridges from about 1890, when Hoosiers begin to explore the potential purposes and experiment with the possible forms of concrete spans, to the early years of the Second World War, by which point design and construction of highway bridges had become largely consolidated in public agencies and significantly standardized. While the narrative focuses on Indiana, the context quite often reaches beyond the state to the broader Midwest and to the nation as a whole. Hoosiers may sometimes have disliked it, but their lives have been caught up within broader regional and national currents. The development of professions and of government bureaucracy, the reform of highways into a federal-state system, and the changing forms of transport and life-style significantly affected twentieth-century bridge design and construction in every part of the nation including Indiana. State boundaries have only sometimes stopped the promoters of new structural materials or of new designs.

The Era of Experiment and Expression, 1890-1917

Raised in Chicago to celebrate the four-hundredth anniversary of Christopher Columbus's first voyage to the Western hemisphere, the 1893 World Columbian Exposition provides our beginning (Chapter I). Members of two of the nation's emerging professions – architecture and engineering – planned an ensemble of buildings and bridges for the Exposition. This imperial White City and its spans captured and enhanced citizens' pride in the world's largest industrial nation. The Exposition turned out to be a dual celebration, first, for the European conquest of the Americas and, second, for the arts, manufactures, and architecture of its conquerors.

Inspiring imitation across the country, the White City gave impetus to the City-Beautiful movement. Parks, boulevards, artistic bridges, and monumental public buildings that reflected this shared, public memory sprang up in American cities. For a while, architects and their allies triumphed with concrete-steel bridges, designed with metal frames to carry most of the structural load and with concrete sheathing that in part celebrated community achievement. Indiana cities had their share of City-Beautiful bridge advocates. J. Clyde Power and George Kessler, Alonzo Hammond and William S. Moore, Charles M. Robinson and A. W. Grosvenor, as well as Otto O. Clayton and E. A. Gast successfully insisted on design that dressed a bridge's transportational usefulness in symbols of elevated civic purpose.

Engineers soon freed themselves from the control of the architects who led the City-Beautiful movement. Unfettered, engineers embraced simple and unalloyed utility, which, in the case of bridge design, promised the public more structural efficiency and less symbolic significance. Reducing decoration cut the cost of design and construction. But the concerns of the engineers went beyond the elimination of what most regarded as decorative frills. Seeking the maximum amount of safe carrying-capacity from the materials used for a span, some engineers who designed metal and concrete bridges began to explore more fully how each material contributed to load bearing. By 1900, avant-garde engineers had begun to conceive of steel and concrete as a team of nearly equivalent materials when designed into what they increasingly called reinforced concrete bridges. Shortly thereafter, a few of the most conceptually daring engineers moved beyond the notion of steel and concrete as yoked materials to the consideration of reinforced concrete as a single, integrated substance.

Much as engineers agreed on the importance of structural efficiency, British and French practitioners took divergent paths to the goal. The British transferred to North America the oldest European tradition of invention and innovation, one based on intuition followed by a process of trial and error to identify an acceptable solution to a problem. To be acceptable to the empirical engineer, a solution did not need to have some theoretical explanation nor be situated within some existing system of knowledge; it just needed to work or function. Empirically-based invention and innovation underlay most of the world's oldest and greatest structures. The tradition fitted well in Britain, where most engineers worked for private individuals and firms in an environment often congenial to pragmatic and parsimonious approaches to problem-solving. The same entrepreneurial environment for invention and innovation prevailed in North America through the nineteenth century. Until after the First World War, American bridge engineers largely worked within this empirical tradition, including Indiana's Daniel B. Luten, one of the nation's most inventive and prolific entrepreneurial designers of reinforced concrete bridges (Chapter II, III).

By no means, however, were all engineers committed to empirically-based bridge design. An increasing number of engineers – probably a majority of practicing ones after the First World War – owed a primary allegiance to the second European tradition of invention and innovation, a more rigorously scientific approach which the French promoted. The French insisted that engineers take the systems of knowledge which scientists created for the physical world and apply them when creating substances, machines, or structures. Engineering preoccupied with the prevailing systems of scientific knowledge and theory about the natural order required the establishment and support of institutes or other agencies for giving prospective practitioners adequate training in the canons of science which had to be applied, and scientific engineering worked well where government itself employed a lot of engineers, often for military purposes.

As allies of the colonies in the American War for Independence, the French had influenced engineering in George Washington's army, and French traditions took root in the U. S. Army Corps of Engineers and the Military Academy at West Point. Later, private and land-grant colleges that trained "civilian" or civil engineers borrowed heavily from the Military Academy's curriculum and practices. As the nineteenth century progressed and engineers became increasingly more formally educated and committed to scientific practices, an increasing number of civil and other engineers secured government employment.

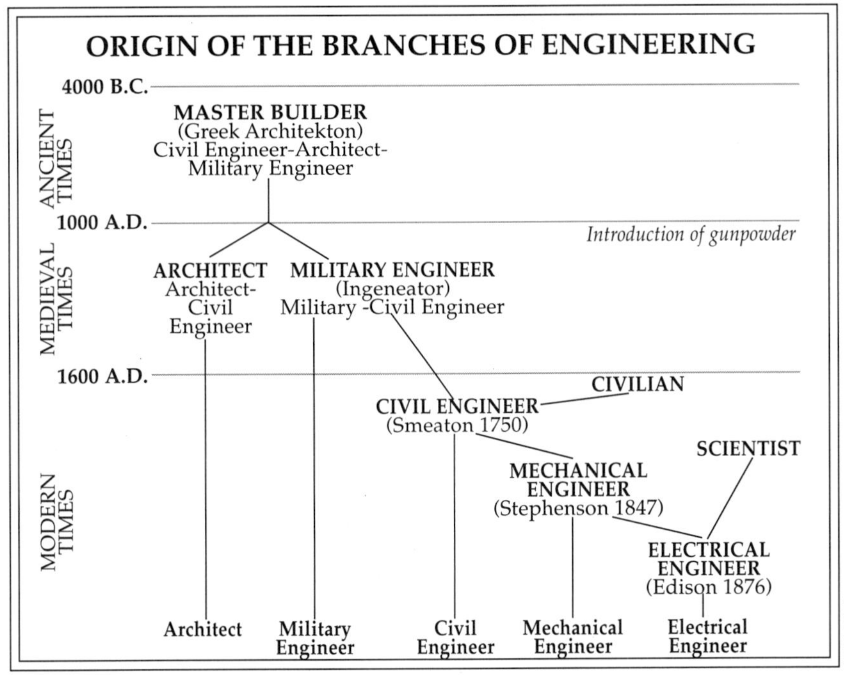

(ENR, 1930)

Towards the end of the nineteenth century a number of graduates from engineering colleges and universities that promoted science-based research and invention joined the movement for Good Roads. At a conference associated with the 1893 Columbian Exposition, the movement's leaders agreed on the need for direct federal participation in highway design and building and promptly lobbied influential national figures (Chapter IV). Consequently the predecessor of the Bureau of Public Roads emerged as an agency within the U. S. Department of Agriculture, from where its cadre of college-educated engineers created an impressive federal-state highway system over time.

By century's end, the idea that disinterested experts should dominate policy-making in local, state, and national government had become a hallmark of Progressivism. These reformers included civil engineers with pertinent educational credentials who headed the new highway system. They wanted to wrest control from politicians, whose expertise was reportedly limited to managing electoral activities, and from the private interests, which were allegedly preoccupied with lining their own pockets. To keep the politicians and the private interests at bay, the federal-state highway engineers not only gradually made road and bridge design more scientific, they also turned it into a government function. Indeed, the orderly, regulated, and incremental attributes of science-based work suited professional life in these government bureaucracies exceptionally well. Such professionalism both promoted meritocratic decision-making and separated it from the public. These Progressive engineers were, in short, more hierarchical than egalitarian, their design more safely conservative than innovative (Chapter V).

Consolidation and Standardization, 1917-1942

The struggles between entrepreneurial and empirical engineers against the government-employed and science-based colleagues highlight only part of the evolution of twentieth-century bridge design. Motorized vehicles, which introduced a revolution in transportation that the engineering profession had only partly anticipated, explain other themes of fundamental change. As soon became clear, motor vehicles took a heavy toll on roads and bridges designed for horse-drawn wagons and carriages. In short order, cars and trucks tore up even the newer "Good Roads" and destroyed many of the older bridges, dramatically increasing demand for a federal-state highway system with roads and bridges of far different design. The government-employed civil engineers in charge of the evolving highway system strengthened their commitment to science-based design and increased their sway over bridge-building as they accommodated the internal combustion revolution (Chapter VI).

Of the states in the union, Indiana joined the federal-state highway alliance last. As a result, however, the Indiana State Highway Commission faced automobility immediately and started with science-based design. Under the leadership of William J. Titus, the state designed more than two-thousand bridges in the Twenties and Thirties as part of an evolving intercity and interstate highway system (Chapter VII).

The dramatic expansion of the federal-state highway system essentially halved the market for private bridge design in Indiana in the Twenties, and the Great Depression caused its virtual collapse in the Thirties. Still, empirically-committed designers outside the system continued to generate the most innovation in bridges constructed on city, county, and park roads in the Twenties, and a few examples of private engineers' innovative designs appeared on the Hoosier landscape in even the hardest of ensuing times (Chapter VIII, IX).

Taken together, the government and private designers had provided Indiana by 1942 with a solid infrastructure of concrete spans that accommodated early effects of motor transport. As automobility evolved, however, it made new demands on the federal-state highway system and on other community services. The car ultimately created a drive-in culture where standardized institutions for lodging, eating, banking, and many other services became commonplace. Even earlier, the horseless carriage had generated standardized highways and bridges for vehicles operated under an essentially uniform code of rules. In that process, the roles which bridges played had changed dramatically, too.

Once waved, the automotive wand rendered most new spans so nearly invisible that few Americans would find even a metaphorical connection between a contemporary bridge and a life passage (Epilogue). Older bridges on Hoosier highways and byways, however, still testify to our forebears' dreams, values, and achievements and provide modern Amandas with life-line connections. They communicate especially clearly and forcefully to the observer who uses informed imagination to know the every-day past.

NOTES

[1]"Bridges," *Midwestern Review* 7 (Fall 1993): 4-5.

[2]Linda Degh, "The Haunted Bridges Near Avon and Danville and Their Role in Legend Formation," *Indiana Folklore* (1969) II: 89.

[3]The bridge is named for the Big Four Railroad, as the Cleveland, Cincinnati, Chicago and St. Louis has been known in Indiana. The railroad became a part of the New York Central system by 1900, although it continued to operate separately until about 1930.

[4]Ronald L. Baker, *Hoosier Folk Legends* (Bloomington, Ind., 1982), 1-6, 63-66.

[5]Degh, "Haunted Bridges," *Indiana Folklore* II: 81.

PART ONE

Experiment and Expression, 1890 - 1917

(Indiana Historical Society Library, Bass Photo Co., neg. #20643)

(Parke #7056)

Left – The Emrichsville Bridge (1905-1947) once provided stately access via Sixteenth Street to Thomas Taggart Riverside Park in Indianapolis.
Above – Originally on a county road, now resides inside Turkey Run State Park.

(Marion #2501)

Above – Cartouche from Capitol Avenue Bridge over Fall Creek, Indianapolis.

I

Spanning the American Heartlands with City-Beautiful Bridges

Many Hoosiers applauded the selection of Chicago as the site for the celebration in 1893 of the four-hundredth anniversary of Christopher Columbus' landing in the Western Hemisphere. Old Glory flew over "waves of grain" and astride smokestacks in the "heartlands" of the greatest industrial nation on earth. Indiana was contributing to the imperial sway that extended from the Atlantic to the Pacific Ocean and beyond. Hoosier corn fattened livestock to energize meat-eating Yanks, and Indiana industries helped well-fed midwesterners flex their manufacturing muscles. Americans from all regions would display their productive prowess in the World Columbian Exposition's nearly two hundred buildings and exhibition halls.[1]

Such accomplishments needed to be presented in a properly planned and expertly designed context, one which would weave thousands of parts into a celebratory ensemble capable of being heard across and beyond the continent. Frederick Law Olmstead, the nation's best-known landscape architect, transformed the Exposition's site from a swamp into a lagooned park. Daniel H. Burnham, a rising Chicago architect and planner whose rule was "to make no little plans," organized a committee of the nation's leading architects to design the buildings that enclosed the acres of exhibition space that twenty-seven-and-a-half million visitors, including tens of thousands of Hoosiers, viewed.

Although architects led the Exposition planning efforts, engineers also contributed significantly, adding their knowledge of recent industrial technology and its structural implications to the architects' expertise with design. Since antiquity buildings and bridges had been built largely of stone and wood. But, midway through the nineteenth century, designers and builders learned that iron or steel framing allowed for larger and more flexible structures at lower cost. Experienced in using metal to frame factories, train sheds, and skyscrapers, the Exposition's planners naturally chose metal trusses to support exhibition buildings and metal arches to span lagoons and canals.

Architects dressed the Exposition in neoclassical garb fit for a modern metropolis. The landscape, the statuary, and the structures reinforced a metaphoric association of the Roman with the Euramerican empire. As the Romans had demonstrated their genius by mastering the ancient western world, so the

THE WORLD'S FAIR.

Designs Completed for Another Monster Building.

A Palace of Mechanical Arts That Will Cost Half a Million Dollars – The Structure to Be One of the Largest at the Exposition.

Peabody and Stearns, the Boston members of the exposition board of architects, have finished their designs for the machinery building [which] ... will cover a space of 850 by 500 feet. It is spanned by three arched trusses and the interior will present the appearance of three railroad train houses side by side, surrounded on all the four exterior sides by a fifty-foot gallery. The trusses are to be built separately so that they can be taken down and sold for use as railroad train houses ...

The two exterior sides adjoining the grand court are to be rich and palatial in appearance. All the buildings on this grand plaza are designed with a view to making a grand background for displays, and in order to conform to the general richness of the court and add to the festal appearance, the two facades of the machinery hall on the court are rich with colonnades and other features. The design follows classical models throughout, the details being borrowed from the renaissance of Seville and other Spanish towns as being appropriate to a Columbian celebration. An arcade on the first story admits passage around the building under cover. As in all the other buildings the front will be formed of 'staff' colored to an ivory tone. The ceiling will be enriched with strong color.

A colonnade, with a cafe at either end, forms the length between machinery and agricultural hall. In the center of this colonnade is an archway leading to the cattle exhibit. From this portico there will be a view nearly a mile in length down the lagoon. The obelisk and fountain in the lagoon will form the southern point of this vista.

– Logansport *Daily Journal* (1891)[2]

Metal framing of Machinery Hall being prepared (below) for sheathing in "staff." Right – Machinery Hall, cost $1,200,000.

European sons and daughters of Columbus showed their prowess by conquering and cultivating North America in modern times. The architects shaped the buildings of their White City like oversized Greco-Roman structures and sheathed their metal frames in "staff," a mixture of cement, plaster, and jute fibers developed in France in the 1870s. The architectural imitation was more imaginative than literal, more a combination of Greek, Roman, and Renaissance elements than a copy of any, more a superimposition of ancient appearance over modern structural engineering than over classical technology – in sum, a contemporarily invented neoclassical amalgam.

I.A Artificial Stone

Over one hundred miles of paved walkways and concrete bridges connected the exhibition halls and other buildings. The oldest of all synthetic materials used for building, concrete "is the artificial stone formed by the hydration of a hydraulic cement around a matrix of inert aggregate."[3] To produce "artificial stone" requires the interaction of two active materials, cement and water. Nature provided the second; the first was manufactured. "Quicklime," as cement in its oldest and most natural form has often been called, starts from limestone (calcium carbonate) which is kiln-burned into lime (calcium oxide) and then slaked with water (into calcium hydroxide). Exposed to the air, the calcium hydroxide of the quicklime combines with the carbon dioxide in the air to form again the calcium carbonate of which it was originally composed. When combined with inert elements such as sand and gravel, the hydrated quicklime and the aggregates can be shaped at will before setting in the air into a solid and durable concrete mass.

The ancients produced concrete. Lime-burning kilns date to about 2450 B. C. The Greeks used lime with sand and water to form a hardening mortar. And the Romans discovered that mixing lime with pozzalana, a sandy volcanic ash, yielded a hydraulic cement, one that hardened under water as well as in the air. The Romans combined their inheritance with their discovery for the foundations of the Temple of Concord in 121 B. C., and, in the early imperial period, erected walls, domes, vaults, and breakwaters of concrete, too.[4]

The knowledge of concrete-making disappeared from Europe with the disintegration of Roman power and influence, and its recovery came in phases in modern times. Sixteenth-century English builders rediscovered hydraulic-lime mortar when they imported Italian mixtures of volcanic ash and lime. Roman-style concrete, however, was not revived until the eighteenth century when George Semple mixed hydraulic lime with gravel for the foundations of the Essex Bridge in Dublin.[5]

But neither Europeans nor Euramericans would use concrete extensively until they had developed a steady supply of reliable hydraulic cement. For guaranteed supply and consistent quality, an artificial product edged out natural cements. Joseph Aspdin of Leeds, England, in 1848 patented a cement he had created by calcinating a mixture of chalk and clay. He named it Portland cement after a limestone found in the region. It was an American, however, who benefited most from Aspdin's creation. In 1871, David O. Saylor patented an American equivalent of Portland cement, built a mill to manufacture the product, and quickly exploited the great and growing American common market.

The first concrete bridges and buildings began to appear on the American landscape as sales of American Portland cement spread across the country. At least two decades before the Columbian Exposition, Frederick Law Olmstead had watched the successful erection in a Brooklyn park of the nation's first concrete-arch bridge. Shortly thereafter the first concrete house was built in Port Chester, New York, shortly thereafter. Widely manufactured and distributed throughout the country by 1890, Portland cement had made it possible to exercise considerable control over the quality of concrete before the Exhibition opened.[6]

Arches and vaults as well as concrete fit well with the classical imperial theme of the Exposition. The Romans no more invented the arch and the vault than they had discovered cement. The Egyptians and the Mesopotamians had originated the arch centuries before. But their Roman successors made the most of arches and vaults in the ancient world, enough to keep a number of Roman

ARCH TERMS

Ancient and trustworthy, the arch remains an extraordinarily useful and stable structural form. Once properly erected, arches can carry extraordinarily large loads. Neither abuse nor neglect, furthermore, keeps most arches from fulfilling their appointed functional duties over long periods of time.

The arch rests at its ends against lateral supports called **abutments**. If several arches are placed side by side, **piers** provide their intermediate support. The arch **rib** or **ring** rises from the **springing** at a pier or abutment in one or more arcs to the springing on the next pier or abutment. The distance between the adjacent springings provides the ring's **span**, and the height a ring reaches from its springings to its **crown** at midspan denotes the ring's **rise**.

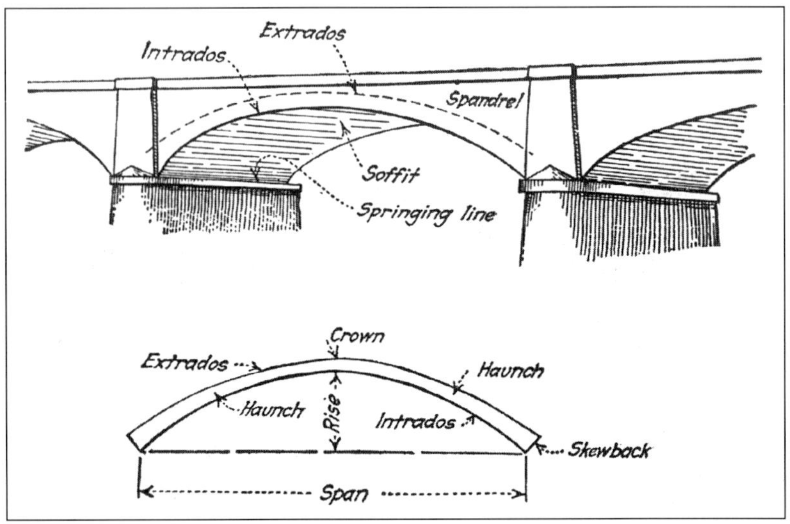

The space between the arch ring and the roadway or **deck** over which traffic travels is called the **spandrel**. To contain the earth-fill between the ring and the deck, many designers have **spandrel walls** erected at the sides of the ring. Sometimes the spandrel is not walled and filled. In **open-spandrel** arches, columns typically support the deck above the ring.

(Parke #156)

A filled-spandrel arch. Dunn's Ford Bridge.

(Putnam #159)

An open-spandrel arch. Reelsville Bridge.

construction records unsurpassed until the modern era.[7] These were among the kinds of substantial achievements to which the Exposition's leaders implicitly laid claim for the United States by imitatively steeping the White City in neoclassical forms and garb.

In addition to the Exposition's buildings, statuary, and landscaping, its bridges played a part – even if a minor one – in the metaphoric association of modern America with ancient Rome. The architects shaped modern metal ribs to the classical arch form and then enclosed most of the ribs in staff or in concrete. The architects decorated the piers and walls of their more important bridges to look like natural stone, and they used rails with neoclassical balusters to bound the roadways and walks. Bridges which spanned the canals adjoining the lagoon around the Court of Honor received this imitation-classical sheathing. Bridges crossing the canals some distance from the Court of Honor did not merit as much finery, although the arch remained the form of choice. Here the metal ribs which served as the skeleton of all the arches were not always fleshed-out with a staff or concrete sheathing.

To achieve beauty in the eyes of leading architects, metal-framed buildings and bridges had to be naturalistically enclosed in cut or artificial stone. Industrial giants need not stand in their steel skeletons. They could – and, landscape architects thought, should – exist in harmony with nature or at least appear to do so. In qualitative if not quantitative terms, less was not more. Industrialization generated power and prosperity, but not beauty, for beauty involved integration with the natural order even if the integration were manufactured. The industrial skeleton needed a "natural" integument.

American pride blazed through the windows of the Columbian Exposition's halls, and reflections of the electrically-lighted White City shimmered in the lagoon at night. This concentrated show lasted for only six months. But its powerful image lingered indelibly, and it inspired other construction – some as ingenious, some as exhibitionistic as their model – across the land.

Lagoon bridge, World Columbian Exposition.

Canal bridge with Woman's Building in background, World Columbian Exposition.

ARCH FORCES AND SHAPES

Gravity, a force of attraction by which bodies tend to fall towards the center of the earth, gives weight to matter which, in turn, exerts pressure on other matter in its field. Take, for example, three stone blocks placed vertically upon one another (see drawing to left). The **V**ertical pressure arising from the weight of the blocks accumulates (1>3) and, if uninterrupted, would flow directly downward. In the case shown here, an externally imposed **L**ateral pressure on the top block (1) is large enough to move the otherwise vertical line of pressure sideways and to turn it on a diagonal. When the second block's weight and consequent vertical pressure (2) is added in, the role of the external lateral pressure makes up less of the whole, and the accumulated pressure becomes less diagonal and more vertical although still displaced from where the vertical pressure line alone would have been. Adding the third block's weight and resulting vertical pressure (3) to the total pressure from the two blocks above turns the line of pressure to a nearly vertical but still displaced position.

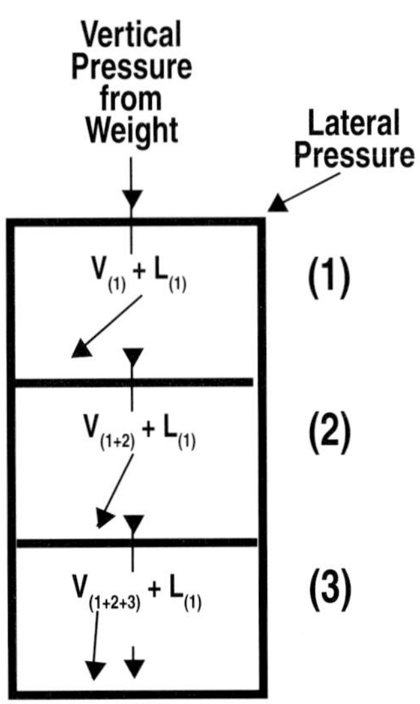

To turn the vertical stone blocks considered above into an arced ring secured at each end and then to cover the ring with incased fill and a roadway is to form the essential elements of an arch bridge. Like the stone column, the weight and consequent vertical pressures of the arch ring, fill, deck, and whatever travels across the bridge are not equally distributed over all blocks in the ring. Since the ring is arced downward towards the earth from each side of the crown, the ring's weight is lowest at its vertex where the spandrel area above also happens to be smallest. Weight accumulates towards the lower part or haunches from the ring's blocks and from the greater spandrel area above. Unlike the stone column where lateral pressure must be externally imposed, the blocks in the arch ring pass some of their vertical pressure on to adjacent areas as lateral thrust, again with the crown region having the least and the area around the springings the most. The lateral pressure which runs around the ring compresses its sections against one another. The longer the span, the greater the accumulated pressures grow.[8a]

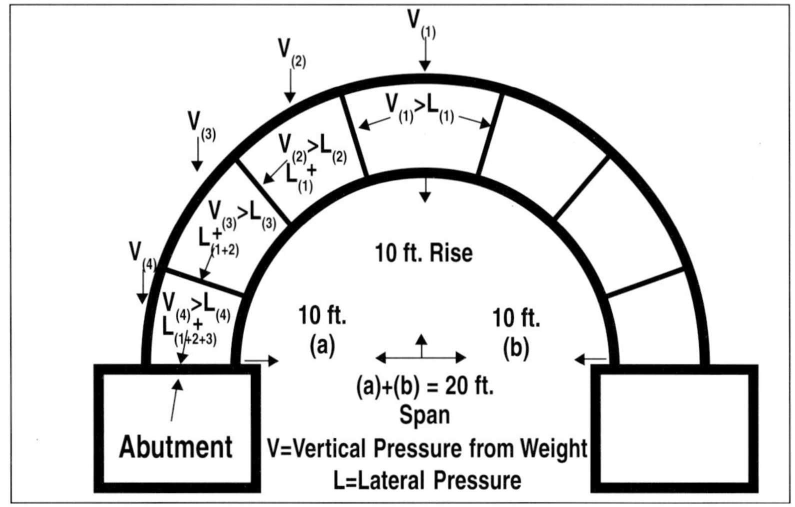

Line of pressure or thrust in the semicircular (above) and the segmental (below) arch ring.

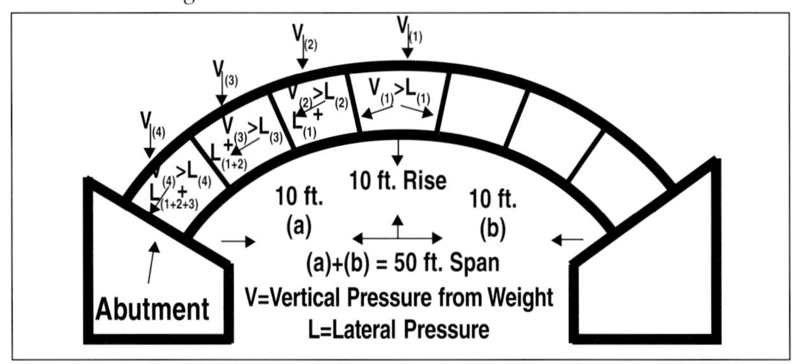

(Continued on next page)

The abutments sustain both the vertical pressure arising from the structural and traffic loads as well as the lateral pressure generated within the arch ring. Piers, needed where a bridge consists of more than one arch span, typically address only the vertical pressure arising from the structural and traffic weight of the arches supported. If the arches are of equal size and their springing lines are in the same horizontal plane, then the accumulated lateral pressure transmitted through one ring to a pier is countered by an equivalent amount of horizontal thrust carried to the opposite side of the pier from the adjacent ring. In sum, to insure that a bridge stays intact or in equilibrium, the substructure must be strong enough either by itself or with the assistance of other elements to mount equal and opposite amounts of pressure to resist the accumulated ones delivered through the arch ring(s). Above all, the substructure must not allow the ring to move a lot.

To assist in carrying the imposed pressures safely, the arch designer can vary the depth of the ring. The more combined pressures mount, the more material that may be used to share the carrying and thus avoid crushing or fracturing the ring. At the crown where the combined pressures are least, the ring is likely to be its slimmest. At the haunches where the combined pressures are greatest, the ring will probably be at its thickest.

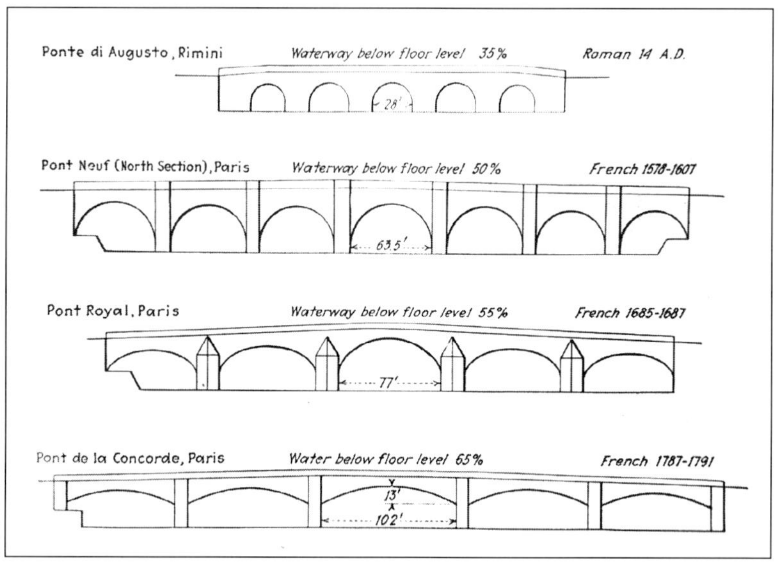

(ENR, 1930)

"Progress in bridge design from the first to the eighteenth century." From Roman and French semicircular arches in stone to segmental ones.

The shape of the arch ring also affects the amount and nature of the pressures upon its different parts. From the Romans onward, designers of cut-stone arches – along with those of many plain or unreinforced concrete spans – have favored semicircular rings. Compared with other shapes for a given span length, semicircular rings develop relatively little pressure at the crown and about four times as much at their haunches. Although the overall range is great, the increments of change in pressure are rather uniform over the sections of the ring.[11] By delivering their resultant forces almost vertically into the piers and abutments, such rings have the considerable advantage of simplifying substructure design and construction. Bedrock helps a modest but well-grounded substructure stand firm against the ring's accumulated pressures.

Segmental rings, the other traditional form of choice, are also circular, although with an included angle of less than 180 degrees. They generate about twice as much pressure on the ring around the crown as a semicircular one of equivalent span length, but the segmental ring's increase in pressure at the haunches is only about a third over that at its crown. While the semicircular ring's rise-to-span ratio is set by definition at 1 in 2, arcs are variable, and the segmental ring's span length can grow much more rapidly than its rise. This, in turn, allows for a reduced distance between the substructure and the roadway without resorting to subdividing the bridge into more spans in order to retain a given vertical distance.

Segmental rings have an Achilles heel, however. Their springings are inclined, and therefore their rings deliver considerable lateral force to the substructure. Without assistance from the force of the earth, their abutments lack natural aid in resisting these lateral pressures, increasing the danger that segmental arch rings will shift, flatten, and ultimately collapse unless buttressed to forestall movement with larger and more costly substructures than bridges with semicircular rings require. The Romans used segmental arches on special occasions for their cut-stone arches as did later builders for some unreinforced concrete bridges.

I.1 TO CELEBRATE HOOSIER PROWESS

(St. Joseph #3690)

In Indiana, as in the nation as a whole, late-nineteenth-century economic growth occurred largely in the industrial and trading sectors which in turn fueled a surge in urban development. While almost two-fifths of the state's citizens remained in agriculture, manufacture and mining supported one-fifth, and another tenth engaged in trade. A third of Hoosiers lived in urban areas.[8b] By 1890, the population of Indianapolis had passed one-hundred thousand. By 1900, Evansville, Fort Wayne, South Bend, and Terre Haute each claimed more than thirty-five thousand residents.

Increased production and urban growth caused the transportation system to expand. Counterparts of the well-established steam-rail network and the later electric street-railway or trolley systems that carried crowds to the Columbian Exposition on the south side of Chicago could be found in Indiana. When city dwellers did not walk, horse-drawn and electric railways shuttled many of them from home to work, shop, or play.

Urban growth increased the demand for better as well as more bridges. The advent and then the expansion of electric street railways multiplied the weight and moving-load forces which local roads and bridges had to accommodate, making starkly obvious the obsolescence of a number of the spans built before the age of mass transit. While it would seldom be easy to find the necessary funds for new bridges from public sources, the design, materials, and fabrication techniques needed for bridges capable of accommodating the new and increased demands of transportation already existed.

Hoosiers had a good supply of architects and engineers trained like those who designed and built the White City's buildings and bridges. Indeed, Indiana directly contributed to the nation's growing number of college-educated engineers. Purdue University provided graduate as well as undergraduate programs in mechanical, civil, and electrical engineering, and Rose Polytechnic Institute (now Rose-Hulman Institute of Technology) trained undergraduates in engineering by 1890. As vacancies occurred or new positions emerged around 1900, many Indiana counties and cities replaced surveyors with engineers, an increasing number of whom were college- or university-educated to address local structural needs.

Hoosiers, furthermore, did not need to ask engineers and architects to make concrete bridges out of straw. Indiana rests upon large deposits of limestone suitable for cement and so had an abundance of cement manufacturers as well as an ample supply of metal fabricators. By 1900, plants in one county alone were producing a million barrels of cement a year. While the production numbers and profits of individual entrepreneurs and areas fluctuated, Indiana had become by 1920 the second largest state manufacturing Portland Cement and annually shipped over ten million barrels beyond its borders.[9]

I.1A The Melan System of Concrete-Steel Arches

The staff- or concrete-encased steel arch ribs which the Columbian Exposition's designers used for their 1893 arched bridges were not novel. Indeed, the earliest combinations of iron and concrete antedated the Exposition by decades both in Europe and in the United States, perhaps beginning with a French gardener's flower pots. In 1867, Joseph Monier patented a wire network to strengthen large planters. Jean Monier quickly applied his father's idea to more complex structures and in 1873 patented an iron reinforcing system for concrete bridges. The award of a First Class Diploma of Honor at the 1878 Paris Exposition to the Moniers for their containers caught the attention of Europeans more generally. The French called the Monier system "Breton Arme" or "armed concrete."[12]

In 1880, Gustav A. Wayss of Berlin purchased the Monier patent rights for Germany and Austria, where erection of unreinforced concrete bridges had flourished for some time. The German Royal Architect, A. Koenen, developed a method for calculating stress in Monier construction which, Wayss then published in an 1887 promotional volume that spurred the building of many Monier arches throughout western Europe.

Early Monier arches were slender and typically made from mortar – a combination of cement and sand – with a network of thin rods placed midway between the upper and lower edges of the ring. After Wayss and Koenen – and especially as span-lengths and loads grew – Monier designers began to run a network of rods just above the lower edge of the arch ring supplemented either with short sections of meshed rods along the upper ring area near each abutment or across the whole upper area. They also increasingly relied on concrete by adding gravel or stone to mortar.

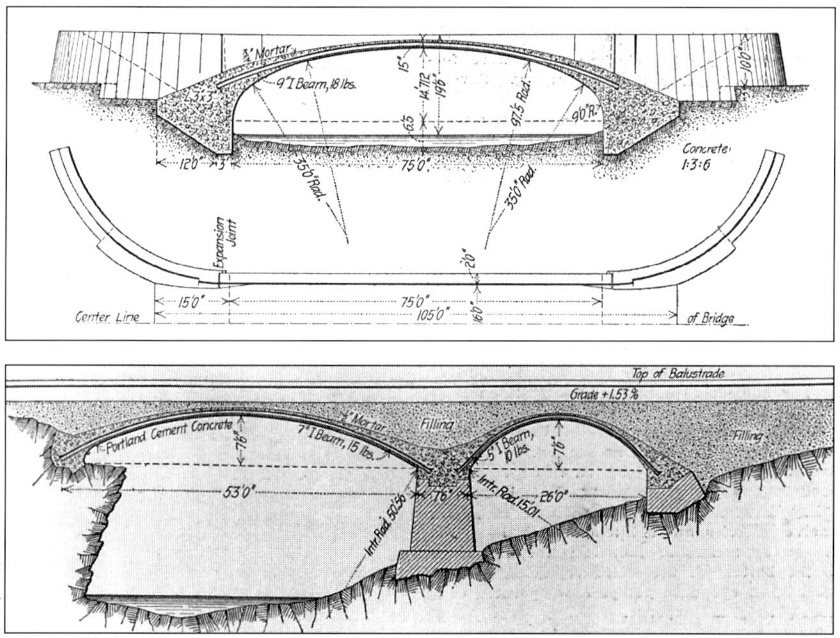

(ENR, 1898)

Plans showing Melan's arced I-beam reinforcing for single and double-span bridges constructed on Frederick W. Vanderbilt's Hyde-Park-on-the-Hudson estate.

After learning how the French gardener-craftsman and his German promoters used metal with concrete, a noted Viennese engineer subjected Monier's finding to engineering science and developed a hybrid system out of it around 1890. Professor Joseph Melan's system reordered the importance of iron and concrete for load-bearing in bridge arches and gave priority to metal. Melan transformed "armed concrete" into "concrete-iron/ steel," with "iron/steel" describing and "concrete" qualifying the relative structural significance of the two materials.[13] As European engineers understood, Melan designed "a series of parallel metallic arches placed from distance to distance and embedded in concrete." In essence, his ribs become "girders" or "a structural skeleton ... enveloped in concrete." They "safely carry alone the load intended."[14]

In a well-publicized test of the "existing system of arches" in 1892, the Austrian Society of Engineers and Architects loaded a series of arches they had built for the occasion and "broke them down."[15] The tests revealed that a Melan arch could handle four times the load that a Monier arch sustained before breaking. With these results triumphantly in hand, Melan secured a patent for his system in the United States in 1893, and the Melan Arch Construction Company soon opened in New York city.

Despite the vigorous efforts of Eyvind Lee Heidenreich to promote the Monier system without Melan's transformations of it, American engineers shied away from the unamended original. Used as they were to metal trusses and plate girders and inexperienced with concrete, Americans considered Monier's netting to be dangerously light. Leading engineers routinely dismissed the Monier system in the 1890s. Carl Gayler predicted in 1894 that "Monier system" arches "will probably never be used for spans exceeding 50 feet, as it admits in nowise of a calculation of its strength, the engineer having to be guided entirely by the results of tests made on test arches." Edwin Thacher in 1899 gently noted that Monier arches built in Europe had "exceedingly delicate dimensions." As late as 1900, H. W. Parkhurst still thought of the Monier system as *"experimental."* Heidenreich chafed at the objection "that it *looks* too light" which he heard "every day from laymen and engineers." He pleaded unsuccessfully with American engineers to critique the Monier system as professionals "on its merits, and not alone on its looks."[16]

While Americans continued to keep Monier design at arm's length, Fritz von Emperger, an Austrian engineer, both experimented with and effectively promoted Melan's system in the United States in the mid-1890s. He secured an American patent in 1897 for a modification of Melan. In place of a single row of I-beams, von Emperger substituted fabricated ribs consisting of a pair of parallel, curved, rolled I- or T-beams stabilized with "distance rods." Bent into W-shaped diagonals and rivetted to the upper and lower beams where they met, the distance rods formed a web designed "to stiffen" the outer beams "and prevent their sagging under pressure."[17] The von Emperger modification of Melan underwent further change and soon turned in practice into an open-webbed arched girder of a pair of angles above and below laced together with bars.

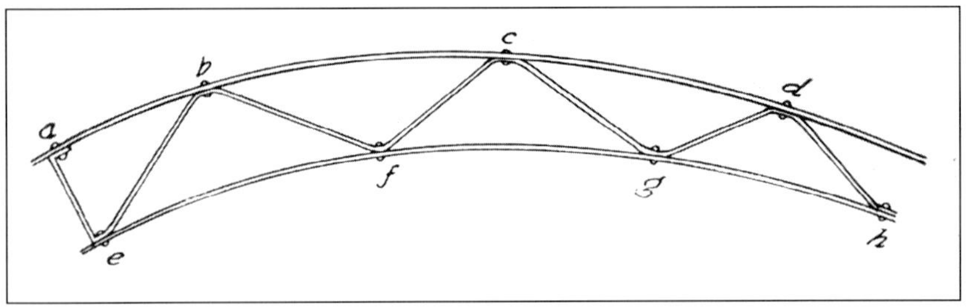

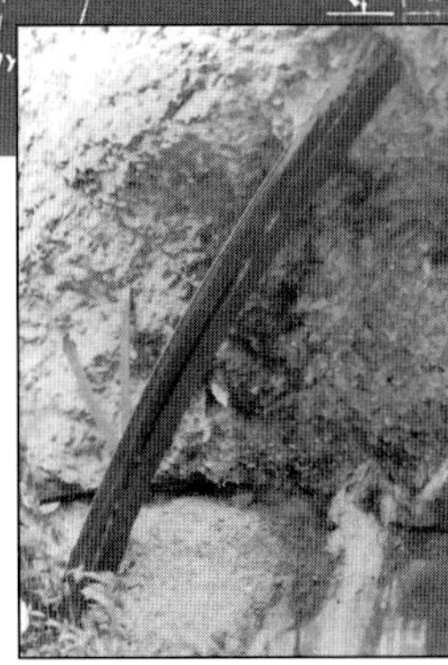

Von Emperger's patented reinforcing rib (top diagram), and its application to the segmental ring of the Burlington Bridge over Wildcat Creek in Carroll County (lower plans). This exposed von Emperger rib on the Minnesota Street Bridge over Eagle Creek in Indianapolis probably encouraged the structure's removal (right).

He who has the best intuition may not prosper most. If, for example, the Monier-system advocates took two steps forward in their thinking about metal and concrete working together, Melan and von Emperger backed off a step when they conceptualized their system. The latter relied mostly upon the familiar – trusted-and-tried metal – and experimented with the new – concrete sheathing. They expected the iron or steel structural frames in the system to bear most of the span's load. In the American marketplace, however, the more daring thinking failed, and those who stepped back from the conceptual cutting edge executed a brilliant strategic retreat. By falling back into the more conservative domain of American architects and engineers, the later European design advocates planted Melan-system bridges across the American landscape.

While Melan and von Emperger designers seemed to underestimate the amount of carrying which the concrete in their arch ring achieved, they did understand concrete's contributions to protection and stability. The concrete reduced rusting of the metal ribs, provided support for the structure's earth fill and roadway, and allowed greater decoration than would metal alone. Concrete between the I-beams or ribs also added considerable structural stability, a regular matter of concern to iron or steel-arch or truss designers and one for which von Emperger took the added precaution of latticing his metal ribs to each other.[18]

The shapes typically used for Melan-Emperger system's arch rings and the way the rings related to the piers and abutments remained essentially traditional. I-beams and fabricated metal ribs allowed designers to move with ease and assurance beyond the semi-circular rings favored since ancient times for cut-stone arches. But Melan-system designers did not move very far beyond the segmental rings that had come into fashion during the Renaissance. While the need to bend I-beams encouraged the use of a single wide arc, the building of ribs from a collection of smaller pieces of steel stock made more variegated shaping easier.

Von Emperger's patented ribs started with two different segmental arcs, the top flange coming from a circle with a wider radius than the bottom one. Since the segment which the lower flange followed included a larger portion of a semicircle than would the upper one, the thrust it delivered to the substructure was somewhat less lateral than that of the upper one. Over time, concrete-steel designers shifted the shape of the rib's lower flange from a large, tight segment towards a semi-ellipse. The ellipse provided a curve similar in appearance to an egg. Since the ends of a semi-elliptical arch rib dropped more rapidly than its center, the thrust the rib delivered to the substructure became less lateral and somewhat more vertical.

Whether I-beams bent into segments, or ribs fabricated with semi-elliptical lower flanges, Melan-system rings nonetheless delivered a lot of lateral thrust to the substructure. This thrust required Melan-system designers to place their sheathed rib-rings upon substantial pier and abutment anchors. To balance out lateral thrust over the piers, these designers also typically planned the springings of all their spans' rings for the same horizontal plane, and they kept as many of a structure's spans as feasible of equal length.

I.2 BRIDGES BEFITTING A CAPITAL CITY

(Marion #2501)

The civic leaders of late nineteenth-century Indianapolis embraced the city's growth andwanted to embellish its status as the state capital. The White City of 1893 – and its evolution in the next five years into the City-Beautiful movement – caught the imagination of many urban promoters.[19] The movement developed in Indianapolis upon the base which the period's foremost landscape architect, Frederick Law Olmstead, provided. Among his major achievements as a designer, Olmstead could boast of the community of Riverside built outside of Chicago at the close of the Civil War, Central Park in New York City, and the White City's landscape.

Shortly after the White City moved from material reality into memory, landscape architects joined forces with the broader municipal improvement movement, which insisted on combining beauty and utility. The movement united civic leaders (mainly businessmen who would take risks in the name of collective, enlightened class and community interest) with experts, notably architects to plan urban spaces which combined beauty and usefulness.

Olmstead and other landscape architects promoted naturalistically constructed park and boulevard systems in urban areas to produce a form of civic art. Parks, they believed, could provide areas for city dwellers to bathe the restorative powers of nature which, among other things, would encourage social-class reconciliation. While urban growth seemed inevitably intertwined with an increasingly industrial economy and its material dividends, leading Americans had not forgotten the Founding Fathers' warnings that the city would threaten the soul of the republic. Settlement in American cities in the late decades of the nineteenth century of tens of thousands of non-English immigrants with different languages, religions, and political beliefs – some radical – underlined the arguments for various forms of social control. By ameliorating class tensions, parks provided social insurance which, Olmstead argued, more than offset their cost in part by increasing property values.

The landscape architects and municipal reformers who made parks a centerpiece of City-Beautiful programming wanted parks dispersed throughout the city, especially along rivers and streams, and connected by boulevards. Building a park system around waterways, though, inevitably included bridges needed to get to, through, and around the parks. City bridges thus became a concern of landscape architects locally and nationally.

A leading national advocate of civic improvement, Charles M. Robinson, formally incorporated bridges within the City-Beautiful canon. With "monumental" structures or spans, he stated, "we should not be satisfied merely with durability and strength, but should demand that to these be added fitness, grace, and beauty."[20] Bridges, like the parks in which they were located, needed to be naturalistically constructed, that is, to fit in with – rather than stand out against – the natural environment.

City bridges carried vehicles, some with fairly large numbers of passengers, as well as pedestrians. However, a bridge's functions, Robinson believed, did not have to be narrowly construed. Drivers and riders should be able to view the streams and adjacent parks as they cross. Proper urban spans should include wide sidewalks for strolling and refuge bays where pedestrians might tarry for conversation and might observe the beauty around them. Olmstead, Robinson, and other City-Beautiful advocates articulated standards for community development that local leaders soon applied in Indianapolis.

I.2A The Park Board's Merritt and Power

Business leaders in the Commercial Club of Indianapolis began the state capital's City-Beautiful movement. "Father of the park idea," George Merritt, served on the committee that drafted the first park law, which the Indiana legislature passed in 1895 especially for Indianapolis. He redrafted the act in 1899, after a state court dismissed the earlier version, and he served on the city's park Board from 1899 to 1906. A successful woolen manufacturer and among Indiana National Bank's founders, Merritt concentrated much of his considerable civic participation on city schools. His commitment to free kindergartens with ample playgrounds fueled his interest in public parks.[21]

Thomas Taggart, manager of the Union Railroad Station hotel and restaurant, boosted Merritt's law in the capital city. A Democrat elected mayor in 1895, he was imbued with this "spirit of progress." Taggart appointed an Indianapolis Board of Park Commissioners to take "the initial steps ... for a system of parks which, when completed, will make Indianapolis the `park city' of the continent." On receipt of the first of the Board's annual reports, Mayor Taggart repeated his call for a "park system" to "assist materially in beautifying the city and [to] offer the residents a place for healthy recreation."[22]

From the start, the park commissioners recognized an inextricable link between urban parks and transportation. While walking played a large role in urban life, the commissioners knew they needed to address mass transit immediately. The Board's members became "urgent in their demands" for an extension of street-railway service to Garfield Park, for example, in order that "the masses ... from the densely populated portions of the city" could reach the park quickly and inexpensively. Street railways increasingly connected neighborhoods within the city and also tied regional urban areas together. Lafayette had the first fully electrified city railway system in Indiana by 1888. Trolleys operated in all the other major Indiana cities by 1895, and interurban service covered about 2,500 miles within the state by 1920.[23]

At their Board's genesis, the park commissioners turned to two firms of "expert landscape architects" for advice. The Board brought John C. Olmstead, son of the venerable Frederick Law Olmstead, to town for his advice and blessing. Following City-Beautiful tenets, John Olmstead declared that "the best and almost the only park sites worth considering are those including some portion of the rivers or runs which pass through or close to the city."[24]

To turn plans into action, the Board selected J. Clyde Power as engineer and superintendent of Indianapolis' developing system. Power brought experience and credentials in his own right to his new job. His father and maternal grandfather had been civil engineers, and Power had received a lot of "practical training" as a youth which his formal education in Pennsylvania supplemented. He had graduated from Beaver College and had received training in engineering at Lafayette College. Bridge-building and park construction in Washington, D. C., and Philadelphia were among the many jobs young Power held after college and before coming to Indianapolis in 1895.[25] Power's family background, education, and practical park experience prepared him to be a local leader and not just a subordinate of out-of-state consulting landscape architects. He treasured his own ideas, and with unabated will-power he gradually exercised a great deal of influence over park policy.

Power shared the national concern of City-Beautiful planners with bridges. His undeviating preference for stone arches was more narrowly derivative than most, however, and it consequently both limited park-board influence and increased local resistance to park development in Indianapolis. Artificial stone or concrete-steel, for example, was cheaper, more flexible, and easier to decorate than cut-stone masonry. If the White City's bridges could be framed in steel and sheathed in staff or in concrete, then reliance on concrete arches would generally not exclude a city from the City-Beautiful fold. But Power took exception.

After focusing attention on "the very unsightly bridges over Fall Creek," Power declared that "only masonry arches of artistic design should be built over streams within or near a growing city. Masonry is there for ages, while iron or steel bridges require repairs, paint, etc., constantly, and in a few years must be replaced by another structure. It is true that the first cost is somewhat greater in the masonry arch, but in a few years the steel bridges will have exceeded the masonry in total expense."[26]

Power's claim that cut-stone arches provided permanent and maintenance-free structures paralleled the claims of those who pressed for artificial stone or concrete-steel bridges. City-Beautiful proponents argued that officials who purchased bridges of wood or steel should consider the costs of periodic upkeep (painting and repairs) on such structures and add them to the price of construction. Here, advocates of stone and concrete bridges had a point although they tended to underestimate the amount of maintenance that structures of their preferred materials required, even in days before salt was used widely on the roads.

Power kept repeating the landscape architect's concern with bridge-design and style. "There is nothing which will do so much to improve the appearance of a parkway, street or road as beautiful bridges. No one style of bridge is proper in every location, and the things to be studied are beauty of design and appropriateness to the location in which it is to be built. Handsome bridges have their effect upon the people, and often houses in the vicinity of these bridges will show some of the details of the bridge design."[27]

The Indianapolis park board administered only a small portion of the monies budgeted to build city and county bridges. By 1900, they did not yet control all the lands adjacent to rivers and streams that they planned to incorporate within their system. The city's civil engineer, who worked for the Board of Public Works, handled a larger bridge construction budget than did the park's engineer, and the Marion County surveyor had a larger budget than those of all the city's engineers together. To make their ideas about bridge design prevail throughout the city and county, the Indianapolis City-Beautiful advocates needed to persuade other city and county officers. The city civil engineer was within the easiest reach, since he and the park superintendent served under the mayor. The county surveyor, however, reported to a more independent third party, the Board of Commissioners.

The park superintendent kept up a steady drumbeat for City-Beautiful bridges and, when opportunity appeared, achieved some considerable success. The city civil engineer responded affirmatively to some but not all of his park department colleagues' aesthetic bridge-design standards. The city erected fewer metal spans, and it "ornamented" some of its more important new bridges. On occasion the county's officials also accommodated that influential minority among their constituents who lobbied for a City Beautiful.

Although the city and the county accepted Power's insistence on cut-stone arches only peripherally, they sometimes disguised their waywardness in City-Beautiful garb. When, for example, they built concrete-steel spans, they generally faced them with stone and on occasion decorated the concrete to look like stone. Perhaps as a result of Power's influence, the stone facing tended to be rough cut, the pier pilasters rounded and extended to support refuge bays, and the bridge rail-posts and panels made entirely from cut stone.

Feeding on itself, transportation development heightened the demand for new bridges and provided some of the numerous opportunities for the dialogue that City-Beautiful people exploited to discuss their form and style. The electric street-railways, for example, on which the landscape architects depended to get "the masses" into their park systems, not only allowed for more effective transportation within the growing and developing city but also rendered some bridges obsolete and destroyed others. While streets could easily carry trolleys, most bridges had been designed for animal-drawn vehicles and pedestrians. They were built for weights and moving-load stresses that were dramatically lower than those electric railways generated. The predictable results followed.

In Indianapolis in May 1899, a motor car pulling two trailers loaded with gravel fell through the north span of an 1872 metal-truss bridge on Central Avenue over Fall Creek. Patrons of the Central-Avenue line recalled "that during the times when the street cars were crowded by visitors to the military camp or the state fair, the bridge creaked and swayed in a manner that seemed to indicate it was unsafe."[28]

To replace the partially collapsed metal-truss bridge on Central Avenue in 1899, both the county surveyor and the city civil engineer submitted separate sets of plans for steel plate-girder spans, much to the unhappiness of the park board's members who encouraged bridge contractors to bid for a City-Beautiful structure of their own design. The county commissioners were far less sympathetic to the landscape architect's ideal than to the engineer's concern for functional efficiency. Above all, they were responsive to constituents' interest in holding down taxes. In advance of the bidding, two of the three commissioners had declared themselves in favor of metal spans because "of the probable tremendous cost" of a City-Beautiful structure. Only after the commissioners received the lowest of all bids for cut-stone arches reinforced with Melan style I-beams and complete with refuge bays did the two original metal-span advocates join their colleague in a unanimous vote for a City-Beautiful structure.[29]

The Board of Public Works also contracted for its first "concrete and steel arches" on Meridian and Illinois Streets over Fall Creek (Marion #2502) in 1899. They were "permanent" structures – not made of wood or steel – and they were "ornamented" – faced with Bedford limestone. Of more moment to the engineer, the bridges followed the Melan system of reinforcing: Ten-inch I-beams, spaced about three-feet apart and weighing twenty-five pounds to the lineal foot, formed a steel arch inside the concrete. These bridges, Mayor Thomas Taggart thought, "represent modern ideas as to bridge architecture and are added ornaments to our city." Taggart also favored the erection "of bridges of similar construction, or of

(Marion #2502)

Illinois Street Bridge over Fall Creek in Indianapolis.

Illinois Street Bridge stone railing.

REFUGE BAYS

Bridges built in medieval Europe typically carried rather narrow roadways, partly to contain construction costs and sometimes to defend a crossing from unwelcome intrusions. The Romans had also fortified their spans against natural forces by building upstream structures or cutwaters ahead of piers to prevent structural undercutting. Once late medieval Europeans learned the virtue of building protective cutwaters downstream as well as upstream, they found they had a structural foundation extended ahead of the piers that, if raised to roadway level, widened the deck for a short distance. Such recesses or bays provided space for sentries or lookouts keeping watch up and down the river as well as across the bridge. More typically, they provided pedestrians refuge from horsemen and carts on the narrow roadway.[30]

Well before Europeans came to North America, their bridge builders had largely abandoned construction of refuge bays although not cutwaters. Nineteenth-century Americans, furthermore, did not usually plan their bridges with military defense in mind, and they routinely built urban spans wide enough to accommodate pedestrians on sidewalks. The reappearance of refuge bays with City-Beautiful design in the United States, then, was no mindless continuation of established tradition. Here was a conscious return to the past, and – as is almost always the case with such "returns" – in a somewhat different form for quite different reasons.

Semicircular cutwaters and pilasters carry the refuge bays on the Thirtieth Street Bridge over the White River in Indianapolis.

(Marion #1705)

Refuge bays recalled that same melange of ancient and Renaissance European authority that the White City architects symbolized in their structures. For landscape architects and municipal-improvement advocates, refuge bays provided pedestrians a place to tarry and to view the waterway enclosed in a naturalistically-constructed park. The design as well as the function of City-Beautiful refuge bays had also changed. While their European predecessors had mostly raised triangular cutwaters into triangular recesses, City-Beautiful architects and engineers typically designed semicircular pier columns and bays.

stone," at Washington Street and River Avenue over the White River. The city civil engineer noted in 1901 "that in the future it will be better to build more substantial and ornamental bridges, even though it be necessary to build fewer of them."[31]

The Central Avenue Bridge story was in part repeated in January 1902. The 1871 metal trusses of the West Washington Street Bridge's center span collapsed under the weight of a motor car with two trailers filled with cinders on one track and a team of horses pulling four empty cars on the other track. Both trains, eleven men, and eight horses fell into the White River without loss of life. Because "the citizens, especially those residing west of White River, were very much wrought up over. . .the probable unsafe condition of the other bridges spanning the river," the Indianapolis Board of Public Works hired a consulting engineer from Chicago to examine all the bridges over the river.[32]

The county commissioners had established an informal rule for themselves by which they would normally accommodate City-Beautiful design only where a bridge was near or adjoined a park. The West Washington Street Bridge over the White River fell outside a park area. City officials judged the location differently and began a two-pronged campaign for replacement. They pressed the county to pay for a new bridge at the same time that they prepared plans for a City-Beautiful structure to their liking – one of stone arches. The city's plans for the stone-arch bridge were "well advanced" when the county commissioners finally agreed to build a new structure.[33]

The county commissioners used their own surveyor, John V. Coyner, to prepare a set of plans and specifications for the Washington Street Bridge. Coyner proposed steel plate-girder spans rather than either stone or concrete arches, and the commissioners duly approved the plans and authorized them for bidding. At the bidding, William Fife and Son submitted two: one for a stone substructure on Coyner's plans, and another for a stone-arch structure on the city's plans.

The county commissioners bowed to city pressure and voted unanimously to award a contract to Fife for a complete stone-arch bridge, if the county council would appropriate the extra sixty-five thousand dollars needed for the more expensive structure. As a fallback, the commissioners voted to accept Fife's bid for a stone substructure plus the Brown-Ketcham Iron Works' bid for a steel plate-girder superstructure.[34]

At its meeting a week later, the Marion County Council split along party lines on the requested additional appropriation, with Republicans supporting and Democrats opposing it. With the council's pro-arch president absent because of illness, the opponents of the stone bridge carried the day. The commissioners, in turn, moved quickly to their fallback position for a plate-girder superstructure on a cut-stone substructure.[35] Engineers and fiscal conservatives had narrowly won a round to contain the landscape architects and their urban supporters.

The commissioners could not easily duck the park aesthetic when they addressed bridges across the White River close to the two ends – one on Sixteenth and the other on Thirtieth Street – of the Thomas Taggart Riverside Park. They agreed in 1905 to build the most extraordinary of these structures at Emrichsville (Sixteenth Street) – a three span, Melan-system, concrete structure faced with smoothly cut stone. A large archway arose between towers over the roadway at the park entrance on the west. The bridge's piers were rounded and extended to accommodate refuge bays, and each carried an elaborately ornamented frieze. The stone facing panelled the spandrels, outlined the arch rings, and framed a substantial cartouche over each crown. The rails behind quite wide sidewalks carried neoclassical balusters. Power complimented the county commissioners for building such a "handsome" bridge.[36]

The county commissioners also ultimately responded to Power's long litany of attacks on the existing metal Thirtieth Street Bridge over the White River along Riverside Park, a bridge the park superintendent found "inadequate" and "dangerous." "Having no

(Marion #1705)

Thirtieth Street Bridge, Indianapolis.

(Marion #1705)

Grand stairway and plaza between the Thirtieth Street Bridge and Riverside Park in Indianapolis.

sidewalks, pedestrians have to take the roadway, and on several Sundays I could not see how accidents were averted, the bridge being crowded and teams becoming frightened at the steamboat running under it. This bridge was built about twenty-four years ago and as this is about the life of an iron or steel bridge, a new one should be built, which should be a handsome and ample masonry bridge, for Thirtieth Street is one of the principal approaches to the park."[37]

The commissioners agreed in 1907 to replace the Thirtieth Street Bridge. H. W. Klausmann, the county surveyor, designed a concrete-steel, arched structure with rough-hewn limestone facing, refuge bays, stone railings, fancy light standards, and wide, curved stairways at the ends of the bridge's sidewalks and down into Riverside Park (Marion #1705). The new structure won the park board's praise as "one of the handsomest bridges in the country."[38]

While determined to be in charge of their own fate, the Marion County commissioners and their surveyors nonetheless sometimes supported City-Beautiful style outside park areas. The county built a considerable number of small and moderately sized stone-arch spans in the first decade of the twentieth century. Structural and economic reasons also led the county to rely increasingly on concrete rather than steel structures, again largely for small and moderately long single-span bridges. To placate the White City enthusiasts, the county specified that some concrete arches be stone-faced, it ordered that the walls and rails of others be decorated like stone, and it routinely called for parapet rails with neoclassical balusters. While Marion County pushed its design ahead with the building of open-spandrel, concrete arch spans on Thirtieth Street over Fall Creek in 1905, it had the structure stone-faced and contracted for stone railings.[39]

I.2B George E. Kessler to the Fore

Changes in membership and staff underlined the park board's growing importance in the first decade of the new century. Its patriarch, George Merritt, retired in 1906, and its executive officer, J. Clyde Power, resigned at the end of 1907. The new mayor, Charles Bookwalter, seemed even more committed to the importance of parks than had been his predecessors. In his first annual message,

Bookwalter stated that "public parks are a public necessity. Public parks are the lungs of a city."[40] Dr. Henry Jameson, whom the new mayor appointed to the park board, "wished his Indianapolis to be a city beautiful." To get to that end, he thought the city needed a more "definite, organized, forward-looking plan." Jameson asked Hugh McGowan, leader of the Indianapolis street car interests and a major figure in interurban companies, to recommend some possible consultants, and McGowan proposed George E. Kessler, a nationally known landscape architect.[41]

The park board, the mayor, and some of their weighty associates first interviewed and then enthusiastically hired Kessler to inaugurate a wider and more systematic effort to beautify Indianapolis. Kessler came to his post as "Landscape Architect" with impressive credentials. Born in Germany and reared in Missouri, Wisconsin, and Texas, Kessler returned to Germany for advanced study in botany, landscape design, and civil engineering. On reentering the United States, he worked briefly for New York City's Central Park project before moving off to Kansas City, where he designed the city's park and boulevard system. Cincinnati, Dallas, Denver, Memphis, Salt Lake City, and Terre Haute also profited – sometimes considerably – from Kessler's advice during the years in which he consulted for Indianapolis.[42]

Kessler's planning dramatically extended and systematized what the park board under Merritt and Power had started. Kessler used the White River, Fall Creek, Pogue's Run, and Pleasant Run as the system's skeleton, proposing boulevards along each bank like sinews connecting the members and adding movement to the urban body.[43] Such initiatives brought substantially more river-edge and stream-front property as well as the bridges which spanned the waterways of Marion County within the landscape architect's purview.

Rather than focus almost exclusively on cut-stone arches as Power had preferred, Kessler embraced well-dressed concrete-steel structures. Indeed, in the last year of his service, Power had also accepted the construction of three concrete spans within the park

Spades Park footbridge, Indianapolis.

system (Marion #2615, #3215, Spades Park footbridge), although he had his concrete structures decorated like cut stone.[44] Kessler's first chance to show his prowess came with the design of the Capitol Avenue Bridge over Fall Creek, a significant crossing as it intersected two projected park boulevards.

The political road to a Kessler-designed bridge was more tortuous than the planning for intersecting boulevards might suggest since the project inevitably involved the Marion County commissioners. The commissioners had their surveyor draw up plans for a stone-faced, concrete-arch structure for which bidding was advertised in December 1908, but a temporary injunction by a county-court judge prevented the commissioners from immediately letting a contract.[45]

Discussions about the Capitol Street Bridge must have occurred behind the scenes during 1909 and 1910, since commissioner and park board records are silent about the project. By 1911, the city and county authorities had reached an understanding in which Kessler designed the bridge, the county paid for the basic structure, and the city underwrote the costs of beautification – roadway pavement, sidewalks, balustraded rails, and light standards.

(Marion #2501)

Above – Capitol Avenue Bridge over Fall Creek, Indianapolis. Two of the bridge's cartouches are shown on pages 7 and 17; right – Detail of Capitol Avenue Bridge cutwater-pilaster.

The careful commissioners must have thought the county council too generous with their constituents' monies, for in May they asked that $15,000 of the council's budget for the bridge be transferred to culvert construction and bridge repairs instead. The commissioners awarded a contract for the bridge in June with park-board officials supervising construction (Marion #2501).[46] Still, the Kessler-design bridge cost the commissioners $70,000 as against $40,000 for the lowest bid by a reputable firm on the county's original plan. The city, in turn, put another $40,000 into the project, bringing the final bill to almost three times the original estimate.

The City-Beautiful leaders were delighted to see the "eyesore" – the old iron girder bridge – replaced by "the handsomest bridge across Fall Creek in this city."[47] In it Kessler had given Indianapolis a neoclassical monument that exuded the kind of civic pride and

(Marion #4395)

Top – Meridian Street or Joseph W. Summers Memorial Bridge; above left – detail of elaborately decorated spandrels and pilasters; above right – detail of neoclassical rail balusters.

authority which the White City had represented. Although the concrete arches were faced with Bedford limestone, the stone was smooth-cut and light, not rough-hewn and dark. The arch rings were carefully framed, the pier pilasters substantially panelled, and both decorated with cartouches.

The park board persuaded the county commissioners to build a smaller but no less imperial version of the Kessler design for a span over Crooked Creek on West Riverside Drive in 1913 (Marion #1715). Once more the county set an upper limit to its contribution, this time of $10,000, apparently leaving it to the city to provide for the "stone trimmings." The commissioners' resolution of a few months later to accept only bids based on plans drafted by their own county engineer may have revealed their continuing irritation with the City-Beautiful cadre.[48]

The city contracted in 1915 for the Meridian Street Bridge over Fall Creek, the last and most monumental Kessler-designed structure in Indianapolis (Marion #4395). The concrete arches carry Kessler's stylistic imprint of smooth-cut limestone facing and balustraded rails with framed arch rings, panelled pier pilasters, and cartouches. Emblazoned stone caps raised ornate, light standards above the pier-rail columns.[49]

I.3 BEAUTIFUL SPANS ALONG THE BENDS OF THE SAINT JOE

(St. Joseph #3690)

Indianapolis was not the only Indiana city touched by the City-Beautiful movement. Although no other city had a nucleus for civic improvement as large and as weighty as the state capital, the South Bend-Mishawaka area also responded. Living within Chicago's gravitational field, St. Joseph County residents and leaders felt the pull of the symbolically-powerful White City.

The landscape architect's impact on bridge design showed up in South Bend about a decade later than in Indianapolis. South Bend's city engineer, Alonzo J. Hammond, and his younger associate, William S. Moore, began designing concrete-steel arches with largely neoclassical exteriors by at least 1906. Unlike their Indianapolis peers, Hammond and Moore did not cover their concrete arches with cut stone. Instead, they decoratively molded the concrete.

In 1906, Hammond designed a four-span arch with Melan-system reinforcing to handle a two-track street-railway, vehicular traffic, and pedestrians. It would replace the metal trusses of the Jefferson Street crossing of the St. Joseph River in South Bend (St. Joseph #209). Hammond shaped the pier pilasters to look like classical goblets which opened at their tops into refuge bays. Behind wide sidewalks, the deck carried open-arch, not neoclassically balustraded, parapet rails and ornate light standards.

Alonzo Hammond planned the east approach of the Jefferson Street Bridge "to harmonize with the improvements in Howard Park," including "winding concrete steps" which wove the bridge into the park. "As it is believed that this bridge will be one of the orna-

(St. Joseph #209)

Jefferson Street Bridge over the St. Joseph River as exhibited on the cover of the Annual Report of Departments *for South Bend for 1907.*

(St. Joseph #210)

LaSalle Street Bridge over the St. Joseph River, South Bend; (left) detail of rectangular pier pilaster.

mental structures of the city, the character of the surroundings at each approach should be carefully considered and harmonize as much as possible with the structure, which will also add an additional degree of beauty and interest to the adjacent park."[50]

The next year Hammond supervised the design of "a modern bridge" for the LaSalle Avenue crossing of the St. Joseph River (St. Joseph #210), and Moore planned a look-alike structure for Bridge Street in Mishawaka. The county hired two consultants for the designing of these bridges. First, they paid the Concrete-Steel Engineering Company of New York for the use of their patented, Melan-system, reinforcing designs. Second, they contracted with Professor Malverd A. Howe of Rose Polytechnic Institute to inspect the plans and specifications for the two bridges. "Permanent bridges of this character should hereafter be erected," Hammond concluded, "where the demands of travel demand new bridges or the replacement of old ones."[51]

Although adjusted to the natural surroundings, the decor of the LaSalle and Bridge Streets' arches was somewhat more restrained than that of the Jefferson Street Bridge of the year before. Hammond and Moore used panelled and rectangular pier pilasters without refuge bays above corbelled cones. Hammond decorated his rings and piers like stone, while Moore preferred panelled spandrel walls with raised ring borders and a cartouche at the crown. They both retained wide sidewalks and balustraded rails, again with some modest differences. Although neither used neoclassical balusters, Hammond repeated the open-arch pattern of the Jefferson Street Bridge while Moore designed intersecting circles. Both employed ornamental lampposts. Each of the eight light standards on the LaSalle Street bridge, for example, carried five incandescent lamps.

With the Leeper or Michigan Street Bridge of 1914 (St. Joseph #3690), Charles W. Cole, Mishawaka's city engineer, designed the last of the City-Beautiful arches built in St. Joseph County. Hammond had started the push to replace the metal-truss structure on Michigan Street at least as early as 1908, and Moore kept up the pressure when he became South Bend's city engineer by 1912. Moore favored "a permanent concrete bridge" wide enough to accommodate all the vehicular and pedestrian traffic on the thoroughfare and beautiful enough to complement "the development of Leeper Park and the surroundings."[52]

Cole chose a City-Beautiful decor far more akin to Kessler's designs for Indianapolis than an extension of the homegrown and distinctive styles of Hammond and Moore. Indeed, the Cleary Kuert Company of Indianapolis, which constructed most of Kessler's bridges in the state capital, won the contract to build the new Leeper bridge.[53] Cole sheathed his concrete in smooth-cut, Bedford limestone. He panelled the spandrel walls, raised the borders of the arch rings, and cartouche at the crown. His pier pilasters rise from a pedestal base like panelled rectangles with decorative scrolls at the parapets. The deck carried wide sidewalks and ornamented light standards between open-arch balustraded rails.

Moore praised Cole's work as "one of the most artistic structures in the state." The next South Bend city engineer joined in the accolade: "This bridge adds greatly to the attractiveness of the surroundings of Leeper Park, and is an improvement in which the whole city can take pride." In sum, "all of the bridges now spanning the river in South Bend are of modern type of construction" with the exception of the metal spans moved earlier to make way for the Jefferson Street arches.[54] Thus the rebridging of South Bend and Mishawaka in the City-Beautiful mold was almost complete by 1915.

(St. Joseph #3690)

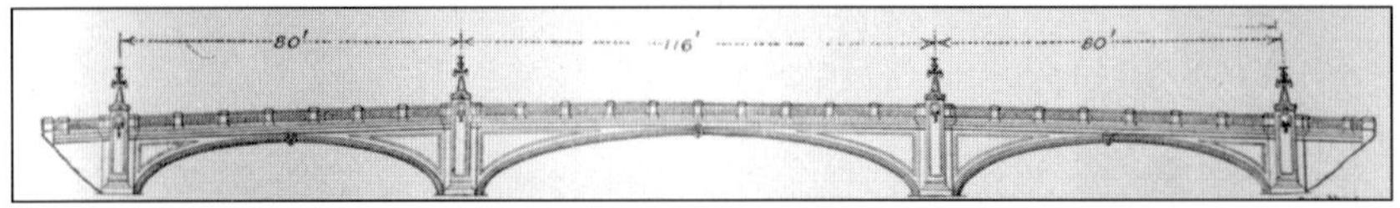

Above – Leeper or Michigan Street Bridge over the St. Joseph River in South Bend with a drawing provided the ENR *(1914) below. Two of the bridge's cartouches are shown on pages 14 and 26; right – detail of Leeper Bridge pier pilaster.*

I.4 FORT WAYNE'S MOMENT "WITH MIGHT AND MAIN"

(Allen #539)

Here and there across the state other bridges survive from the City-Beautiful era. In Fort Wayne, the Commercial Club engaged a University of Chicago professor to lecture on municipal improvement in 1910. The lectures inspired organization of the Fort Wayne Civic Association, which promptly hired the national City-Beautiful advocate, Charles M. Robinson, to draft an improvement plan for the city. He proposed a park and boulevard system centered on the city's rivers and streams and thereby included a significant portion of its bridges. Robinson apparently "came on pretty strong to the ultraconservative community of 1910 Fort Wayne." Nonetheless, Fort Wayne adopted some of Robinson's proposals, especially those "quite apparent and easy to accomplish" including some naturalistically constructed bridges.[55]

A. W. Grosvenor, a Fort Wayne engineer who designed a number of bridges in the area, adapted to the City-Beautiful movement in an even more special way than had Hammond and Moore in South Bend. The difference was not in his engineering, for indeed he followed the standard preference for arches with Melan-system reinforcement. Grosvenor depended upon the von Emperger modifications of steel ribs running near the upper and lower edges of the ring and consisting of pairs of angles riveted together with lacing bars.[56]

(Allen #537)

Grosvenor's creative adaptation of the City-Beautiful aesthetic distinguished his design. His decor abandoned references to the classical and accentuated the angular. In 1912, he designed a bridge to cross the St. Joseph River at Tennessee Avenue between two parks

(Allen #539)

Above – Tennessee Avenue Bridge over the St. Joseph River, Fort Wayne as printed in the ENR *(1915); below – drawing showing the von Emperger reinforcing system. Detail of the bridge's pier and pilaster shown on page 28.*

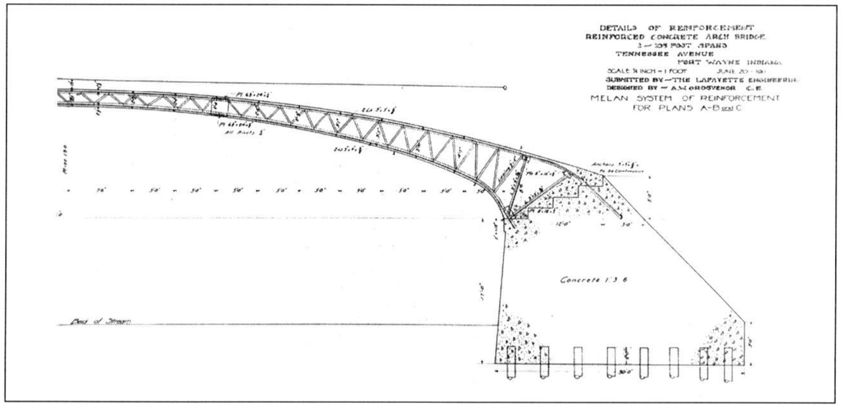

(Allen #537)

Combs Street Bridge over the Maumee River, Fort Wayne with its neoclassical rail balusters. Detail of the bridge's pier and pilaster shown on page 28.

(Allen #539). Grosvenor sheathed the spandrels and pier in Flemish-bond brick, used brick rails, and framed the whole with light and smooth-cut limestone. He suggested a mosaic finish more than a curvilinear or rough-hewn stone imitation of nature, omitted carvings, avoided refuge bays, and substituted solid-brick rail panels for neoclassical or other balusters. In short, the structural system was standard or even old-fashioned, but the decoration was novel.

Grosvenor's bridge emerged nationally when C. E. Drayer compared the Tennessee Avenue spans with those built on Combs Street (Allen #537) in the *Engineering News-Record* [ENR].[57] George Jaap, who had been foreman of the carpenters on the Tennessee Avenue project, apparently traced Grosvenor's plans and used them for the Combs Street Bridge.

Drayer ridiculed Jaap for behaving like an artisan rather than as an engineer. The engineer insisted on a different standard of conduct when he criticized the carpenter for doing what carpenters and other mechanics had done for centuries – build empirically from the designs of others. Jaap reportedly copied Grosvenor's ring shape and his use of the von Emperger version of Melan-system reinforcing. Grosvenor had balanced the lateral thrust over the pier of the Tennessee Avenue Bridge by designing two spans of equal length. Jaap, however, flanked a 100-foot center span with 90-foot outer ones. Whether Jaap adjusted his substructure to accommodate the imbalance of the thrust on the piers remains a mystery.

According to Drayer, Jaap underestimated the length of the ribs needed for the center span. The building contractor discovered the ribs' shortfall after the bridge's substructure had been constructed, so by eye the contractor figured out what was needed and had Melan-type ribs bent in Fort Wayne shops to reach from one pier to the other. Not surprisingly, the adjusted rib design did not meet the "proper elliptical curves" that Grosvenor had used and somewhat reduced waterway clearance. Furthermore, Drayer reported, the carpenter's "ignorance" of science led him to omit plates over the top and bottom flanges where rib sections abutted.

Not surprisingly, the engineer had less to say about the appearance of the carpenter's Combs Street Bridge which was remarkably different than that of the Tennessee Avenue spans. But, even if brief, Drayer's comments remained critical: "The aesthetic effect depends largely on the emphasis of the arch barrel by the quarry-faced ring stones of the facing, although the effect is somewhat reduced by the detail of the railing." Drayer may not have known that Grosvenor had submitted a "Plan B" to the Allen County commissioners for the Tennessee Avenue Bridge, one with stone facing. Indeed, some of the very matters of decor Drayer criticized Jaap for, the carpenter may have also copied from Grosvenor's second-choice plan.

In place of Grosvenor's preferred linearity, Jaap curved his piers and their pilasters into refuge bays. The carpenter finished his spandrels with rough-hewn sandstone rather than with smooth-cut limestone and brick. His rails consisted of neoclassical balustrades, not solid brick panels with limestone borders. Jaap, in sum, followed the early conventions of City-Beautiful bridge style, while Grosvenor considerably extended them.

Drayer's public criticism of Jaap only added insult to his injury, for the carpenter may not have profited from his tracings anyway. Having for some time regularly won contracts from the Allen County commissioners for bridge repair or for the construction of small spans, George Jaap did bid for the important Combs Street Bridge contract. After accepting Jaap's plan and specifications, the county commissioners then gave the contract to the Herman Tapp Construction Company which underbid Jaap by less than a mere seventy-five dollars on proposals of more than fifty-six thousand dollars.[58]

I.5 A "BETTER AND NEATER" BRIDGE FOR A SMALL CITY: THE PORTLAND ARCH

The leaders of some small cities also became infected with the municipal improvement bug, even when they did not contract a full case of City-Beautiful mania. Enough of the city fathers of Portland, the seat of Jay County, fell under the movement's spell in 1913 to document their urban progress through monuments like the South Meridian Street Bridge (Jay #6182).[59] Under the leadership of Mayor C. A. Paddock, Portland had been building sewer lines, paving its streets and sidewalks – including the working-class neighborhoods south of the Salamonia River – and electrically-lighting the city center. In addition, the Jay County commissioners planned to build a new court house.

An elegant bridge over the best-travelled route between the south side and the center city would convey Mayor Paddock's separate-but-equal message to the different classes of Portland's citizens. The responsibility for paving, lighting, and bridge-planning fell to Otto O. Clayton, the city civil engineer, who was determined not to replace the perfectly serviceable iron bowstring span then in service on South Meridian Street with another metal bridge. To stay in step with progress as defined by the larger metropolitan areas required either a cut stone or a concrete bridge.

Clayton probably decided on concrete for two reasons: concrete bridges usually were cheaper to build than stone structures, and

(Jay #6182)

Meridian Street Bridge over the Salamonie River in Portland.

he knew where to obtain the special design he wanted. Having observed flooding on the Salamonia for several years, Clayton concluded that the new bridge should not restrict the flow of water more than the existing structure had done, and so to resort to a spandrel arch would not do. To the rescue came E. A. Gast of Cincinnati, Ohio, who earlier designed and had constructed what Clayton now needed – a tied-arch or bowstring-girder in concrete. Gast described his design in the *ENR* two years before, and the editors praised it. Gast had also stumped the region around Cincinnati to promote his design.[60]

With the Clayton-Gast plans in hand, the members of the Portland City Council confronted the county commissioners in December 1913. The commissioners were prepared for the councilmen, and after much talk the former agreed to pay for "a better and neater bridge than they would [build] in the country," thereby implicitly agreeing to adopt the city's proposed plans. On one matter, however, they were adamant. They did not build sidewalks in the countryside, and they would not pay for them in Portland. If the city wanted sidewalks, it would have to pay for them.

While the commissioners generally sought construction bids on approved plans and specifications which the county engineer had prepared, an experienced Indiana engineer also knows that nothing is settled until a contract is let. Commissioners sometimes succumb to temptation and overrule their in-house professional expert when presented with lower offers made on alternate designs submitted at the bidding. To avoid this, Clayton brought in "an engineer from Cincinnati" – probably Gast – as an outside consultant. He was to keep the commissioners committed to the Clayton-Gast plan so they would not stray to some other concrete or steel design. The stitch may have been in time, for Daniel B. Luten of Indianapolis submitted the lowest bid for a concrete bridge of his own design – probably for a filled-spandrel arch. No metal bridge fabricators bothered to bid. In the end, the commissioners honored "the Clayton plan" by awarding the contract to the lowest bidder on it, Israel E. Smith of Richmond.[61]

The Meridian Street Bridge appears to be a copy of the Benson Street span over Mill Creek in Hamilton County near Cincinnati, and the procedures used for the earlier construction seem to have been followed in Portland. Out of anxiety for building what was for him a new form, I. E. Smith added, at his own expense, more steel reinforcement to the bridge than the plans and contract required.[62]

When completed with the extra steel, this copy of a nationally advertised bridge provided a remarkable monument for central Portland and a very special City-Beautiful structure for Indiana. The Portland Arch – "being one of the first bridges of this type and design" – was no imitation of something built in Indianapolis, South Bend, or Fort Wayne. Panelled, sidewalked, and electrically-lighted, it stood out as the only structure of its particular kind ever to be erected in the state.[63]

BUILDING THE PORTLAND ARCH

A segmental concrete rib rises above and on each side of the roadway of Portland's Meridian Street Bridge. If this were a spandrel arch, the ribs or ring would deliver a great deal of lateral pressure or thrust to the abutments which would in turn have to be built quite wide and heavy to resist significant movement and consequent flattening of the ribs. The Portland arch ribs are tied, however, and thus deliver largely vertical pressure to the substructure.

The tieing is all done by steel rods placed inside the concrete structure. Probably four rows of rod are spaced at intervals longitudinally within each rib. As these rib rods extend into the haunches and approach the substructure, they are hooked by steel rods which run through the bridge's floor from one rib end to the other. Thus arises the name, "Bowstring" girder. The rod in the arch ribs is like the hunter's bow. The rod in the flooring acts like the string attached from one end of the bow (rib) to another to hold it in an assigned arc. The bridge floor is suspended from the arch ribs by several lines of concrete-encased steel rod hooked around the upper row of rib rod and extended into the floor beams below. In all, the floor and hangers contain forty of the fifty-two tons of steel laid in the bridge.[64]

The building of Portland's arch reversed the usual order of construction where the ring or ribs are typically the first parts of the superstructure erected. In Portland as in Cincinnati, they were the last to be constructed. The builder started by fabricating a very heavy scaffolding or false-work which consumed much of the four railroad-car loads of lumber ultimately used for the bridge. Then the contractor laid the floor and rib haunches and next erected the hangers. It took nearly three weeks to construct the false-work and forty hours working day and night to mix and pour the concrete for the floor. Once laid and allowed to set for ten days, the concrete and steel in the floor, rib haunches, and hangers provided a rigid base for the completion of the arch ribs to specification. The finished floor and rib haunches secured the ribs' crucially tied areas against movement, and the set hangers provided rigid support for the arch-rib forms.[65]

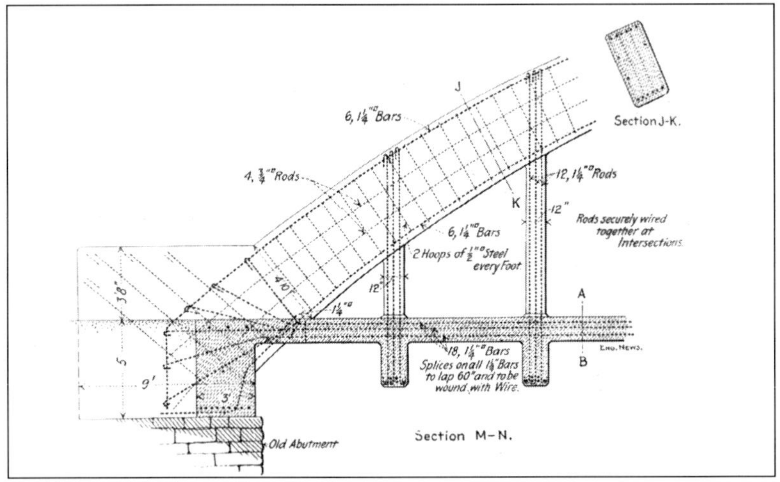

Right – Plans of the Benson Street Bridge, Hamilton County, Ohio, from which Portland's Meridian Street Bridge was probably copied. The construction photo for the Benson Street Bridge (right, bottom) shows that the deck was built before the arched ribs. The reinforcing for the rib extends from a timber frame in the upper center of the photo as well as clusters of rods sitting upright at intervals for the hangers between the deck and the rib. Note how the reinforcing of the arched ribs are tied into that of the deck. The plans and photo were published partly in the ENR *(1911) and partly in George Hool,* Reinforced Concrete Construction *(1917).*

I.6 SAVING THE WORLD FOR DEMOCRACY AND AMERICA FROM BEAUTIFICATION

The First World War soon brought closure to the City-Beautiful movement. At the end of 1915, Kessler resigned as Indianapolis' landscape architect. The Indianapolis park board's second great surge of activity had passed its peak by the time Kessler resigned, although he continued to advise the park board and the city planning commission until his death.[66]

By the close of 1915, South Bend had also spanned almost all its park-side rivers with concrete, and Charles Robinson's mark on Fort Wayne was about as prominent as it ever would be. The war concluded Otto Clayton's tour of engineering duty in Portland, too. In all, "Saving the World for Democracy" soon took precedence over making America beautiful.

NOTES

[1]There are many good studies of the World Columbian Exposition. For the most easily readable one, see Robert Muccigrasso, *Celebrating the New World* (Chicago, 1993). For a good photographic record, see Stanley Appelbaum, *The Chicago World's Fair of 1893* (New York, 1980). For the standard scholarly treatment, see David F. Burg, *Chicago's White City of 1893* (Lexington, Ky., 1976). The summary in the text owes much to all three.

[2]Logansport *Daily Journal* (13 June 1891): 3.

[3]"A Definition of Concrete," *Engineering News-Record* [*ENR*] (6 March 1924) 92: 387-88.

[4]The Romans are not known to have built a concrete bridge. They did, however, often use concrete as a core in the upper part of the substructure and the lower reaches of arch rings. The concrete cores were invariably enclosed within squared stone walls and a squared stone vault. Colin O'Connor, *Roman Bridges* (Cambridge, U. K., 1993), 57-58.

[5]Carl W. Condit, *American Building* (Chicago, 1968), 155-56.

[6]William P. Chamberlin, "The Cleft-Ridge Span: America's First Concrete Arch, *IA* 9: 29-44; Carl W. Condit, *American Building Art: the Nineteenth Century* (New York, 1960), 247-254, 340-341.

[7]O'Connor, *Roman Bridges*, 187-88.

[8a]See, for example, Frederick P. Spaulding, A. Lincoln Hyde, and Ernest F. Robinson, *Masonry Structures* (2nd ed., New York, 1926), 374-77; O'Connor, *Roman Bridges*, 163-86.

[8b]For topics discussed in this and immediately ensuing paragraphs, see Clifton J. Phillips, *Indiana In Transition: The Emergence of an Industrial Commonwealth, 1880-1920* (Indianapolis, 1968), 323, 363-71, 416-18, 567-74.

[9]Phillips, *Indiana in Transition*, 207-8.

[10]See, for example, Frederick P. Spaulding, A. Lincoln Hyde, and Ernest F. Robinson, *Masonry Structures* (2nd ed., New York, 1926), 374-77; O'Connor, *Roman Bridges*, 163-86.

[11]John C. Trautwine, *The Civil Engineer's Pocket-Book* (20th ed., Philadelphia, 1919), 614.

[12]The following discussion on the Monier and Melan systems owes much to "The Monier Method of Constructing Arches," *ENR* (23 May 1891) 25: 499-500; "Some Recent Examples of Monier Arch Construction," *ENR* (16 February 1893) 29: 148-49; Fritz von Emperger, "The Development and Recent Improvement of Concrete-Iron Highway Bridges," American Society of Civil Engineers [ASCE] *Transactions* (April 1894) 31: 438-88; Paul Christophe, "Le Breton Arme et ses Applications," *Annales de Traaux Publics de Belgique* (1899): 452-57; Edwin Thacher, "Concrete-Steel Bridge Construction," *ENR* (21 September 1899) 42: 179-85; Eyvind Lee Heidenreich, "Monier Constructions," *Journal of the Western Society of Engineers* [*JWSE*] (June, October 1900) 5: 208-37, 329-39; "The Monier Exhibition at the Paris Exposition of 1900," *Cement and Engineering News* (October 1900) 9: 54; Eyvind Lee Heidenreich, deposition (20 September 1916) for Luten v Marsh, Attorney General Cases S.II (State Historical Society of Iowa Library and Archives, DesMoines); A. Marston, transcript of testimony in Thacher v Polk County, Attorney General Cases S.II (State Historical Society of Iowa Library and Archives, DesMoines), 25-30, 36-38; "Melan Honored by His Colleagues," *ENR* (7 February 1924) 92: 242; Condit, *American Building*, 169, 173-75.

[13]In a letter to the editor, Fritz von Emperger, who represented Melan interests in the United States in the mid-1890s, addressed the issue of which word should come first for a bridge that combined two materials such as concrete and iron/steel. He emphatically wanted to reserve the second spot in a hyphenated label for the interior, load-bearing material. Thus Melan arches were known as "concrete-iron/steel" and not the other way around. But, given his system of labelling, von Emperger admitted to "a good deal of trouble" over what to call a stone-faced Melan arch which combined three materials. *ENR* (9 May 1895) 33: 306-7. For examples of the "concrete-steel" designation for Melan-design structures, see "Two Recent Melan Arch Bridges," and "Melan Arch Bridges over Fall Creek, Indianapolis, Ind.," *ENR* (10 November 1898, 11 April 1901) 40, 45: 290-91, 258-59.

[14]Christophe, "Le Breton Arme," *Annales de Traaux Publics de Belgique* (1899): 452-57; Heidenreich, deposition (20 September 1916) for Luten v Marsh, 7; von Emperger, "Concrete-Iron Highway Bridges," ASCE *Transactions* (April 1894) 31: 444.

[15]Von Emperger, "Concrete-Iron Highway Bridges," ASCE *Transactions* (April 1894) 31: 446-47.

[16]Von Emperger, "Concrete-Iron Highway Bridges," ASCE *Transactions* (April 1894) 31: 468; Thacher, "Concrete-Steel Bridge Construction," *ENR* (21 September 1899) 42: 181; Heidenreich, "Monier Constructions," *JWSE* (June 1900) 5: 225, 235.

[17]For Melan's U. S. patent of 12 September 1893, see #505,054; for von Emperger's of 1 June 1897, see #583,464. For a brief description and discussion of these patent developments, see Condit, *American Building*, 174-75, fig. 63.

[18]In practice, the concrete in many Melan and von Emperger arch rings has carried considerably more structural burden than their designers probably expected. Concrete often became an equal load-bearing partner with steel. Looking back, one can identify some elements of "overdesign" in these systems. Today's proposals for removing redundance in these earlier designs would likely include greater proportional reductions in steel than in concrete.

[19]The following summary concerning the City-Beautiful Movement and its roots owes much to William H. Wilson, *The City Beautiful Movement* (Baltimore, 1989).

[20]Charles M. Robinson, *The Improvement of Towns and Cities* (1901), 35.

[21]For many years Merritt underwrote a kindergarten which maintained playgrounds in Military Park, an area he had procured for park purposes from the state. *Encyclopedia of Biography of Indiana*, ed. George Read (2 vols., Chicago, 1899), II: 141-43; Jacob P. Dunn, *History of Greater Indianapolis* (2 vols., Chicago, 1910), II: 1198-99; "Twelfth Annual Report of the Board of Park Commissioners," *Indianapolis City Reports* (1906), 200.

[22]"Thomas Taggart," *Men of Progress: Indiana*, ed. William Cumbach and J. B. Maynard (Indianapolis, 1899), 372-76; "Mayor's Report", *Indianapolis City Reports* (1895), 9.

[23]"First Annual Report of the Board of Park Commissioners," *Indianapolis City Reports* (1895), 332; Phillips, *Indiana in Transition*, 251-58.

[24]As quoted in J. Clyde Power, "Third Annual Park Report," *Indianapolis City Reports* (1897), 245.

[25]"James Clyde Power," *Men of Progress: Indiana*, ed. Cumbach and Maynard, 324-25.

[26]"Second Annual Report of J. Clyde Power, Engineer and Superintendent," *Indianapolis City Reports* (1896), 256.

[27]J. Clyde Power, "Fourth Annual Park Report," *Indianapolis City Reports* (1898), 135.

[28]"Through a Bridge," *Indianapolis Journal* (12 May 1899): 8.

[29]"Contracts Not Awarded", "It Will Be of Stone," *Indianapolis Journal* (1, 31 August 1899): 8; Marion Co., "Commissioner's Record," 34: 13, 37; 36: 146.

[30]Jeffrey A. Hess and Robert M. Frame III, *Historic Highway Bridges in Wisconsin: Stone and Concrete Bridges* (2 vols., Wisconsin Department of Transportation, 1986), I: 286-93; Derrick Beckett, *Great Buildings of the World: Bridges* (London, 1969), 29; Martin Hayden, *The Book of Bridges* (New York, 1976), 33.

[31]"Fifth Annual Message of Thomas Taggart, Mayor," *Indianapolis City Reports* (1899); "Sixth annual Message of Thomas Taggart, Mayor", "Tenth Annual Report of the City Civil Engineer," *Indianapolis City Reports* (1900), 6, 58; "Report of the City Civil Engineer," *Indianapolis City Reports* (1901), 57.

[32]"Report of the City Civil Engineer," *Indianapolis City Reports* (1902), 53-54.

[33]"Report of the City Civil Engineer," *Indianapolis City Reports* (1902), 54-55.

[34]"Stone Bridge Contract," *Indianapolis Journal* (7 May 1902): 10.

[35]"New Bridge Not Stone," *Indianapolis Journal* (15 May 1902): 11; Marion Co., "Commissioners' Record," 35: 343.

[36]The Central States Bridge Company of Indianapolis, largely a builder of steel spans, won the contract for the Emrichsville Bridge with a bid of $141,293. "Indianapolis - Bridge" in Construction News, *ENR* (22 June, 6 July 1905) 53, 54: 237, 3; Marion Co., "Commissioners' Record," 37: 296, 409, 465, 524; 38: 17, 37, 119, 151, 206, 247, 280, 345, 398, 477; 39: 35; 40: 463; "Eleventh Annual Report of the Board of Park Commissioners," *Indianapolis City Reports* (1905), 158.

[37]J. Clyde Power, "Sixth/Seventh Annual Report of the Board of Park Commissioners," *Indianapolis City Reports* (1900), 161; (1901), 153.

[38]"Indianapolis - Bridge" in Construction News, *ENR* 56 (20 December 1906): 211; "Indianapolis - Bridge" in Construction News, *ENR* (14 March 1907) 57: 92; Marion Co., "Commissioners' Record," 38: 382; "Bridge Open for Traffic," *Indianapolis News* (21 October 1908): 1; "Department of Public Parks," *Indianapolis City Reports* (1908), 213-14, 218.

[39]Marion Co., "Commissioners' Record," 37: 411.

[40]"First Annual Message of Charles A. Bookwalter, Mayor," *Indianapolis City Reports* (1906), 9.

[41]"Genius of George Kessler, Who Planned Beautification of Indianapolis, Reflected in Kansas City's Move to Make a Park His Resting Place," *Indianapolis News* (31 March 1923): 21. On McGowan's role in mass transit, see Phillips, *Indiana in Transition*, 154, 253, 257.

[42]While Kessler's work is highlighted in the best study to date on the City Beautiful, his contributions in Indiana are ignored. On Kessler's national role, see Wilson, *City Beautiful*, especially chapters 5, 9, 12.

[43]Map of "Present and Proposed Park and Boulevard System of Indianapolis," *Indianapolis City Reports* (1908).

[44]"Thirteenth/Fourteenth Annual Report of the Board of Park Commissioners," *Indianapolis City Reports* (1907), 208; (1908), 212-13.

[45]"Capitol-Ave. Bridge Contract is Held Up," *Indianapolis News* (30 December 1908): 1, 14; Marion Co., "Commissioners' Record," 39: 555; 40: 27.

[46]Marion Co., "Commissioners' Record," 42: 179, 213, 284; 43: 52.

[47]"Report of the Department of Public Parks," *Indianapolis City Reports* (1911), 348-49.

[48]The commissioners stuck to their bottom line, paying out in all $9,998.56 for the bridge over Crooked Creek. Marion Co., "Commissioners' Record," 44: 19, 303; 45: 506; "Report of the Board of Park Commissioners," *Indianapolis City Reports* (1912), 268, 272.

[49]"Indianapolis - Bridge" in Construction News, *ENR* (4 November 1915) 74: 283; "Report of the City Civil Engineer," *Indianapolis City Reports* (1915), 101.

[50]Alonzo J. Hammond, "Report of the City Civil Engineer," *Annual Reports of the Heads of Departments of the City of South Bend, Indiana* (1905), 61-62; St. Joseph Co., "Commissioners' Record," 18: 171, 479, 260; "South Bend -Bridge" in Construction News, *ENR* (28 July, 1 September 1904) 52: 38, 102; "The Reinforced Concrete Bridge," *ENR* (21 June 1906) 55: 704.

[51]Alonzo J. Hammond, "Report of the City Civil Engineer," *Annual Reports of the Heads of Departments of the City of South Bend, Indiana* (1906), 67; (1907), 70; St. Joseph Co., "Commissioners' Record," 19: 262, 271; "South Bend - Bridges" in Construction News, *ENR* (24 May, 21, 28 June 1906) 55: 189, 231, 243; "South Bend - Bridge" in Construction News, *ENR* (21 March, 11 April 1907) 57: 105, 133.

[52]"Report of the City Civil Engineer," *Annual Reports of the Heads of Departments of the City of South Bend, Indiana* (1908), 95; (1913), 98.

[53]St. Joseph Co., "Commissioner's Record," 21: 109, 118, 475, 493; "South Bend - Bridge" in Construction News, *ENR* (21 May, 25 June 1914) 71: 309, 399.

[54]"Report of the City Civil Engineer," *Annual Reports of the Heads of Departments of the City of South Bend, Indiana* (1914), 106; (1915), 122.

[55]Robert D. Parker, "Fort Wayne with Might and Main," *Old Fort News* (1974) 37: #2.

[56]The author thanks Michael H. Wenning of American Consulting Engineers of Indianapolis for providing access to the Tennessee Avenue Bridge plans. See also Allen Co., "Commissioners' Record," 7: 90, 118-19, 220, 275; 8: 142.

[57]C. E. Drayer, "Concrete Bridges with Stone and Brick Facing," *ENR* (24 June 1915) 73: 1214-15.

[58]Allen Co., "Commissioners' Record," 7: 43-44, 80, 373; 8: 138, 171, 319, 387, 404, 434-35, 439, 464, 468, 493.

[59]The factual bases for this and the ensuing paragraphs can be found in the following city and county records: Portland *Commercial-Review* (13 November, 2 December 1913; 3 February, 2, 3 March, 1, 18 June 1914); "City Council Proceedings, 1912-1917," (Portland City Hall), 101-102, 109, 128, 131; Jay Co., "Commissioners' Record," V: 148, 172, 248-52; Jay Co., "Commissioners' Docket," 9: 117; Jay Co., "Record of Appropriations and Disbursements: Auditor's #3," 310-11, 314-15, 318-19.

[60]Gast acknowledged and Hool confirms that "several like" his Benson Street bridge "have been built in Europe." E. A. Gast, "A Through Reinforced-Concrete Arch Bridge," *ENR* (16 February 1911) 65: 196-97, 199; George A. Hool, *Reinforced Concrete Construction* (2nd ed., 3 vols., New York, 1917), III: 306, 370-72.

[61]The commissioners had the old iron bowstring moved and re-erected over Limberlost Creek, where it served the county for many additional years. Insisting on their full due, the commissioners were officially unhappy to learn that in the replacement process Portland officials had the wooden deck of the old bridge carried away for the city's use.

[62]Steve Macy of the Hamilton County Engineer's Office in Cincinnati, Ohio, graciously provided photocopies of the Benson Street Bridge plans for this comparison. (Plans for the Meridian Street Bridge appear to have been lost.) On the Benson Street Bridge, see also David A. Simmons, "Hamilton County Bridge Believed To Be First of its Kind in Ohio," *Ohio County Engineer* (November 1985): 21; "New Bridge Is All Complete," *Commercial-Review* (12 November 1914): 1.

[63]"New Bridge Is All Complete," *Commercial-Review* (12 November 1914): 1.

[64]Estimates of steel apportionment varied some over time. "Bending Steel for the Bridge," "Big Task Is Nearly Done," *Commercial-Review* (6 October 1914): 1; (13 October 1914): 5.

[65]"Forms Ready to Pour Concrete," "Big Task is Nearly Done," "Building Forms for the Arches," "South Bridge Is Completed," *Commercial-Review* (9 October 1914): 1; (13 October 1914): 1, 5; (23 October 1914): 1; (4 December 1914): 1.

[66]"George E. Kessler Dies at St Vincents", "Genius of George Kessler, Who Planned Beautification of Indianapolis, Reflected in Kansas City's Move to Make a Park His Resting Place," *Indianapolis News* (20 March 1923): 19; (31 March 1923): 21.

II

Design Prophet for a Practical Country: Daniel B. Luten

Let us have the city useful, the city practical, the city livable, the city sensible, the city anything but the city beautiful.
– Cass Gilbert, address to the American Institute of Architects (1909)[1]

Landscape architects and their allies capitalized on the opportunities which the World Columbian Exposition of 1893 offered. With the ideal of a City Beautiful, they captured much of the informed public's imagination for a decade or two. They dominated the practices of other specialists and professionals employed in a number of urban bureaus, commissions, and departments in Indiana. But architects gradually lost their superintending role as specialties and professions both developed and exploded through subdivision, each protecting its own rituals and honoring its own special expertise. Having established at best a beachhead in bridge design, architects soon retreated from what engineers claimed as a part of their professional turf.

Engineers have always been a practical lot, if not always a highly professional one. In 1882, the *Railroad Gazette*'s editors prescribed a training "for the occupation of engineer" that fit within the tradition that had long prevailed in North America, but which was seriously challenged by the late nineteenth century. The editors viewed engineers as closely akin to craftsmen or mechanics who needed some secondary education and apprenticeship. The editors believed that after mastering geometry, trigonometry and algebra in the era's equivalent of high school, the prospective bridge engineer "can learn how to design and build bridges very much better if he enters the employ of some firm in that line of business than he can in a technical school." "There is not a branch of civil or mechanical engineering now practiced in this country in which the men who have achieved either the greatest distinction or pecuniary success have had a liberal education either classical or technical."[2]

The *Gazette*'s editors placed themselves firmly in the engineering tradition inherited from the British and widely practiced in the United States after independence.[3] The British tradition emphasized an observational and experimental approach to design and construction that worked well with individual, group, or corporate entrepreneurs. Civil (in effect, "civilian") engineering blossomed in this context of apprenticeship training and empirical strategies for problem-solving (i.e., using what works without necessarily being able to explain how or why).

While not ending British design and construction practices, the very act of securing political independence introduced challenges to them. Indeed, French soldiers who fought alongside George

Washington's continental troops brought a different engineering tradition to North America – one that required a higher level of formal training in mathematics and science. Even as the British empirical tradition of engineering continued to flourish in North American private sectors, the French tradition took root in one of the new nation's few permanent military establishments, the U. S. Military Academy at West Point. French engineering – having risen within a tightly monarchical state that had a centralized tax-collection systems and other bureaucracies – responded to the needs of a permanent royal military establishment and provided trained personnel for government agencies. The handful of private engineering colleges established in the first half of the nineteenth century, in turn, borrowed heavily from West Point's Francophile curriculum. The new land-grant and research universities established in the second half of the century also modelled their engineering programs on West Point's partly imported traditions.

After the *Railroad Gazette*'s editors in 1882 advised the young to tread the old and hallowed steps to empirical engineering, the editors of *Engineering News-Record* [*ENR*] raised the banner for the professionalization of engineering following the French tradition. The *ENR* first published several letters from engineers who insisted on the importance of formal technical education. One civil engineer even concluded "from the tone" of the *Gazette* editorial that "it was not written by one in the active pursuit of the profession."[4] After publishing readers' salvos, the *ENR*'s editors weighed in on behalf of increasing professionalization. They argued that four years of study in a technical institute or college "developed the mind and character" in scientific and general thinking for a more flexible and wider range of problem solving. "The indications are, that technical education, as a qualification for technical pursuits, will grow every year more and more important, until finally it will become. . .indispensible."[5]

As the *ENR*'s editors had predicted, engineering would soon be professionalized. In 1850 the United States contained only "four small struggling schools of engineering," but by early in the twentieth century there were "more than fifty times that number, and many of these schools number their students by hundreds and even by thousands." "Training men for the engineering profession in college is no longer an experiment. The engineering graduate is a staple product."[6] Challenges to the landscape architects' control of bridge design, then, came increasingly from the bumper crops of graduates feeding the engineering profession.

II.1 AN EMPIRICAL AND ENTREPRENEURIAL PROFESSIONAL ENGINEER

An engineer is very properly defined as one who *best* adapts the forces of nature to human needs at the least expenditure of dollars.
– *Daniel B. Luten*[7]

Daniel B. Luten did more than any other single person to advance the movement from concrete-steel to reinforced concrete bridge design and to bring the latter to maturity in Indiana during the first decades of the twentieth century. Countless engineers throughout North America incorporated one or more of Luten's ideas into their projects, and across the country many contractors built Luten-designed spans. Dominant for some decades in Indiana, Luten's considerable influence reached towards the continent's corners from Maine to California and from Canada to Mexico. Probably a thousand out of approximately twelve thousand of his structures remain to bear witness to his once ubiquitous presence.

Daniel B. Luten early in his entreprenurial career.

Like many an engineer, Luten scrambled upward from a humble beginning. Born (1869) and raised on the family farm of Dutch immigrants near Grand Rapids, Michigan, the young man sought escape from manual labor through education. To

(*Debris*, 1899)

Daniel B. Luten at a Purdue University blackboard.

go to college, however, he "not merely had to pay his way, but also had to pay the wages of men hired to take his place" on the family farm.[8] Still, Luten felt privileged to be at the University of Michigan from 1890 to 1894 and to study engineering, an occupation which promised upward social and economic mobility.

With a bachelor of science degree in civil engineering, Luten served as Instructor of Civil Engineering and Surveying at Michigan for the 1894-1895 academic year. He then moved to Purdue University as Instructor of Architectural and Sanitary Engineering, where among other things he taught arch design. He also married Edith Hull, daughter of an established and prosperous Lafayette, Indiana, family.[9]

Others in Luten's position might have settled comfortably into an academic career. But neither teaching nor scientific research in engineering excited Luten.[10] He found engineering professors too caught up in institutional politics and too preoccupied with scientific theory unchecked by practical experience. In their search for scientific principles, academic engineers became so obsessed with abstracting structure and trying to solve specific problems through general mathematical equations that they did not check their generalized abstractions against physical reality often and fully enough to discern even the modifications to theory that experience would suggest. As a consequence, academic engineers "are almost always years behind the practical men of the profession." "College professors," Luten concluded, "are not leaders in engineering."[11]

Luten strode between the two engineering traditions inherited from Europe. Trained in the French tradition of mathematically-expressed mechanical theory, Luten tilted toward the British and the empirical in his practice. He would have endorsed the 1882 *ENR* editors' argument for college training and professionalization of engineering over the *Railroad Gazette*'s insistence that common school education was enough. But he wanted the trained professional's mettle tested in the marketplace, where utility outweighed scientifically certified correctness.

Luten's affinity for the empirical resonated with his commitment to the independent owner-engineer operating outside a large or bureaucratic organization. Both the academic engineer and his government-employed students lacked the necessary incentives to contribute their utmost, Luten believed. Without "the usual spur of competition," the only incentive for academic engineers to invent came from celebratory status achieved by publicizing improvements through technical journals and societies. At most, this could help one "stand high in the profession," a decidedly modest and largely inconsequential achievement in Luten's eyes when compared with the material good one might do for self and society.

Typically on a fixed salary and thus underpaid as well as underappreciated, the government-employed civil engineer had even less incentive than the academic one to invent, for he was "rarely forced into competition except in securing employment." Worse yet, by being freed "entirely from competitive incentives," government engineers would too often not perform at their best. Indeed, they might well have acquiesced in poor proposals or themselves carelessly suggested inefficiencies.[12] Luten would intertwine professional engineering with commerce. The design of concrete bridges, he argued, was "largely a commercial proposition" which "makes it in its highest sense an engineering proposition. For any mechanic can build a bridge; but it requires an *engineer* to build a bridge that is best suited to its requirements and *at the lowest cost*."[13]

II.2 THEORY AT WORK

The lineage of Luten's design does not flow directly back through the von Emperger-Melan line of concrete-steel to which the City Beautiful advocates largely owed allegiance: "I consider them simply steel structures encased in concrete, and not reinforced concrete in its proper sense, which is that the steel is used to reinforce the concrete and better enable it to sustain stress."[14]

If Luten had acknowledged a conceptual lineage, it probably would have been to the same source as Joseph Melan's – the Moniers. But, unlike the Melan-system developers who backed away from some of the implications of Monier, Luten extended the Moniers' ideas. He took the message that the leading Monierist prophet, E. L. Heidenreich, had preached unsuccessfully in the 1890s to professional engineers in the United States and transformed it into practical and eminently salable design after 1900.

II.2A A Forward Lateral: From Melan to Von Emperger to Thacher

By Luten's day, the Melan system interests in the United States had come to rely increasingly on a third designer – Edwin Thacher – who, like Fritz von Emperger, modified the system he had inherited. Thacher brought a wealth of knowledge and experience in bridge engineering to concrete-steel design.[15] He held a degree in civil engineering from the Rensselaer Polytechnic Institute of Troy, New York, had practiced as a "calculating engineer" for a major bridge company, originated the multiscale cylindrical slide-rule,

(ENR, 1909)

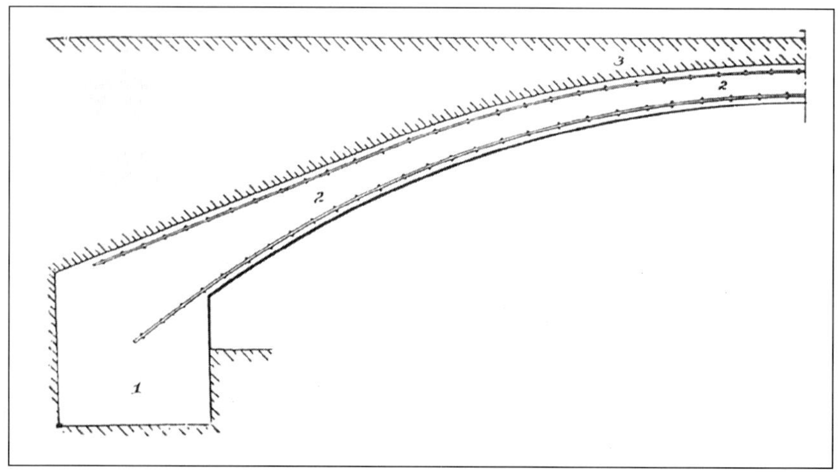

Thacher's patented ring reinforcing.

and invented a metal truss which carried his name. The currency and thoroughness of Thacher's knowledge of the stresses and strains in spans and his ability to calculate them seemed assured.

Thacher's promotion of the Melan system in the West in the mid-1890s must have seemed like a natural extension of his interests in iron trussing; his growing experience with the system inevitably played into the arch design he started through the patent process in 1896. That he and a partner in 1901 formed "the Concrete-Steel Engineering Company to develop jointly the Melan, von Emperger and Thacher methods" underscores their design compatibility.[16]

The patented Thacher system consisted of two tiers or lines of longitudinally-placed "metal bars," one line situated above the lower edge of the arch ring and the other under the upper edge. By being placed above one another, bars from the two tiers were paired without being connected with one another. As such the tiers operated "independent of the other."[17]

Although Thacher, like Melan and von Emperger before him, still relied on steel to address most of the stresses confronting the arch ring, he made his inherited system more flexible. For example, Thacher could easily move both of his tiers of reinforcing close to the arch ring's edges. Melan, in contrast, could place his single set of bent I-beams close to only one edge of the arch ring. He favored the lower area, so the top flange of his "I" approached the upper edge of the ring only at the crown and then grew increasingly distant from it towards the springing. By fabricating his arched girders, von Emperger could adjust the depth of his webbing over the course of a rib to allow each flange to parallel the top as well as the bottom edge of the ring more closely, thereby addressing somewhat more effectively the stress that was present near the haunches. Still, as Thacher noted, "it is not practicable" in long spans to have the arched ribs "follow throughout the lines of greatest strength, as the beams become too deep and unwieldy."[18]

Thacher bars or rods also used less steel than Melan's mill-rolled I-beams or von Emperger's ribs made from structural stock. In place of

THACHER PRACTICE AND FUTURE REINFORCING PATTERNS

In practice Thacher reinforcing came closer to von Emperger's system than the Thacher patent taken alone suggests, for Thacher frequently used radial rods or stirrups to connect bars paired from the tiers of his longitudinal reinforcing. Finding that, as practiced, the Thacher and the von Emperger systems were alike, a federal judge in 1916 quoted Thacher directly against himself: "The stiff ribs built according to the von Emperger system are much superior to any bar system, not excepting even my own." From this the judge concluded:

> These appeals by Thacher for use of the Melan and von Emperger types in preference to his own, and the failure to use the strictly independent bar system in any construction, so far as the record discloses, convinces me that Thacher himself recognizes that cross-bars and ties and lattice bars used with longitudinal bars, have a function and a purpose in concrete construction; in fact it is quite apparent from the record in this case, that limited strictly to the bars independent of each other. . , the Thacher invention is not, and has not been, of practical use.[19]

Once the era of concrete bridges with embedded steel structural frames had passed, most bridge designers followed Edwin Thacher's lead and used a two-tier system of reinforcement, one row of rods running longitudinally close to the bottom and another close to the top of the arch ring. In most cases, double-row reinforcers followed Thacher's practice more than his patent. They used a third set of rods as stirrups.

standard structural steel stock, Thacher patented a far more efficient deformed bar specifically for use with concrete. Omitting radial connections or stirrups between the tiers of longitudinal bars saved still more metal.

In addition, the Thacher system took a step towards shifting more of the ring's thrust from the horizontal to the vertical by carrying the bars "well into the abutment." This allowed for a more compact substructure than the previous versions of the Melan system.[20]

II.2B Monier with Elastic Theory

When Luten discussed his engineering roots, he spoke as a professional about his education, not about Melan, Monier, or other designers of concrete-bridge systems. Luten had spent an extra year at Michigan beyond his bachelor's degree to work under Charles E. Greene's tutelage, one of several professors who taught Luten how to think like an engineer. Nationally known for his use of elastic theory for arch analysis, Professor Greene provided his bright and eager student with a conceptually advanced apparatus and

ELASTIC THEORY: STRAINING WITHIN LIMITS

A seventeenth-century contemporary of Isaac Newton and a contributor to his identification of the law of gravity, Robert Hooke had articulated a natural law about elasticity—the ability of a material to recover its original size and shape after force has been applied to it.[21] Hooke had seen that weights hung from an iron wire caused the wire to stretch—its particles pulled further apart. As he studied the effect of force or stress upon a material like an iron wire, he discerned that the amount of stretch or strain depended both upon the amount of weight used and the area of the wire. Force, though, is not always applied to a material in a way that causes it to stretch. Sometimes force compacts or compresses the material—pushes its particles closer together—much as a press hanger can hold a skirt or pair of trousers by squeezing the cloth together. Hooke's law posited a linear relationship between stress and strain in a given material under load.

The ease with which a body can be deformed varies from material to material: rubber, for example, stretches more than does plastic. But the elasticity of a material—the ability of particles either to separate or to be pushed together and then for the material to return to its original size and shape—has its limits, too. A load can become so great that the material passes its so-called yield point (its elastic limit); its particles no longer return to their original state even with the load removed. The rubber band becomes permanently stretched; the steel paper clip fails to spring back. At even higher levels of stress the material fails completely. The rubber band or paper clip snaps; the concrete column shatters during an earthquake.

Before the middle of the nineteenth century, some bridge designers began to take account of the elasticity of the materials they used and the forms in which they used them. Elasticity found its earliest and widest acceptance among the designers of timber and metal trusses and arch ribs. Here the strains of the structure's own weight and the vehicles passing over it could be assessed in a focused manner. Consequently, bridge members could be designed specifically within safe limits (short of failure) for the resulting stretching out (tension) or pushing together (compression) of the particles in timber, wrought iron, or steel components.

Blocks of wedge-shaped stone provided the materials for masonry arches built in the United States before the 1890s, and their designers continued routinely to calculate the stability of cut-stone arches according to line-of-thrust rather than elastic theory. Designers altered the arc of the ring to keep the center of compression within the middle third of an arch ring's stone blocks, making the assumption that from the center third the work of carrying the load would most likely be shared reasonably equitably with both outer thirds.

Line-of-thrust theory does not provide results as accurate as those derived from elastic theory. Still, it allowed for some simply-derived calculations that generally sufficed given the reliance on a narrow range of material (specified kinds and grades of stone) that was cut deeply enough to allow a wide area for load bearing and then placed within a limited range of arcs.

vocabulary for thinking about bridges. Luten acknowledged a primary debt to Greene and his work on *Trusses and Arches* (1879) and a supplementary one to Professor Malverd A. Howe's *Treatise on Arches* (1897).[22] Howe had also considered elastic theory "the most rational of all for the design of masonry arches, and particularly arches composed of reinforced concrete," despite the difficulties involved in applying it.[23]

Luten was not the first engineer to use elastic theory to critique the concrete-bridge reinforcing systems which the particular advocates of Monier and Melan proposed. When von Emperger argued for the Melan system at a meeting of the prestigious American Society of Civil Engineers [ASCE] in New York in 1894, he described Monier design with its hybrid offspring unfavorably.[24] Most of the sixteen illustrious engineers selected to comment on von Emperger's paper probed the presenter's endorsement of the Melan system in a spirit of cautious support. Only one compared the Melan system unfavorably with Monierism.

ON STRESS IN SIMPLE, TIED, AND ARCHED CONCRETE BEAMS

The vertical pressure exerted upon a beam by its own weight and of any load placed upon or passing over it causes stress within the beam. The stress in a horizontal beam resting on supports at each of its ends is neither of the same character everywhere nor does it affect all parts in the same way.

The regions of a beam most distant from a support are subject to the greatest amount of stress. Midway between the supports, the particles of the upper areas of a horizontal beam get pushed together or compressed the most. At the bottom of the midsection, the particles of the beam are pulled apart or subjected to the greatest amount of tension.

Since all materials are elastic to some extent, they can have their particles compressed or stretched to a degree before they deform or rupture. To break a stick, for example, a person may hold the ends of the stick and push her foot against the middle. The particles of the stick directly under the foot will be increasingly compressed and the ones on the side away from the foot will be stretched. At first the stick will bend. As more pressure is added, it will crack open on the underside where its fibers, under the greatest amount of tension, pull apart first.

The force that vertical pressure causes can lead to another kind of rupture – shear. The stretching out of a material's particles increases the tendency of one particle to slide on and over another. In this case, the forces are also distributed over the cross-section of the material. But they are parallel to the plane of the cross-section instead of being perpendicular to it as in tensile or compressive stress. Faced with force beyond its elastic limits, a beam supported at its ends may rip apart horizontally, vertically, or in some combination of the two. A beam so situated and circumstanced is likely to shear within the third closest to its supported areas.

The ability of different materials to handle particular kinds and amounts of induced stress varies considerably. Concrete, for example, can sustain a great deal of compression without buckling or crumbling. But concrete has much less ability to withstand tension. Steel, on the other hand, can sustain a great deal of tension or shear without tearing apart. The two materials properly combined produce a powerful and efficient load-bearing machine – the cheap concrete carrying compressive stresses and the more expensive steel positioned to carry tension and

(continued on next page)

shearing stresses. In the simplest reinforced concrete beam, steel rods run longitudinally inside the concrete near the bottom of the beam where tension is greatest. To counter shear, the designer might place additional rods – sometimes diagonally, in recognition of shear's horizontal as well as its vertical dimensions – between the beam's top and bottom in that third of the beam closest to a support.

If the ends of a horizontal beam are tied down to its supports and cannot bounce or deflect when loads cross the structure, the regions of greatest tension necessarily change. When compared with the untethered ends of the simple beam, the high-tension zone along the bottom of the tied beam moves further away from the supports towards midspan, and new tension zones appear towards the top, close to the supports. For the tied beam, the steel, longitudinal reinforcing, which in the simple beam was concentrated along the bottom, now needs to be placed towards the bottom near midspan and towards the top near the supports.

If one changes the horizontal tied beam into an arced beam, it becomes an arch. For a shallow arch, the zones of tension are about the same as they are in a horizontal, tied beam but with reduced intensity. If one increases the arch's "hump" or curvature, the bending stresses decrease almost to nothing, and the vertical pressure is expressed as pure compression. Since almost all materials can handle more compression than tension, forms that translate a higher proportion of their load into compression can withstand a larger amount of stress which, in turn, permits longer spans. Thus, arches can span more space than can beams.

Beams do have some advantages over arches. They generally require simpler and smaller supports, for example. While a beam delivers force vertically to its supports, even the most rounded arch – a semicircular one – delivers some of its thrust to its supports laterally, tending to push piers sideways and thus to apply extra force to abutments. To address the lateral thrust typically requires a more complex and substantial substructure. Beams are also easier to construct than arches.

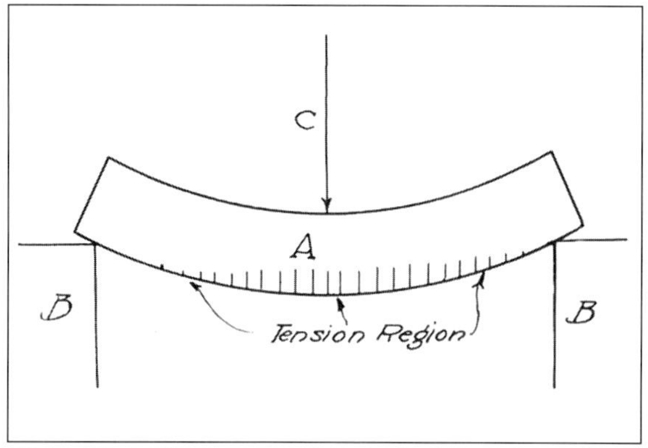

The region of greatest tension which weight or vertical pressure (C) produces in a simple beam (A) resting upon supports (B). Note how the ends of the beam respond.

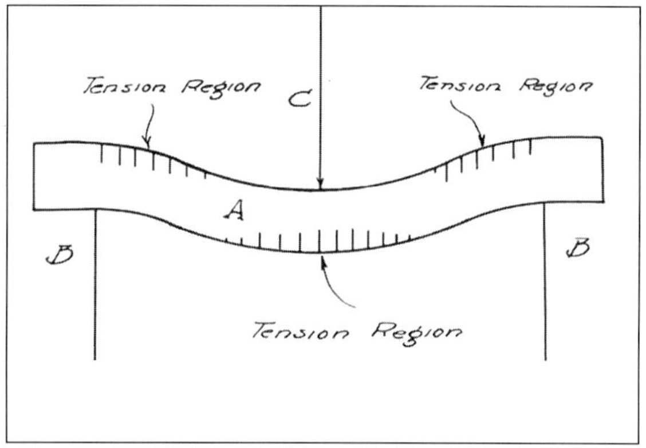

When the ends of the beam (A) are tethered to the supports (B), the regions of greatest tension differ in size and place from those which weight or vertical pressure (C) produce in the simple beam.

Zones of tension in a concrete arch.

"As to the so-called Monier system," W. R. Hutton opened, "the principle is correct to build iron bars or straps into the concrete at points where tensile stresses come, with the condition that the metal shall be used in such a form and manner that its value can be ascertained." In contrast, "the Melan system appears to be an entirely different thing. In it no attempt is made to supply directly with metal the deficiency in tensile strength of concrete."[25]

If proper attention were paid to the elasticity of each material, Hutton argued, concrete and metal combinations could result in structures "much more slender than could be adopted with either material used alone." While Hutton was not prepared to commit the "labor" required for the "theoretical investigation" of particular cases, he "concluded that in the hands of one familiar with the theory of the elastic arch, and thoroughly acquainted with the materials used, some economies may be realized – economies which in unskillful hands may be too dearly bought."

While Luten recognized a role for the Moniers at the genesis "of the art in reinforced concrete bridges," he admitted that he "had not … made any study of the system at the time that I entered the field." To Luten, the Monier system was "reinforced concrete so-called." Since the Monierists used iron/steel with the intent of helping concrete sustain stress, their combination – unlike the von Emperger-Melan system derived from it – was "reinforced concrete." But Luten qualified his endorsement of Monierism with "so-called" for several reasons. First, the Monierists used mortar – cement and sand – rather than concrete – mortar plus gravel or stone. Second, the Monierists did not seem to place metal in bridges with theoretical consistency. They "at first and principally" embedded "reinforcing mesh along the center line of the concrete structure, on the theory that the steel would carry the compression and the concrete prevent it from buckling."[26] Later, they placed their metal networks in arch rings where tension was prevalent but extended the nets into zones of compression, too.

Luten used the kind of theoretical analysis that W. R. Hutton had called for in 1894 to bring a rigor to design that Monierism lacked. By using the elastic theory that he had learned from Professor Greene and found enhanced by Professor Howe, Luten became the spiritual heir of Monierism and transformed it into a system where reinforcing provided "a maximum of tension reinforcement and a minimum of compression reinforcement" and concrete served "as the main structural member resisting compressive forces."[27] The reinforcing in a beam needed to address the tension generated by the weight of the structure itself as well as of loads moving over it. An arch, on the other hand, should require metal reinforcing only for the tension that moving loads produced.

Luten never claimed to be the first to place "metal in concrete to resist tension." That idea, he acknowledged, was "old" by 1900.[28] But he did lay claim, as an inventive engineer for whom economic efficiency ranked almost as high as safety, to a rigorous application of the ideas at the core of elastic theory.

II.2C Theory Applied in Context: Qualifications for "A Practical and Successful Method of Reinforcing Concrete Structures"

Good theory, Luten insisted, generated intuitive insights which needed to be woven into the details of design and through the processes of construction. In short, design had to be "economic and efficient," reducing cost, increasing strength and durability, and improving appearance.[29] "Otherwise it will find but an indifferent market."

The principle that steel and concrete "form a composite whole and act in harmony" provides keys to economic and efficient design. Luten had not matured in the artisan's craft of cut-stone arches nor worked as a metal bridge designer so he found it easier than Thacher, for example, to avoid practices "laid down in text-books for stone arches or steel bridges" and to start afresh with a new material.[30] Formulas for a layered structure of concrete and steel did not provide a truly reinforced concrete product. Ideally, the two ingredients would meld into a new material distinctively designed.

Harmony between the two ingredients in the composite material included the proper distribution of metal in the concrete to avoid "excessive expansion or contraction of the steel" from temperature changes. Since metal conducts heat more quickly than concrete, Luten criticized the use of "steel in heavy masses," as in von Emperger-Melan ribs, which tended "to split the concrete" along the lower edge of the arch ring.

After harmony came simplicity of design so the reinforcing system could be easily analyzed, plainly represented on drawings, readily interpreted "by practical men who build such structures," quickly inspected by an engineer supervising construction, and easily explained to laymen, "for engineering structures such as bridges are often sold to laymen having no engineering advice."

Facility with the purchase, transportation, shaping, and placing of materials generated economies, too. Reinforcement should be designed from commercially available sizes and shapes of steel, purchased and transported in carload lots, shaped with hand-tools at the bridge site, and placed into position easily by hand. These Luten precepts contrasted with Melan-system practice where ribs had to be framed in a metal fabricating shop, specially shipped in sections to the bridge-site to be bolted or riveted together, and then raised into position by a derrick. On the other hand, Melan ribs were easier than light bars or rods to hold in position while concreting. They were not easier than bars or rods, however, to encase in concrete, for getting concrete fully under the Melan ribs' heavy lower flanges often proved difficult.

II.2D Design and Analysis: The Luten Arch

Few issues illustrate Luten's empirical bent better than the way in which he addressed the inextricably linked functions of design and analysis. While physical scientists study a reality outside humanity, Luten as engineer made his own structural form. The scientist takes the forms of the natural world as givens; the engineer creates his own. Instead of the scientist's search for a general theory buttressed with mathematical analysis to explain the given natural world, the engineer designs and builds his own structure and then uses it. Theory helped Luten think about the form he would recommend for a specific setting, but he invested no time in the search for some general form in nature. Indeed, he thought theory should be the servant of the designer, not his master. Analysis must supplement, not hinder the design function.[31]

Luten understood the process of design and analysis as an iterative one between a designer who proposed the size and shape of the members as well as their reinforcement system and a mathematician who computed the strength of the parts and of the whole. If after analysis the mathematician found any element deficient, the designer needed to redesign the entire system and then submit the revised design for reanalysis – "and this process must be repeated until the design approximates perfection. The design of any engineering structure is thus a cut-and-try process."[32]

To Luten, the inventive took priority over the double-checking. "Since engineering structures must submit to engineering analysis, it has sometimes been concluded that invention in engineering structures was nothing but engineering analysis or mathematics. In reality the analysis must follow invention as one of the essentials of a successful structural improvement." "Mathematics and engineering analysis of structures has always followed invention." In his search for maximum efficiency – that is, to get the design process satisfactorily completed with the least amount of the engineer's labor – sensible short-cuts needed to be found in both design and analysis, but especially in the latter.

For his filled-spandrel arches – those with walls which rose above the ring at each edge and contained an earth-fill over which the roadway ran – Luten favored the "linear"-shaped or semi-elliptical ring with "thickness. . .proportional to thrust. In such an arch the stress would be uniform, every particle of material being subjected to compression of the same intensity, with consequent economy of material." After laboriously calculating each of the eight formulas used for formal elastic-theory analysis on many examples of his preferred linear arch form, Luten began to collapse the analytical part of the process into a few empirically obtained, rule-of-thumb formulas. He thought his formulas provided in a few hours results within five to ten percent accuracy of those he would have

MONIER AND LUTEN ARCH-RING REINFORCING

Monier-system designers used nets composed of two sets of round rods, the largest in diameter (.39 in.) ran longitudinally from abutment to abutment and were placed a few inches apart; a smaller set (.28 in. in diameter) ran transversely just above the longitudinal rods to which they were wired.[40] In their very earliest arches, the Monierists placed a net midway between the lower and upper edges of the ring. Soon, however, they ran the net close to the lower ring edge. As they experimented with longer spans by 1890, the Monierists added netting towards the top of the ring, sometimes over the haunches only and sometimes across the whole top.

Luten's single line of longitudinal rods shifted in thirds from the bottom of the ring at midspan to the top areas towards the springing. The first shifted at one-third of the distance from midspan to each springing, the second at half way between the two, and the third at two-thirds of the way towards the supports. The system of longitudinal displacement served specified purposes. First, a full set of rods addressed tensile stress where it was generally greatest—towards the ring's bottom over the crown and towards the top over the haunches. Second, the areas of significant tensile stress can shift some in a ring as vehicular loads move over the span. To accommodate shifts in the boundaries of stress, the rods pealed off from one region to the other in three waves, each time transferring a third of the tensile resistance from one area to the other. Third, as displaced, the rods travelled across the ring diagonally. In so doing, they coincidentally helped to address shearing stress.[41]

Luten also ran one or more straps transversely under the longitudinal rods at the crown to prevent the rods from "whipping out" of the concrete.[42] When under tension, the rods tended to straighten and to pull through the concrete on the lower edge of the ring. Looped at each end around a longitudinal rod in the spandrel, the strap or continuous stirrup was accordioned across the ring, passing under the longitudinal rods and weaving up into the body of the ring's concrete between the tension members.

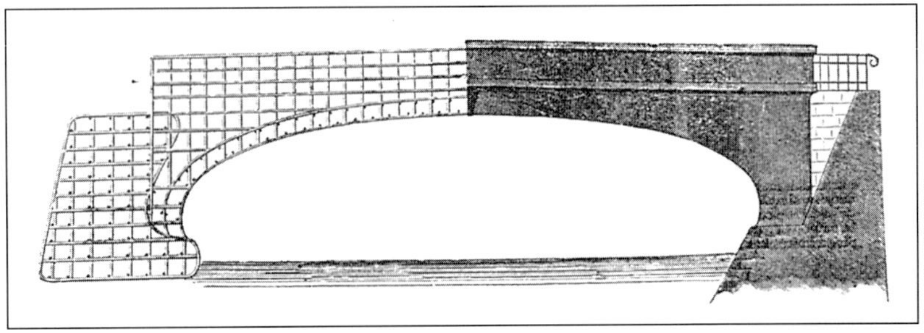

(Cement & EN,1900)

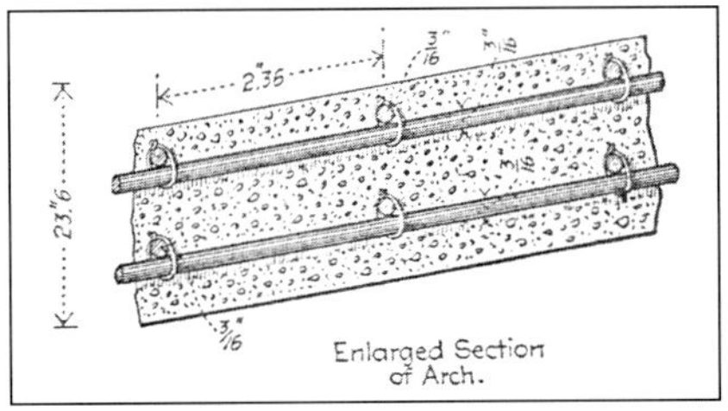

(ASCE, Transactions, 1894).

Above – Drawing of Monier arch exhibited at the Paris Exposition. Left – Von Emperger reported that German Monierists typically used an upper and a lower net in arch rings by 1890.

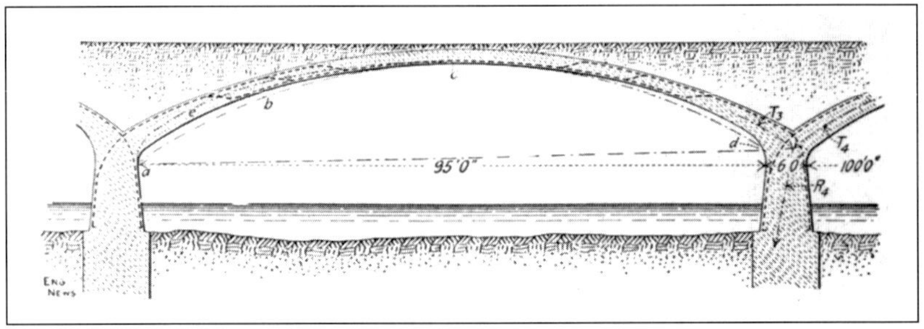

Drawing of Luten's reinforcing for one span of the Wayne Street Bridge, Peru. All the longitudinal rods are placed towards the lower edge of the ring at the crown (c). They shift by thirds towards the upper edge in the quarter section (b) of the ring, until all run along the upper edge over the haunch (e).

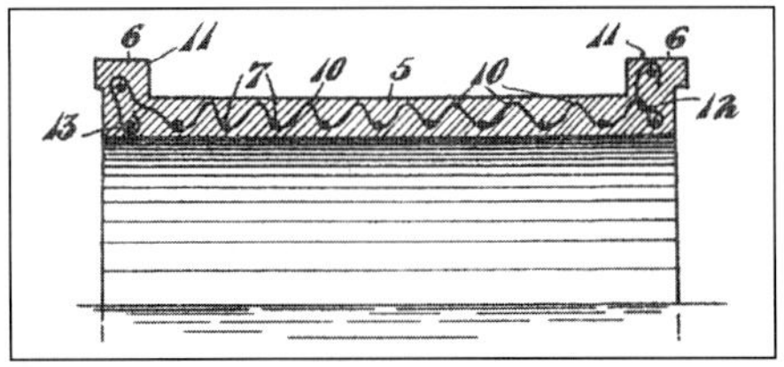

Drawing of Luten's patented transverse reinforcing system at the crown. The transverse strap helped to integrate the spandrel wall with the ring (6).

of its height. Luten believed he "was the first to point out any other function" for the spandrels than their "retaining wall feature."[43]

Luten assigned the spandrel walls two additional roles. First, he used them to stiffen the arch ring and to strengthen it against distortion or buckling. The failure of most arches results from the settlement of one part of the ring, causing a corresponding rise in another section of the ring, and finally a rupture arising out of the new, unanticipated arrangement of stress. By helping the ring hold its originally assigned shape, the walls contribute to the structure's continued integrity. Second, Luten designed the spandrel walls in a way which reduced the load on the arch ring. The walls themselves typically added considerable weight to the ring.

To achieve the two additional functions for the spandrel walls, Luten used both concrete and steel in different ways. Since the ring arced earthward from its crown to its springing, the spandrel walls, which carry the fill for the roadway above it, necessarily grow deeper toward span end. When Luten extended the spandrel walls not just over the abutments but behind them, he added considerable weight at the ends – enough to use the principle of the lever with the abutments as the fulcrum. The extension cantilevered each half of the spandrel wall from midspan to its end and consequently reduced the load it would otherwise add to the ring.[44]

Luten used two pairs of steel rod to help accomplish the new objectives. Both sets of rods addressed tension. The first set united the ring at the crown with the wall: it arced from the lower area of the ring at the crown to the top of the spandrel at its end and then extended downward to an anchor in the abutment. The second set of rods worked with the tensile stress that the cantilevering of the wall-halves generated: it followed the top of the wall to span-end and then reached downward to the same area of anchorage in the abutment as the first set.

The harvest from the integration of the spandrel walls with the ring proved bountiful. At a secondary level, it allowed Luten to decrease the size of the walls at their base and it reduced the tendency of the spandrels and ring to separate from one another over time. At a more fundamental level, the stiffening of the ring – "upwards of 100%" in exceptional cases – allowed Luten to reduce the ring's depth. He attributed "a large part" of his ultimate success "in introducing concrete bridges. . .to a recognition of the stiffening principle of the spandrel."[45]

Piers and abutments also became for the first time extensions of the rings, forming "a continuous elastic body," in Luten design.[46] He tied the reinforcing of the rings together with that of the piers and abutments, a strategy which reduced the burden on the rings and at the same time allowed for the use of thinner substructures.

Luten's extensive use of "unusually thin piers" – where structural leanness in the substructure showed – generated a lot of comment and some national recognition. A leading authority on concrete construction acknowledged his cleverness: "thin piers not only save concrete in most cases but increase the area of waterway, and on this account, it is quite likely that elastic piers will be much used in the future where architectural appearance is satisfied with this type of construction."[47]

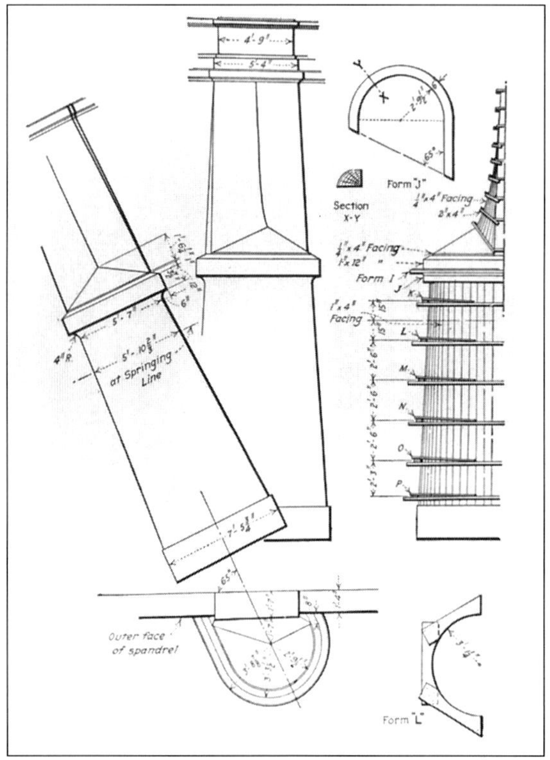

Left – Typical pier forms used in bridges of Luten design.

II.3 TO BECOME A "DESIGNING AND CONSULTING ENGINEER"

When Daniel Luten spurned the academy in 1900, he wanted to become a "designing and consulting engineer" who tested his ideas in the marketplace, not a bridge builder or contractor. He wanted to achieve with reinforced concrete what Edwin Thacher had managed with concrete-steel – establish a firm to sell a distinctive form of design. While Thacher took over the von Emperger-Melan interests in the United States, Luten struck out on his own. As Luten soon learned, though, setting up shop was more easily imagined than achieved. He already possessed a capacity for invention disciplined by professional training in engineering. To sell one's design profitably required far more. He knew at the outset that design needed to be capitalized. It took a little time, though, for Luten to realize that he also needed to develop a strategy for promoting his design to both producers and consumers.

II.3A Capitalizing Design through the Patent System

Luten selected a quite traditional way to capitalize his intellectual property: he patented a great deal of it. A patent provides the originator of an idea with control over who uses it and at what price. Inventors of timber and metal trusses had secured dozens of patents on their designs throughout the nineteenth century, and the concrete-steel bridge interests had followed the same course in the 1890s.

In his last year as an instructor at Purdue University, Luten crossed an unspoken line between an academic and an entrepreneurial engineer when he applied for and secured his first patent. Nineteenth-century engineering professors typically protected their intellectual property through copyrights and by being credited for one's ideas, not in patenting inventions. Instead of being rewarded by salary for one's books and articles in the university and through professional awards and prizes, Luten selected a far more materially competitive outlet for his talents with a much riskier form of return, the royalty.

Patents, Luten proclaimed, produced public and professional progress. "Engineers who hope for any material progress of the profession of civil engineering as a whole," he concluded, "will do well to claim the rewards of patent law." Because patents publicized new ideas and encouraged their test in the marketplace, they led to the building of better bridges at less expense, both advancing the general welfare and reaping a profit for the patentee. Royalties on cost-saving devices also provided the inventor with additional capital to invest in experiment with other new ideas. Indeed, even the possibility of royalties stimulated those who had the intelligence, education, and imagination needed to generate better designs.

Compared with all of Luten's rhetoric about progress and profit through inventive risk and his pedigree in cutting-edge structural theory, his initial patent seems deceptively modest. It concerned an arch tied from abutment to abutment with underwater timbers where decay would be reduced.[48] By fixing the position of the substructure, ties allowed for the abutments to be reduced without risking the ring's spread, flattening, and eventual collapse. The ties also limited the undercutting of the abutments' faces by the stream. In some ways, the patented timber-tie design was quite daring: Luten placed no steel reinforcing in his preferred linear-shaped ring. As if to underscore his life-long conviction that too much steel was typically put in a concrete arch, Luten promoted his no-steel timber-tie arches for railroad use where fixed and moving loads were greatest.[49]

To advertise his patent, Luten issued a catalog in June 1900 for the "Timber-Tie Concrete Arch Company" and gave it "wide distribution throughout the State of Indiana, expecting to receive mail orders in reply."[50] While waiting for orders to flow in during the summer, Luten sought and secured a contract from local county commissioners to build one of his patented structures near Lafayette. With a company in name only, the designer doubled as "a concrete laborer" and did "all of the carpenter work in erection of forms."[51] The academic job which Luten left would not likely have built the callouses or muscles needed for the sustained manual labor involved for even this little job. Furthermore, however good a fifty dollar profit on his first four hundred dollar contract may have seemed, it could not support a family for long.

Good news did not come in the mail either. He "never received a single inquiry," in response to his initial catalog, "not even a request for further information."[52] Luten learned from this resounding silence. The message was clear enough: a new design, like other inventions, may go unnoticed until innovated, i.e., integrated with production and marketed. Innovation in bridge-building generally occurs when contractors bid to construct a particular design, either because they choose or are required to do so.

Luten tried several new initiatives in 1901. He continued to develop his design, informally organized and operated the National Bridge Company with money advanced by his wife and her sister, and raised his visibility with letters and articles in the engineering press.[53] In December of 1900, Luten had moved his "headquarters" to Los Angeles, California, where he designed and oversaw the construction of three concrete bridges in the next three months. Back in Lafayette for the Midwest bridge-building season, Luten designed and built a number of structures using Purdue University engineering graduates as construction foremen of his National Bridge Company. The seventeen bridges he built in 1901 – three California bridges, two in Michigan, three for the CCC&StL (Big Four) Railroad, and others in Indiana – yielded a profit of one thousand dollars without salary for Luten.[54]

II.3B Luten's Tied Arch

The 1901 bridges gave Luten opportunity to introduce steel reinforcing in his timber-tied arch design and to experiment with its placement. At base the new design presumed a connected system of reinforcement through the ring, the abutments, and the streambed. The ring could be built to "any desired flatness," and the ring and the abutments could both be made unusually thin. Paving the stream-bed

THE WORLD'S FLATTEST ARCH

Two tied arches which Luten designed in 1902 for foot-bridges in Columbian Park, Lafayette, provided him with a special opportunity to advertise his work. By leaflet and through the *ENR*, he promoted the longest of the two, the South Bridge, as "the flattest concrete-steel arch in the world." His assertion was challenged in a letter to the editor, the occasion of which provided Luten with yet another chance to peddle his design nationally. Indeed, earlier in the year, the *ENR* published a Luten letter hawking one of his "cheap but substantial highway bridge[s]," a $573 tied-arch special erected in Wabash County.[55]

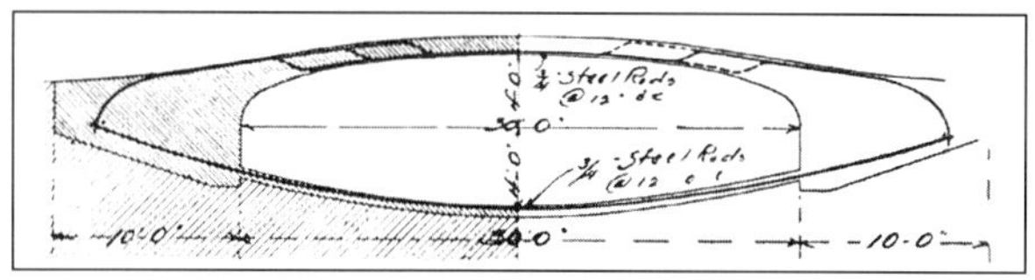

The West Bridge rather than the South one in Columbian Park, Lafayette, is shown here, since the West span survives. Except for span-length, both structures had the same dimensions. Each rose four feet, but the South Bridge spanned forty feet and the West thirty feet, giving the South structure a larger span-to-rise ratio. The West Bridge as it originally looked (top) and as originally planned with integrated Luten-system reinforcing (left, lower). A surviving cast-iron post from the original decorated railing (right) and rare example of Luten's use of metal railings.

protected the steel from rusting, resisted the structural change which settlement in soft bottoms could cause, and "flood-proofed" the bridge by keeping the swift current of a stream from undercutting the inner face of the abutments.[56]

As Luten had learned in 1900, separating design from construction could threaten the life of each. He had weathered the storms of 1901 with makeshifts and a little help from his family. Still determined to turn trial into tribute, the family decided in 1902 to move to Indianapolis, the state capital, and to register the National Bridge Company. In April, three Luten-Hull family members incorporated the National Bridge Company to manufacture, mine, contract, and construct "work of every description."[57] These family members served as directors overseeing twenty-thousand dollars of capital stock. Luten assigned his timber-tie patent "and all subsequent patents" to the corporation for half the capital stock and became president. His wife and her sister purchased the other half of the stock for cash.

Luten matched his inventive talents with his wife's and her sister's money. To capitalize on his talent, the National Bridge Company immediately invested some of its cash in the retention of a Chicago patent attorney, Robert L. Ames, to turn Luten's inventions into protected assets. Being "extremely short of funds," Luten did not travel to Chicago to consult with his attorney. Instead, he "mailed sketches to Mr. Ames for numerous of my inventions and asked him to include as many as possible in one case, to save fees."[58]

Seven patent awards resulted from the several consolidated applications of 1902: one by 1905, another in 1906, three in 1907, and one each in 1910 and 1911. The patents resulting from these early applications included ones on Luten-designed arch-ring reinforcing, the steel-tied arch, the ring-stiffening spandrel, and a "Luten truss" for beam bridges.[59]

Ever the empiricist who applied theory, Luten never claimed that his patented devices "originated any new methods of analysis for arch action or beam action, so far as the theory is concerned. I believe that any improvements that I may have originated have been in the application of the theory, rather than in the theory itself." His central concern, he claimed, was "to reduce the total of material required to produce a given strength," not just "to saving in reinforcing. My methods are intended to produce a more efficient structure; that is, a stronger structure, for a given cost."[60]

II.3C The Luten "Truss" for Beams, Slabs, and Girders

As Luten knew well, the arch was not the only bridge-form used in reinforced concrete any more than in timber or metal. Prosaic, beam and slab spans got tucked away across the countryside, usually over small streams. They did not match arches or trusses for span-length, but simplicity of design and construction made them cheaper when span-length was not a determining factor. They also had a modest superstructure, an especially handy feature where the distance between the waterway and the roadway was necessarily limited: the slimmer the superstructure, the more waterway clearance the structure allowed.

One unmistakable sign of transitional use of a new material or invention is a penchant for shaping it like the old. Integrating the

"GOOD BRIDGES FOR GOOD ROADS"[63]

Luten varied the size of the bars in his "truss" from one-half to three-quarters of an inch and varied their number from three to twelve to address different amounts of tension and shearing as well as to reinforce both deep and shallow members efficiently. He certified his "trusses" for slabs of five- to twenty-foot spans, beams of twenty to thirty feet, and girder spans of twenty to thirty-five feet.

Contractors ordered the trusses from Luten's company, which fabricated them in Indianapolis and shipped them to a designated railroad station near the builder.

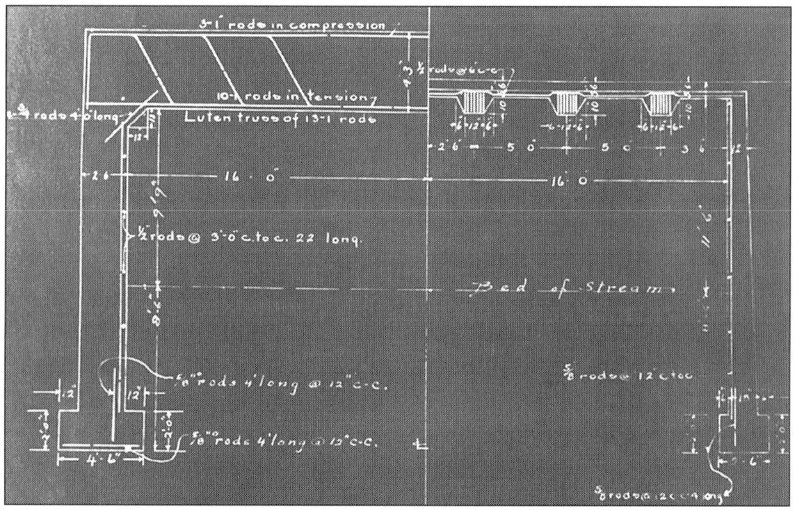

(Morgan #56)

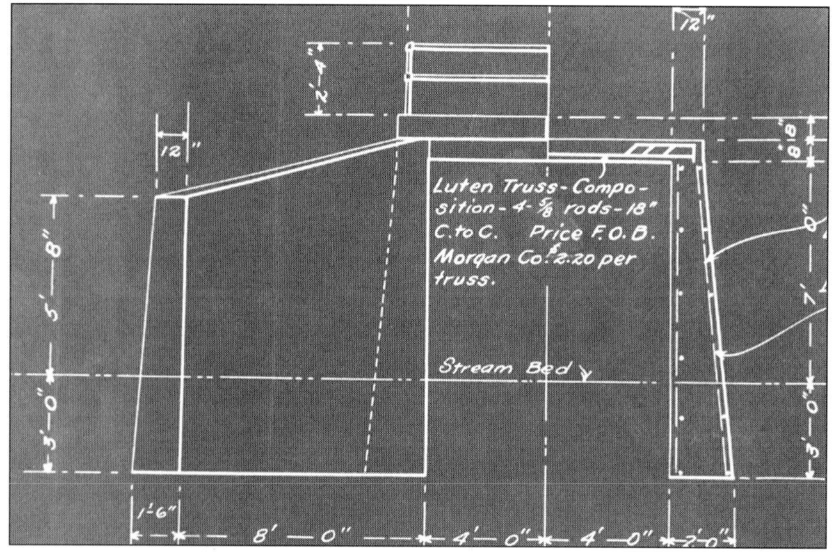

The Stockwell Bridge contains Luten trusses in its girders and its floor beams.

The Thomas Wheeler Bridge in Morgan County consists of a short slab span with Luten trusses embedded. Some of the trusses, with their periodically transposed rods, are now exposed on the underside of the slab (below left).

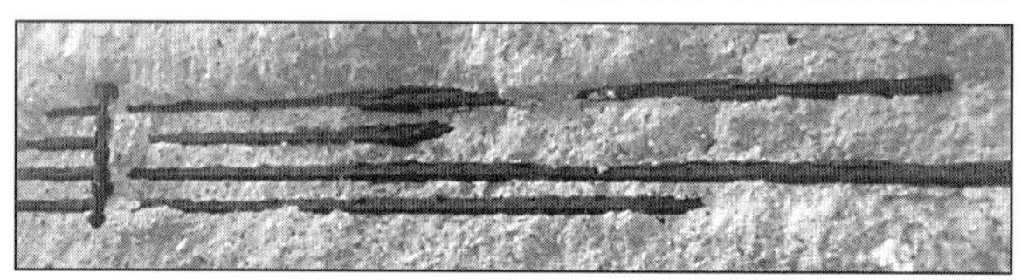

new into the old limits the expression of the new so the innovation is less discomforting. The popularity of the beam- and girder-form of bridges and the shapes they often assumed in concrete, Luten believed, copied steel rather than exploited the particular assets of concrete. While almost always preferring arches to beams in part because the former required far less steel per foot of span, he recognized demand in the marketplace and applied his inventive talents to developing a system of reinforcing especially appropriate for horizontal spans.

Luten designed, assembled, and patented devices that he called "trusses" which could be placed in a concrete beam, slab, or girder. As with his arch reinforcing, Luten turned his "truss" into an integrated system that took elasticity into account and, he claimed, sold for about two-thirds of what it would have cost to purchase mill-rolled I-beams to perform the same amount of carrying.[61] Again, the economy came from using steel only where tensile stress, including shear, became significant.

The "truss" consisted of a number of rigidly-locked steel rods stretched along the lower edge of a slab, beam, or girder with some of its rods bent upwards to cross the structure diagonally at different places – approximately one-fourth, one-half, and three-fourths of the distance between midspan and the abutment. The rods crossing the structure were bent again to parallel the top edge to the beam's end, at which point they were bent downward to adjoin the rods running along the bottom edge.[62]

At first glance, the Luten "truss" looks like his arch-ring reinforcing system. Indeed, for purposes of saving fees, the "truss" and the system of ring reinforcing were included in the same patent application. But the patent office insisted on dividing the application and making two quite separate awards, a distinction that Luten always acknowledged as sound. When asked whether the "truss" bent into the shape of an arch ring would be equivalent to his arch-ring reinforcement, Luten responded that at best "it might involve it in a very inefficient form." The reinforcing in a Luten-design arch extended into the substructure, while the "truss" represented a self-contained system to deliver force vertically to its abutments. Furthermore, "the size and grouping of the members with respect to the thickness of the structure should be entirely different for the arch" where the areas, nature, and intensity of tension would be quite different than in a beam span.[64]

II.3D The Concrete Girder

Of all the steel-design mimicry in concrete Luten found one form too antiquated simply to ignore. He decried the use of concrete girders with load-bearing members placed alongside the roadway. Concrete girders provided only temporary structures, he insisted, for "the time will certainly come when every such bridge will require widening and the only solution . . . is to tear it down" and replace it with an arch or beam "which can be widened without destroying any of the original structure except the railing."[65]

Luten also found "no economy" in concrete girders over a spandrel arch for spans longer than twenty feet. A "concrete girder has upwards of three-fourths of its material subjected to no stress or at best to a very small amount and consequently doing no work. The concrete arch is almost always the more efficient structure since its material may be subjected to more uniform stress throughout." The girder had no earth-filling to protect it "from excessive impact or concentrated loading" as did the filled-spandrel arch. While the girder might offer a bit more waterway clearance than an arch, its shorter spans and more frequent piers effectively limited this advantage by also providing "a very satisfactory straining for collecting drift from a stream."

Luten was not surprised to find all kinds of beams, slabs, and girders appearing in concrete, some hundreds of which used his patented "trusses" for their reinforcement. Despite his railing against their inefficiency and inadaptability, girder spans became almost as common across the Hoosier landscape as redbuds, dogwood, and crabgrass. At least one-hundred, concrete girders built before 1920 survive on Indiana's county roads today, offering mute and undeniable testimony to structural hardiness. Some of these spans reach beyond fifty feet (Owen #143, #148),

(Dearborn #77)

(Dearborn #53)

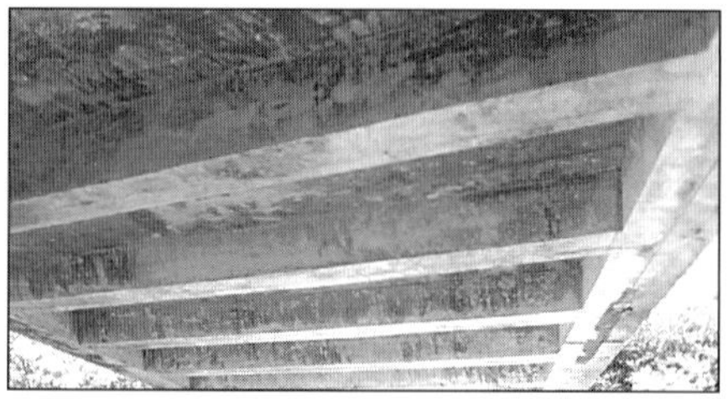

Two of Dearborn County's girder bridges illustrate some of the form's similarities and differences. In all cases, the outer girders carry the basic structural load. Designers have generally selected one of two basic strategies to support the roadway between the load-bearing girders. As with the Wolluing Road Bridge (left), some have opted for floor beams, while others have used a rectangular slab (Kuebel Road Bridge, right).

and most support a slab deck with a roadway occasionally wider than twenty feet (Sullivan #137).

II.3E The Bowstring "Arch" in Concrete

Luten faulted the bowstring in concrete for many of the same reasons that he criticized concrete girders: imitating metal design generated inefficiencies, and structural load-bearing alongside and above the roadway precluded widening. Besides the serious disability concerning widening, Luten faulted the concrete bowstring for possessing many of the general inefficiencies of beams when compared with arches.[66]

A tied structure, the bowstring by definition operated as a beam rather than as an arch: it delivered only vertical pressure upon its abutments. The lateral thrust of the curved members rising alongside the roadway was addressed, not by the substructure as with the arch, but by tying the curved members' reinforcing rods with the rods of deck at the ends of the arcs and through vertical members. As beams, concrete bowstrings inevitably required more steel reinforcing than did a filled-spandrel arch and possibly more concrete, too, and their erection would likely have been more costly because more difficult.

Only on sites where the span's clearance posed a very special problem – where the bridge's roadway was not naturally and could not be conveniently raised far enough above the waterway or rail-bed to allow adequate clearance – could Luten grudgingly find justification for putting working members alongside or above the roadway rather than below it. Very few circumstances induced him to enter a bowstring design into a bidding. No cases apparently oc-

curred in Indiana where only one structure of this kind is known to have been built – the Portland Arch.[67] Luten's representatives, in fact, unsuccessfully submitted the very lowest of all bids, four percent below the lowest bid on the locally preferred, bowstring design for the Portland Arch. But Luten probably proposed a filled-spandrel arch which might not have allowed for as much waterway clearance as the bowstring on which the Portland City Engineer's heart was set for a number of reasons.[68]

II.3F "Warped-End" or "Horse-Shoe" Railroad Arches

No part of the arch was safe from scrutiny and change in Luten's unrelenting search for greater efficiency. In early experiments with small railroad structures, he cut back the ends of the arch ring, eliminated spandrel walls, and modified the depth and reinforcing of the arch ring from side to side rather than from end to end. He had called these structures "warped-end" arches in 1901 but renamed them "horse-shoes" by 1905.[69]

Typically under considerable fill, railroad arches had modest spandrel walling and no rails. Luten took advantage of the railroad's banks to cut the crown of the outer edges of the ring back to the contour of the banks like a horse-shoe. Instead of a spandrel wall, Luten designed a shoulder at the crown's edge to hold the toe of the slope, presenting the edge of the arch ring as "a warped surface

Left – Luten designed concrete bowstrings in extraordinary cases, all of which happened to occur outside Indiana. His Rainbow Arch, for Niagara Falls, New York (left, top), suggests a segmental ring tied through the quarter section. The setting probably dictated a somewhat decorative design. The bowstring span integrated into the Avis Overhead of Hinton, West Virginia, addressed other matters (center and bottom). Rarely can a bridge designer hope to change the established level of a roadway or of railroad tracks in an already-built urban environment. In Hinton, Luten maintained the established level of the roadway with arch spans except where he needed to cross railroad tracks. There, to provide a rectangular clearance for the trains without changing the level of the roadway, he resorted to the bowstring which moved the load-bearing structure from below the roadway to alongside and above it. The ends of the bowstring's arched ribs were tied together through the deck. Since the span's thrust was delivered vertically to the substructure, the piers or bents could be kept slim enough not to obstruct the activity below.

A "horse-shoe" arch under construction for the Big Four Railroad (above). A completed one under fill on the same railroad (left).

of helicoidal form." He calculated that the warped, horse-shoe sides saved from fifteen to twenty-five percent of the total cost of the structure. To make a virtue of a defect, he allowed that "the beauty of this design rests in its common-sense."

Luten designed a second and structurally more significant novelty into his horse-shoe arches. He strengthened the ring's center – where the fill was greatest and over which the trains passed – without adding what would have been redundant concrete towards the edges of the ring. Would the narrowing of the ring's passage at the center by about twenty percent create a bottleneck that limited the effective passage of water through the span? No, Luten reasoned, because of the Venturi principle: funneling the waterway would force an increase in the velocity of flow through the constricted areas, thus moving essentially as much water through the whole as would occur if the span were of uniform width. As with elasticity, Luten made rule-of-thumb use of a principle of scientific theory.

II.3G "Double-Drum" Arches

Experimenting with arch design in yet another way, Luten divided the arch ring longitudinally, separating its lower from its upper half. By putting the concrete in each of the halves at a greater distance from the central axis, he placed it where stress was less intense, much for the same reason that mill-rolled or fabricated I-beams concentrate their material on the flanges and join them with a thin web. With a diminished amount of stress needing to be addressed, the amount of material – both concrete and steel – could be reduced.[70] The reductions came in the ring itself, in the abutments to sustain a lighter superstructure generating less thrust, and in the amount of centering or scaffolding needed to hold the construction, all moves that generated savings over the traditional filled-spandrel arch.

Luten claimed to have designed several hundred double-drums over the course of his career for locations "where there is no danger of the foundations being undermined." According to his calculations, the double-drum often required no more than half of the concrete and only sixty percent of the steel called for in a comparable filled-spandrel structure. With less form-work, the double drum could be built quite profitably for about seventy percent of the price of a comparable, filled-spandrel arch.[71]

Here Luten applied elastic theory to generate an economical and more sophisticated design than the early Monierists had produced but with drums or rings as thin as they had ever used on a bridge. Conservative American engineers almost universally scoffed at Monierist design, and they probably would have criticized Luten's drums, too, had the drums been in plain view rather than hidden behind spandrel walls!

"DOUBLE-DRUM" DESIGN

Luten's double-drum arch typically consisted of two, thin (often four-to-six inches thick) concentric rings each anchored separately on narrow footings with earth sandwiched between.[72] The use of footings both small and separated—the lower and larger one often being only about a foot wide and extending four feet below mean water level – accounted for a considerable portion of the design's special economy over the traditional, single, large abutments. The drums were joined at occasional intervals – about eight feet apart—by vertical and transverse ribs of concrete to provide for shear.

Designed to carry only the lower drum, the centering was much simpler and cost much less than scaffolding for a regular, filled-spandrel ring. Luten believed that with tension thus essentially "eliminated" from the drums, no reinforcing was required. But to protect against contraction or settlement, the lower drum "was reinforced slightly with occasional" rods run longitudinally, three rods being bent, as in regular Luten-style rib-reinforcing, first upwards across areas of shear and, then, horizontally to continue through the upper drum and into its footings.

When adequately cured, the lower drum had earth tamped into place over it as a framing for the upper drum. Except for the three rods carried up from the lower drum and through the haunches of the upper one, the upper drum contained only transverse reinforcing run between the spandrel walls. The upper drum "hangs like a hammock" and pulled the spandrels together, allowing Luten's already slim walls to be designed even thinner than usual. Once the spandrel walls were erected, the upper ring was filled, and the rails built.

A double-drum as designed (below, left) and built (above) on the Holcomb Estate, Indianapolis, in 1919. Now carrying a path between condominiums, the structure retains its urns and neoclassically-balustraded rails (below).

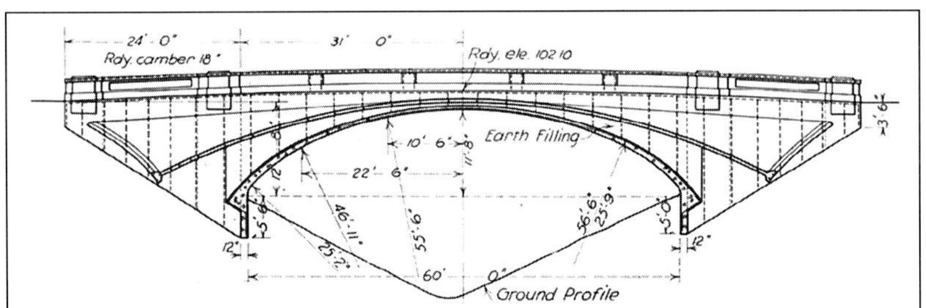

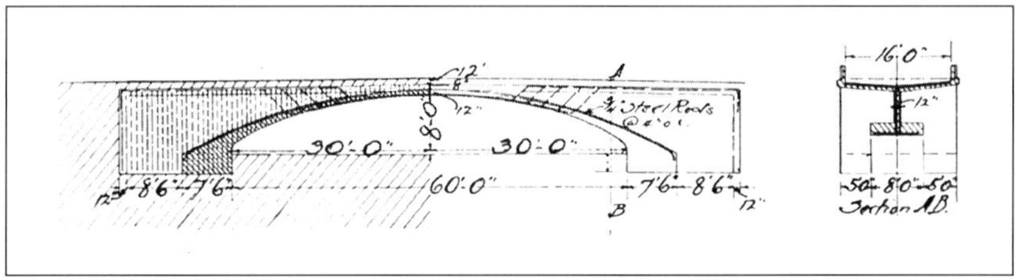

Luten's single T-arch. The Kissinger Bridge in Wabash County.

II.3H The T-Arch: More Experiment than Invention

For Luten, an unusual location seemed as much an opportunity as an obstacle. Opportunities, however, did not always lead Luten to an empirical trial which produced large new efficiencies. Luten once experimented with a T-arch for sites with very high banks where a conventional spandrel structure would have entailed an unusually large amount of fill. Such circumstances would normally have required the design of a heavy ring and spandrel walls to support the extraordinarily large and heavy load of earth. Luten's proposed alternative, the T-arch, carried a quite narrow and slim arch ring with a single spandrel wall placed halfway between the ring's edges. The crown of the arch ring and the spandrel wall, in turn, supported an earth-covered, concrete-slab roadway wider than the stem of the T's spandrel or of its arch ring.

While the T-arch used much less concrete than a filled-spandrel design, its need for more steel reinforcing and especially its added form-work produced only modest savings compared to a filled-spandrel.[73] Luten could not reduce labor costs enough to propose building more than a single T-arch. But he learned from this failure, later integrating his notion into the economically as well as technically efficient spandrel-braced arch.[74]

II.4 DESIGN AND CONSTRUCTION: FISSION AND PROFIT

The incorporated National Bridge Company had provided Luten with an opportunity to develop his design ideas, apply for a series of patents on his inventions, publicize his work, and build a number of bridges. Construction, while an absolutely necessary activity to introduce his designs, remained an Achilles heel. The years from 1902 to 1905 were not altogether good ones for the newly-launched National Bridge Company, a fate Luten shared with most proprietors of firms in their start-up periods.

While he designed and built thirty-three structures and sold a number of "trusses" in 1902, Luten spent many "sleepless nights" as the company lost between four and five thousand dollars.[75] The arches for which he won contracts were small and mostly tied. Contracts with Miami and Wabash Counties and with the Indianapolis and Northwestern Traction Railway did show that Luten had begun to break into the local highway and traction markets. Despite these successes and the promise they carried, Luten's volume was too low to generate enough revenue to cover the annual expenses of his operation.

Electric railway contracts in 1903 played a critical role in wiping out the losses of the previous year. The Indianapolis Northern Traction Railway alone contracted for nearly forty percent of the seventy-five bridges Luten designed and built in 1903. But his market also widened to include the Nowland Avenue Bridge in his new hometown of Indianapolis as well as a structure in Charlotte, North Carolina.

In 1904, Luten suffered more losses despite an increased number of contracts.[76] During the year, Luten designed and constructed about one hundred and fifty bridges. With his eye on design, patents pending, and difficulties with turning a profit at construction, Luten decided to take the next step in his career – to separate design from building with the hope of making both considerably more profitable.

II.4A The National Concrete Company

At Luten's behest, three Indianapolis men – E. H. Lee, J. C. Stuckenberg, and Russell T. MacFall – organized the National Concrete Company in March 1905 "to do manufacturing, mining, contracting, and all kinds of construction work" with a capital stock of $100,000.[77] A few months later the National Bridge Company met to transfer "all the contracting business" it had in Indiana to the Concrete Company which, in turn, agreed "to at all times use its influence to promote in a legal manner the adoption and use" of Luten-designed "improvements", which the Bridge Company continued to control. The Concrete Company agreed to pay a royalty for the use of Luten's improvements and to give $10,000 of its capital stock to the Bridge Company.[78]

Within the month of its organization, the National Concrete Company won its first major contract for a large structure in Indiana – a seven-span, seven-hundred-foot crossing of Wayne Street, Peru, over the Wabash River. This provided Luten with a reasonably large stage on which to practice and thus to promote a number of his ideas. The Concrete Company's $36,900 contract included fifteen percent, or $5,640, for royalties and steel reinforcing for National Bridge.[79]

The spans of the Wayne Street Bridge cleared from one hundred feet at center to seventy-five feet at the ends. In 1905, the rings of the outer and shorter arches would conventionally have been flattened to handle the thrust or push of the longer, inner ones, as had been done with the well-known Y Bridge in Zanesville, Ohio. The conventional procedure, however, inevitably reduced the volume of water which could clear the outer spans – a consequence that concerned Luten, for flooding was considerable along the Wabash. Luten opted for new solutions to the problem of inter-span thrust. He inclined the outer rings upwards towards the center, making them somewhat unsymmetrical but maintaining the volume of water each could clear.

Not surprisingly, Luten used the bridge's substructure to help balance the thrust of rings of unequal span length. To reduce the push against the abutments, he tied each end span through a reinforced streambed pavement – a strategy which added some flood protection at the ends and at the same time permitted a considerable reduction in the size of the abutments.

The National Concrete Company starting the substructure for the Mishawaka Avenue Bridge over the St. Joseph River, Mishawaka.

(St. Joseph #203)

Luten also applied a number of his other design strategies. He treated the piers elastically by tying the reinforcing of the rings together with that of the piers, thereby reducing the burden on the rings and at the same time allowing for the use of very thin piers. Not surprisingly, he used the spandrel walls to stiffen the arch rings and in consequence cut down on the depth and reinforcing in the rings and the walls. He bonded the spandrel walls to the rings through their reinforcement and had the concrete poured for the walls and rings at the same time in part to avoid settlement cracks between them.

With his early construction experience very much in mind, Luten looked for efficiencies in building as well as in structural design. He tried out one of his new styles of centering with the Wayne Street Bridge. He employed flexible uprights where bracing could be easily adjusted to allow for above-average control of settlement at prescribed times.[80]

The National Concrete Company sold many a "truss" for a slab, beam, or girder bridge and designed several dozen arches built in Indiana between 1905 and 1910. But after the Wayne Street Bridge, they did not win another major multiple-span contract within Indiana until 1911, when they triumphed in the bidding for the Waverly Bridge – a five-span arch structure across the White River in Morgan County. In 1912, they secured two contracts to span the Wabash River – a four-span arch at Belden (Wabash #41) and a seven-span arch at Georgetown (Carroll County).

It is an ill flood whose sweep produces no good. Indeed, Luten's firms were well poised for the Spring Flood of 1913, which washed out more than a hundred mostly steel and timber bridges across Indiana. All twenty-two hundred Luten-design bridges then extant in Indiana survived the flood, although two were damaged. Of his major, multiple-span structures, the three Luten arch bridges over the Wabash and the one over the White River remained intact.[81] This success helped Luten advertise his product and enabled him to exploit the failure of others. In Logansport, for example, almost all major bridges washed out. National Concrete won the contracts to replace the Burlington, Cicott, and Third Street Bridges

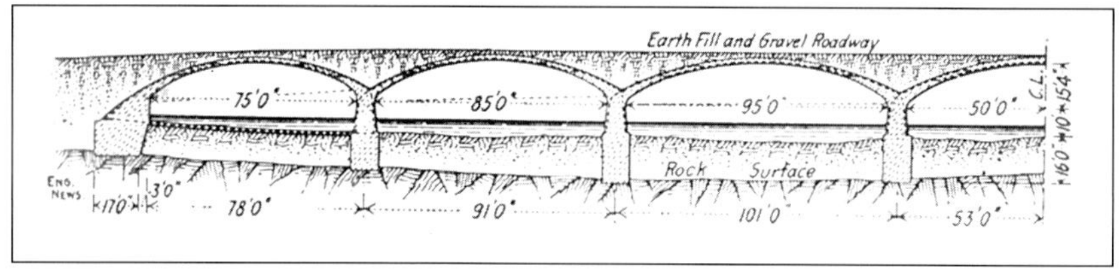

Wayne Street Bridge, Peru, as planned (above) with tied outer spans and somewhat unsymmetrical inner ones, and as under construction (below).

Collapsible framing to support poured concrete until cured.

(Cass #4177). Farther west as well, Luten's firm got the job of replacing the wrecked steel bridge at Lewisburg (Cass #123).

II.4B Bridge Agents and Contractors

Luten's horizons were not limited to Indiana, and he did not give the National Concrete Company a monopoly on his design even in his home state. The separation of design, some of it patented, from construction required operational policy for coordinating the building of Luten-style bridges. From its start, the National Bridge Company followed the recommendation of the International Patent Conference of 1873 to license anyone paying a royalty for use of a patent.[82]

Open licensing seemed like one reasonable way to mitigate the potentially arbitrary use of monopoly inherent in the patent system and to insure something approaching competitive use of new technology. Luten priced his designs in a twofold manner. He offered to supply working drawings and a license for their use to any and all for ten percent of the price a contractor bid to construct the bridge. Or he would furnish agents and builders with drawings, engineering advice, a license to construct using his plans, and the specified reinforcing steel prepared for installation for a set price which usually totalled approximately fifteen percent of the estimated contract price – ten percent for the licenses and five percent for the steel reinforcing.[83]

The National Bridge Company also trained over thirty foremen and fifty assistants whom contractors could hire to build Luten-design structures. While supposedly "the simplest and easiest type of reinforced arch to erect that has yet been devised," Luten bridges nonetheless "demand[ed] foremen possessing a high degree of skill in order to accomplish proper construction with economy of cost and handsomely finished appearance." By 1913, Luten had about one hundred and fifty "trained practical foremen" scattered throughout the United States.[84]

Contractors could bid against one another on equivalent terms using a given Luten design, or they could bid on the basis of an alternative plan. In any case, "the competition is genuine and a public letting of such a structure is legal. When an individual wishes to make a purchase he secures the best at the lowest price, regardless of whether the device is patented; why, then, should the public be denied the same privilege?" Should a given Luten design have resulted in bids that were higher than those on another competent plan, the public would have been best served by building a non-Luten bridge, and the market would have appropriately disciplined Luten. But if, as advertised, Luten-design bridges included a significant number of "cost-reducing devices," savings from the efficiencies should more than offset the royalty which a contractor paid the designer.[85] "A farmer who would cut his grass with a scythe because the mowing-machine is patented, would have to eat grass to live in this progressive age."

(*ENR*, 1907)

As Luten had hoped, the growing demand across the United States for concrete bridges built on competent plans brought a number of building contractors to his door. He publicized his design through many articles and letters in the *ENR* and other publications as well as through the National Bridge Company's catalogs. At least sixteen of his articles had been published in technical journals by 1907.[86] In a 1907 advertisement in the *ENR*, the National Bridge Company named six companies in addition to National Concrete, with which it had a special relationship.[87]

Luten printed and distributed five thousand copies of his 1908 catalog in different parts of the country with different covers and title pages, each emphasizing a particular representative contractor.[88] The 1908 catalog highlighted Luten-design structures built in twelve states beyond Indiana. The other states ranged from Massachusetts and Rhode Island to New Mexico. As the number of states increased and their distribution spread, so the number of Luten designs being built grew gradually from 1905 to 1910 and averaged about five hundred annually.[89]

By 1913, Luten had agents in at least thirty-six states, and his catalogs were now circulating as far afield as Japan. He paid most of his agents half his royalty or five percent of the contract price of a bridge using his design. The agent paid for local advertising, securing information on the bridge site, and his own travel. Each agent was expected to guarantee an annual minimum amount of construction, and the percentage of the agent's commission depended upon the amount of business generated yearly.[90]

N. M. Stark & Company of DesMoines, Iowa, ranked high among the larger agent-builders of Luten designs from 1906 to 1913. Stark erected about one hundred Luten bridges, mostly tied arches, in Polk County, Iowa, alone. Within this same period, about half of Iowa's counties erected several hundred more Luten-designed arches, again largely tied as the deep loam of the prairie produced many stream-beds in which bridge substructures could be easily undercut.[91]

Besides Stark, at least two other agents erected Luten-design bridges in the Plains States. The Topeka Bridge & Iron Company built many from Kansas to Texas and New Mexico. Indeed, Lloyd Smith of Topeka reportedly supervised in person the erection of nearly three hundred. In addition, the Pueblo Bridge Company of Pueblo, Colorado, constructed a considerable number from Colorado to Oklahoma.[92]

The Illinois Bridge Company of Chicago worked the upper Midwest from Michigan to Wisconsin for Luten. In the East, the Ferro-Concrete Company from Pennsylvania erected Luten designs in New Jersey and Pennsylvania, and the Owego Bridge Company built in upstate New York.

Of agents who were not themselves builders, Walter N. Denman of Springfield, Massachusetts, is best known. He secured contracts on Luten-design structures for a variety of different contractors across New England from Maine to Rhode Island, and he also defended Luten interests in federal court on several occasions.[93]

II.4C The Luten Bridge Companies

The most successful of Luten's construction agents – the ones who probably constructed as many Luten-design spans across the eastern U. S. as the National Concrete Company did in Indiana – grew out of the talents of Scots-Irish stone-cutters combined with those of more established, prosperous families of York, Pennsylvania. On his return from the Civil War, A. B. Whittaker went right to work cutting stone, first to build houses and, later, to erect arches for the Pennsylvania Railroad. He plowed his profits back into what became the Whittaker Stone Company and quarry in Curwensville, Pennsylvania. Alex and his older brother, John, spent many an hour learning the trade and business of stone cutting. By 1908, the two brothers had moved to Harrisburg, the state capital, where each shifted from stone to concrete work. John helped to organize the Whittaker and Diehl Bridge Company and Alex worked for the Ferro-Concrete Company, an early Luten agent.[94]

By 1909, the two Whittaker brothers had joined with Lucius G. Brown and G. W. Drury to incorporate the Luten Bridge Company at York with a capital stock of $5,000 for constructing "bridges, buildings and other structures of similar kind out of iron, steel, and reinforced concrete." The company grew quickly and substantially. It tripled its stock to $15,000 in 1910 and multiplied that to $75,000 by 1914. Its other assets grew accordingly.[95]

The company also established a number of branch offices across the eastern United States. The Whittakers spread out: John stayed near the York headquarters, but Alex moved to Clarksburg, West Virginia, and their younger brother, George, was sent to Atlanta, Georgia, where he soon enlisted the aid of V. C. Pickering, a colorful figure from Chatsworth. Other agents who established offices in Syracuse, New York; Concord, New Hampshire; and Palatka, Florida, appear to have come from among Lucius Brown's associates.

Sometime between 1915 and 1925, two Daugherty brothers, who had gone to school and played baseball with the Whittaker boys in Curwensville and who like them had hustled their way up from manual labor, incorporated a second Luten Bridge Company. Scott (C. S.) Daugherty managed the headquarters in Knoxville, Tennessee, and Harold ran a branch office in Little Rock, Arkansas.[96]

As salesmen, the Daughertys and Whittakers hustled successfully for many a bridge construction contract from New Hampshire and New York in the Northeast, through Pennsylvania, Maryland, and West Virginia in the eastern central states, into Tennessee, Arkansas,

(Luten Bridge Company; Luten family)

A gathering of the Luten Bridge Company in York, Pennsylvania, probably about 1925. **Back row** *(left to right): E. R. Patterson, New York; George W. Wolfe, Pennsylvania; V. C. Pickering, Georgia; H. N. Kirkman, Florida; G. A. Whittaker, Ohio; L. H. Brown, York, Pennsylvania; D. H. Daugherty, Arkansas; Roy Daugherty, North Carolina.* **Front row**: *A. B. Whittaker, West Virginia; C. S. Daugherty, Tennessee; L. G. Brown, York, Pennsylvania; Daniel B. Luten, Indiana; J. F. Whittaker, Pennsylvania.*

Georgia, and Florida in the South. In West Virginia alone, A. B. Whittaker sold between fifty and one hundred Luten-design bridges annually by 1913.[97]

II.5 AT THE PEAK OF SUCCESS

The hard selling of Elmer Lee of National Concrete, the Whittaker-Daugherty clans of Luten Bridge, and Luten's many agent companies and individuals were well worth the commissions he paid them. Luten did not feel comfortable with the necessary glad-handing – and sometimes more – that went with selling bridges to county commissioners and city councilmen. He also distanced himself from the rough-and-tumble world of the manual laborers who did a lot of the bridge building.

Once he had successfully separated the siamese twins of design and construction in 1905, Luten devoted his primary attention to what he knew best and cared most about – design protected by the patent system. From 1905 onwards, the pace of patent applications accelerated, reaching seventy-four by the end of 1913 – forty concerning concrete arches and the rest about concrete beams and bridge-construction technique. Somewhat under half of Luten's applications had matured into awards by the end of 1913.[98]

In the decade and a half after the turn of the century, Luten had moved from working with a spade and saw on a four-hundred-dollar contract in Lafayette to becoming a successful "designing and consulting engineer." By 1913, he operated from a downtown Indianapolis office with a staff of a dozen engineers and draftsmen who helped him design bridges for construction contracts worth close to $2,000,000 for that year alone.[99] His bridges were being built across the continent by a number of contractors, some of whom had incorporated Luten's name into their companies' titles.

By 1915, Luten held thirty-nine U. S. patents on concrete bridges, more than all other bridge patentees combined. He had designed about six thousand bridges built in the United States, Mexico, and Canada, and had written over sixty-five articles published in engineering journals.[100] He was not just bragging when in 1915 he entitled one of his pamphlets *Why Indiana Leads the World in Concrete Bridges.* Indeed, Luten was among that select minority of entrepreneurs who succeeded – for awhile.

Luten played a central role in 1912-1913 in the design of a nearly fifteen-hundred-foot bridge to cross the Saskatchewan River on a grade at Saskatoon in Canada. Two-rib, open-spandrel arches seated on spandrel piers provided most of the ten spans. "Longer in span than any of the same kind in Canada," each of the four central spans cleared one hundred and fifty feet. The ENR *printed two reports on the structure's construction in March 1915 issues.*

EXTANT LUTEN-DESIGN SPANS IN INDIANA[101]

	County	Name/Number	Form	Date
1.	Tippecanoe	West Br	tied and filled-spandrel arch	1902
2.	Marion	#2516	tied and filled-spandrel arch	1903
3.	Wabash	#504	filled-spandrel arch	1903
4.	Decatur	#41	filled-spandrel arch	1904
5.	Parke	#156	tied and filled-spandrel arch	1905
6.	Miami	#132	filled-spandrel arch	1907
7.	Vermillion	#76	filled-spandrel arch	1907
8.	Hendricks	Cartersburg Br, THI&E Ry	tied and filled-spandrel arch	1907
9.	Putnam	#111	filled-spandrel arch	1909
10.	Marion	Country Club Br	filled-spandrel arch	1910
11.	Orange	#55	filled-spandrel arch	1910
12.	Wabash	#41	filled-spandrel arch	1912
13.	Cass	#123	filled-spandrel arch	1913
14.	Morgan	Thomas Wheeler Br	slab with Luten trusses	1913
15.	St. Joseph	#205	filled-spandrel arch	1913
16.	Marion	#4396	filled-spandrel arch	1916
17.	Morgan	#30	girder with Luten trusses	[1916]
18.	Morgan	#56	girder with Luten trusses	1916
19.	Morgan	#64	girder with Luten trusses	[1916]
20.	Morgan	#103	girder with Luten trusses	1916
21.	Marshall	#226	filled-spandrel arch	[1917]
22.	Shelby	#8	filled-spandrel arch	1917
23.	Vigo	#321	filled-spandrel arch	1917
24.	Cass	#113*	filled-spandrel arch	1918
25.	Cass	#4177	filled-spandrel arch	1918
26.	Marion	Holcomb Estate Br	"double-drum," filled-spandrel arch	1919
27.	Hamilton	#230	filled-spandrel arch	1920
28.	Marion	#1111	filled-spandrel arch	1920
29.	Jefferson	#2602	filled-spandrel arch	1922
30.	Marion	#3301+	filled-spandrel arch	1922
31.	Marion	#1007	filled-spandrel arch	1924
32.	Franklin	#17+	filled-spandrel arch	1925
33.	Marion	#6190	filled-spandrel arch	1926
34.	St. Joseph	#203	filled-spandrel arch	1926
35.	Carroll	#132	filled-spandrel arch	1927
36.	Jay	#62	spandrel-braced arch	1927
37.	Putnam	#62	spandrel-braced arch	1927
38.	Putnam	#193	spandrel-braced arch	1927
39.	Shelby	#49	filled-spandrel arch	1927
40.	Clinton	#131+	spandrel-braced arch	[1928]
41.	Fountain	#151	spandrel-braced arch	1928
42.	St. Joseph	#206	open-spandrel arch	1928
43.	Fountain	#130	spandrel-braced arch	1929
44.	Marion	#3108	open-spandrel arch	1929
45.	Putnam	#159	open-spandrel arch	1929
46.	Putnam	#229	open-spandrel arch	1929
47.	Fountain	#122	spandrel-braced arch	1930
48.	Fountain	#124	spandrel-braced arch	1930
49.	Fountain	#126	spandrel-braced arch	1930
50.	Fountain	#129	spandrel-braced arch	1930
51.	Switzerland	#59+	spandrel-braced arch	[1930]

+ = probably Luten-design * = bypassed [19xx] = estimated date

NOTES

[1]As quoted in William H. Wilson, *The City Beautiful Movement* (Baltimore, 1989), 287.

[2]"How Shall a Youth Be Trained for the Occupation of an Engineer?," *Engineering News-Record* [ENR] (25 November 1882) 9: 403-4.

[3]For a fuller discussion of the national engineering traditions described in this and the ensuing paragraphs, see Terry S. Reynolds, "The Engineer in 19th-Century America" in *The Engineer in America*, ed. Terry S. Reynolds (Chicago, 1991), 7-26; James K. Finch, "French Pioneers in Engineering," *ENR* (24 April 1930) 104: 676-83. On the role of science in engineering, see David P. Billington, *Robert Maillart's Bridges: The Art of Engineering* (Princeton, N. J., 1979), 47-48, 100-5. For a discussion of the relative contributions of empirical and scientific approaches to invention and innovation, see Abbott P. Usher, *A History of Mechanical Inventions* (2nd ed., Cambridge, Mass., 1954), 56-115.

[4]"How Shall a Youth Be Trained for the Occupation of an Engineer?," "The Education of Engineers," *ENR* (25 November, 2 December 1882) 9: 404-5, 415-16.

[5]"On the Education of Engineers," *ENR* (25 November 1882) 9: 406-7.

[6]Charles W. Baker, "A Broader Training for the Engineer," *ENR* (20 May 1909) 61: 536-37.

[7]Daniel B. Luten, "Concrete Bridges," *Journal of the Engineers Society of Pennsylvania* [*JESPA*] (October 1909) 1: 459.

[8]Daniel B. Luten, Jr., interview with author (audio-taped), August 1988.

[9]See "Luten, Daniel B," John W. Leonard, *Who's Who in Engineering* (New York, 1922-23), 793.

[10]In an interview with the author (August 1988), Daniel B. Luten, Jr., for example, reported that his father discouraged his son from becoming a teacher and on other occasions slighted teaching in passing comments.

[11]Daniel B. Luten, "What's Wrong With Iowa?," *Engineering and Cement World* (15 May 1918): 29.

[12]Daniel B. Luten, *Patented Concrete Bridges* (reprint of address before Ohio Engineering Society in January 1913), 13, 15.

[13]Luten, "Concrete Bridges," *JESPA* (October 1909) 1: 459.

[14]Luten, deposition (28 August 1913), District Court of the U. S. for New Jersey, *Luten vs. The Dover Construction Company: Complainant's Record*, 169.

[15]"Edwin Thacher," *ENR* (30 September 1920) 85: 675.

[16]For sample advertisements of the firm, see Supplement, *ENR* (29 June 1905, 20 December 1906) 53, 56: 61, 66.

[17]U. S. Patent #617,615 (awarded 1899).

[18]Thacher, "Concrete-Steel Bridge Construction," *ENR* (21 September 1899) 42: 181.

[19]Judge Martin J. Wade, Southern District of Iowa, "Edwin Thacher v The Board of Supervisors of Polk County, Iowa " (12 September 1916), *Federal Reporter* 235: 724. E. Lee Heidenreich – a European-trained civil engineer, chairman of reception for foreign engineers at the World Columbian Exposition, promoter of Monier system bridge-design in the United States, and nationally-recognized consulting engineer – also described "Thacher construction" as "more or less of an imitation" of Melan and von Emperger. Heidenreich stated the following in a deposition:

[20]U. S. Patent #617,615; Frederick P. Spaulding, A. Lincoln Hyde, and Ernest F. Robinson, *Masonry Structures* (2nd ed., New York, 1926), 401-2.

Question: "By Thacher bridge you mean the combination of von Emperger and Melan?"

Heidenreich: "Their successors or imitators; but it is called the Thacher bridge in advertising, for advertising purposes, I presume."

Question: "But it embodied the von Emperger and Melan systems?"

Heidenreich: "Absolutely, as far as I can see."

E. Lee Heidenreich, deposition (20 September 1916) for Luten v Marsh, Attorney General Cases S.II (State Historical Society of Iowa Library and Archives, DesMoines).

[21]For a glossary of the terms used in this box, see L. B. Johnson, C. W. Bryan, and F. E. Turneaure, *The Theory and Practice of Modern Framed Structures*, (9th ed., 3 parts, New York, 1910), I: 1-4. For an early discussion of elasticity and civil engineering, see D. H. Mahan, *An Elementary Course of Civil Engineering for the Use of Cadets of the United States' Military Academy*, (6th ed., New York, 1858), 375-401. For discussion of elastic theory as applied to reinforced concrete design, see George A. Hool and Nathan C. Johnson, *Concrete Engineers' Handbook* (New York, 1918), 659-68; John C. Trautwine, *The Civil Engineer's Handbook*, (20th ed., New York, 1919), 454-82.

[22]Since Howe taught at Rose Polytechnic Institute in Terre Haute, Luten did not study under him. He read Howe's work and became acquainted with him only after Luten moved to Indiana. Luten, deposition (9 September, 12 November 1913), *Luten v Dover: CR*, 208, 415.

[23]Malverd A. Howe, *A Treatise on Arches*, (2nd ed., New York, 1906), xvii.

[24]Fritz von Emperger, "The Development and Recent Improvement of Concrete-Iron Highway Bridges," American Society of Civil Engineers [ASCE] *Transactions* (April 1894) 31: 438-88.

[25]In his time for rebuttal, von Emperger ignored rather than responded to Hutton's critique. See von Emperger, "Concrete-Iron Highway Bridges," ASCE *Transactions* (4 April 1894) 31: 472-73, 481.

[26]The Monier representative in America, E. L. Heidenreich, agreed with Luten on the nature of the material used: "I will repeat that the Monier construction is not a concrete construction but a fortified mortar construction." Heidenreich, "Monier Constructions," *Journal of the Western Society of Engineers* [*JWSE*] (October 1894) 5: 331; Luten, deposition (28 August, 18 September 1913), *Luten v Dover: CR*, 169-70, 260-61; Daniel B. Luten, deposition (2, 4 August 1916), District Court of the U. S. for New Jersey, *Daniel B. Luten vs. The Dover Construction Company: Complainant's Record (Supplementary Testimony)*, 3, 21-24.

[27]Luten, deposition (2 August 1916), *Luten v Dover: CR(S)*, 9-12.

[28]Luten, deposition (20 August, 10 September 1913), *Luten v Dover: CR*, 105, 222.

[29]For the list and descriptions of Luten's thirteen qualifications summarized in this section, see Luten, deposition (2 August 1916), *Luten v Dover: CR(S)*, 7-11.

[30]Luten, "Concrete Bridges," *JESPA* (October 1909) 1: 459-60.

[31]For a more extended discussion of the difference between the natural scientist and the bridge-design engineer, see Billington, *Maillart's Bridges*, 91-105.

[32]Luten, deposition (2 August 1916), *Luten v Dover: CR(S)*, 8-9.

[33]Daniel B. Luten, "Design of a Concrete-Steel Arch Bridge," *ENR* (8 May 1902) 47: 377-80. Luten updated his formulas from time to time: see Daniel B.

Luten, *Manual of Design for Luten Reinforced Arches* (Indianapolis, 1905), 3-4; Daniel B. Luten, "Empirical Formulas for Reinforced Arches," *ENR* (28 June 1906) 55: 718-20. Asked about them in 1916, Luten judged his 1902 formulas as "no longer of any value." Luten, deposition (5 August 1916), *Luten v Dover: CR(S)*, 38.

[34]Daniel B. Luten, *Reinforced Concrete Bridges: Luten Patents* (1907), 83.

[35]Luten, deposition (9 September 1916), *Luten v Dover: CR*, 200, 207; Luten, *R C Bridges* (1907), 83.

[36]Luten, "Concrete Bridges," *JESPA* (October 1909) 1: 474.

[37]*ENR* (8 May 1902) 47: 370. At another time, though, the editors chided Luten for protesting too much about the shortcomings of elastic theory. At the end of a Luten letter on "The Elastic Theory and a Faulty Arch," the editors commented: "It is by no means clear, to us at least, that the elastic theory enjoys enough of either use or abuse in arch design to make the question very serious." *ENR* (9 January 1908) 59: 43-44. After listening to a Luten paper on another occasion, *ENR* editor L. J. Mensch again dissented: "To sum up, the figuring of most reinforced concrete arches by the elastic theory is not so difficult a matter as would appear in Mr. Luten's paper." In rebuttal, Luten responded that Mensch's examples of analysis all referred to parabolic arches, a shape which, Luten insisted, "had only limited applications." Daniel B. Luten, "Arch Design: Specialization and Patents" and discussion, *JWSE* 17 (September 1912): 624-32.

[38]Luten, deposition (18 September 1913), *Luten v Dover: CR*, 260; U. S. Patent #818,386 (filed 1902, awarded 1906).

[39]Luten, deposition (2 August 1916), *Luten v Dover: CR(S)*, 12-13.

[40]"The Monier Method of Constructing Arches," *ENR* (23 May 1891) 25: 499-500; "The Monier Exhibition at the Paris Exposition of 1900," *Cement and Engineering News* (October 1900) 9: 54; von Emperger, "Concrete-Iron Highway Bridges," ASCE *Transactions* (4 April 1894) 31: 441-42.

[41]Luten, deposition (11, 18 September 1913), *Luten v Dover: CR*, 228-29, 260-62; U. S. Patent #818,386.

[42]U. S. Patent #853,203 (filed 1902, awarded 1907).

[43]Luten, deposition (2 August 1916), *Luten v Dover: CR(S)*, 13-15; Luten, deposition (22 August 1913), *Luten v Dover: CR*, 124-26.

[44]U. S. Patents #852,202 (filed 1902, awarded 1907), #852,203, #923,058 (filed 1907, awarded 1909).

[45]Luten, deposition (22, 27 August 1913), *Luten v Dover: CR* 126, 148.

[46]Luten, deposition (9, 11 September 1913), *Luten v Dover: CR*, 208, 229-30.

[47]George Hool dedicated a chapter to "Arches with Elastic Piers." While Luten was probably the earliest and likely the most active early designer, he offered little serious written analysis of elastic piers and abutments. A. C. Janni provided the first systematic theoretical analysis in 1913. George Hool, *Reinforced Concrete Construction* (2nd ed., 3 vols., New York, 1916), III: 98-108, 271.

[48]U. S. Patent #649,643 (filed 1899, awarded 1900).

[49]Daniel B. Luten, *Timber-Tie Arches: Design of a Railroad Bridge of Concrete* (Lafayette, 1900).

[50]Luten, deposition (24 October, 12 November 1913), *Luten v Dover: CR*, 373, 421; Daniel B. Luten, *Timber-Tie Arches* (Lafayette, 1900).

[51]Luten, deposition (12 November 1913), *Luten v Dover: CR*, 421.

[52]Luten, deposition (24 October 1913), *Luten v Dover: CR*, 373.

[53]For Luten's published articles and letters in these years, see: "Concrete Arches-I & II," *Railroad Gazette* (11, 18 May 1900); "Expansion in Brick Pavements," *ENR* (14 March 1901) 45: 193; "The Effects of Low Temperature Upon Resilience," *ENR* (11 April 1901) 45: 266; "The Design of Arch Culverts," *ENR* (13 June 1901) 45: 435.

[54]Luten, deposition (12 November 1913), *Luten v Dover: CR*, 422. The Cleveland, Cincinnati, Chicago and St. Louis Railroad became a part of the New York Central Railroad system by 1900 but operated separately until about 1930.

[55]Daniel B. Luten, "A Cheap Concrete Steel Highway Bridge, *ENR* (14 August 1902) 48: 114; Daniel B. Green, "The Flattest Concrete-Steel Arch in the World and Some Other Flat-Arch Bridges," *ENR* (30 Oct 1902) 48: 362; Daniel B. Luten, "The Flattest Concrete-Steel Arch in the World," *ENR* (13 November 1902) 48: 402.

[56]Daniel B. Luten, *The National Bridge Company* (Indianapolis, 1905), 9-11; U. S. Patent #818,386.

[57]Articles of Incorporation, National Bridge Company, 4 April 1902, Indiana State Archives.

[58]Luten added that he "was unable to pay Mr. Ames his fees until months or years later." Luten, deposition (30 September 1913), *Luten v Dover:CR*, 331.

[59]The Luten patents awarded from the 1902 filings included: #785,676 (1905), #818,386 (1906), #853,202 (1907), #853,203 (1907), #853,204 (1907), #979,776 (1910), and #999,663 (1911).

[60]Luten, deposition (11 September 1913), *Luten v Dover: CR*, 228-29.

[61]U. S. Patent #979,776 (filed 1902; awarded 1910); National Concrete Co., *The Luten Truss* (undated pamphlets).

[62]Luten, deposition (4 September 1913), *Luten v Dover: CR*, 190-99.

[63]National Concrete Co., *Luten Truss*, 3.

[64]Luten, deposition (30 September 1913), *Luten v Dover: CR*, 330-34.

[65]Luten, "Concrete Bridges," *JESPA* (October 1909) 1: 460-68.

[66]Luten, deposition (11 September 1913), *Luten v Dover: CR*, 224.

[67]See I.5.

[68]Unfortunately the county has not retained its old bidding records or bridge plans, so surmises are all that remain on the Luten plan. Jay County, "Commissioners Record," 5: 248.

[69]Luten, *National Bridge Co.* (1905), 24-25; Daniel B. Luten, "*Horse-Shoe Concrete Arches*" (as reprinted from the *Railroad Gazette* of 20 April 1906).

[70]Daniel B. Luten, "A `Double-Drum' Reinforced Concrete Arch Highway Bridge," *ENR* (3 May 1906) 55: 496-97.

[71]Daniel B. Luten, *Double-Drum and Cantilever Arches* (Indianapolis, 1921), 5, 7, 9, 11.

[72]The specific information provided here pertains to a bridge with a 38-foot span erected in Miami County. See Luten, "A `Double-Drum' Reinforced Concrete Highway Bridge," *ENR* (3 May 1906) 45: 496-97.

[73]Daniel B. Luten, *Reinforced Concrete Bridges: Luten Patents* (Indianapolis, 1908), 66, 105, 124.

[74]See VII.1.

[75]Luten, deposition (12 November 1913), *Luten v Dover: CR*, 422.

[76]Luten, deposition (23 October 1913), *Luten v Dover: CR*, 366-67.

[77]Articles of Incorporation, National Bridge Co., 2854-4, Indiana State Archives.

[78]National Bridge Co. & National Concrete Co., author's photocopy of loose-leaf notebook. Original in possession of Luten family members.

[79]Daniel B. Luten, "Reinforced Concrete Arch Bridge at Peru, Indiana," *ENR* (29 March 1906) 55: 347-49.

[80]On his "collapsible centers," see also Daniel B. Luten, "Reinforced Concrete Arch Bridge at Yorktown, Ind.," *ENR* (17 May 1905) 53: 477-78; Luten, *R C Bridges* (1907), 9, 81; Luten, *R C Bridges* (1908), 128.

[81]Daniel B. Luten, *Reinforced Concrete Bridges in the Easter Flood of March, 1913.*

[82]John M. Blair, *Economic Concentration: Structure, Behavior and Public Policy* (New York, 1972), 610-11.

[83]Luten claims to have kept to the same terms while his patents had effective force. Luten, deposition (21 August 1913), *Luten v Dover: CR*, 114-15. For his breakdown of fees see Luten, "Reinforced Concrete Arch Bridge at Peru, Indiana," *ENR* (29 March 1906) 55: 347-49.

[84]Luten, *R C Bridges* (1907), 82, 87; Luten, deposition (12 November 1913), *Luten v Dover: CR*, 422-23.

[85]Luten numbered the cost-saving devices "from five to ten."

[86]Luten, *R C Bridges* (1907), 86.

[87]Berlin Construction (Berlin, Connecticut), Ferro-Concrete (Philadelphia, Pennsylvania), Illinois Bridge (Chicago), N. M. Stark (Iowa), Topeka Bridge & Iron (Kansas), and Pueblo Bridge (Colorado). Supplement, *ENR* (27 June 1907): 57: 85.

[88]Luten, *R C Bridges* (1908). The author is grateful for this information to Victor Darnell of Kensington, Connecticut, who owns or has seen copies of catalogs under the names of Berlin Construction Co. (Connecticut), Ferro-Concrete (Harrisburg, Pennsylvania), The Ley Construction Co. (Springfield, Massachusetts), and Owego Concrete (New York).

[89]Luten, deposition (23 October 1913), *Luten v Dover: CR*, 367.

[90]Luten, deposition (24 October 1913), *Luten v Dover: CR*, 372, 374.

[91]Luten claimed 700 structures of his design in Iowa. Daniel B. Luten, "What's Wrong With Iowa?" *Engineering and Cement World* (15 May 1918): 26, 29; Luten, deposition (25 October 1913), *Luten v Dover: CR*, 385-91.

[92]The information on Luten's agents remains quite scattered. For the best single sources, see: Luten, *R C Bridges* (1908), passim; Luten, deposition (25 October 1913), *Luten v Dover: CR*, 193-97; Daniel B. Luten, *Luten Design Bridges Under Construction Spring of 1915* (Indianapolis, 1915), passim.

[93]See, for example, Judge Dodge, U. S. District Court for Massachusetts, "Luten v. Town of Lee" (29 July 1913), *Federal Reporter* 206: 904-6; Judge Hale, U. S. District Court for Maine, "Luten v. Bearce" (22 December 1914), *Federal Reporter* 219: 237-41.

[94]"Whittaker," *Colonial and Revolutionary Families of Pennsylvania*, ed. Wilfred Jordan (New Series), V: 794-96, and interview with Norris F. Whittaker, 17 November 1992 (Woodsfield, Ohio).

[95]Luten Bridge Co., Charter (27 September 1909), 112: 271; Miscellaneous (2 June 1910), 56: 398, 532; 60, 28; Miscellaneous (2 February 1914), 75: 199; 79: 7-8 in Corporation Bureau, Department of State, Harrisburg, Pa.

[96]Information for this and the preceding paragraph comes largely from photocopies of Luten Bridge Co. papers, originals of which are in the possession of Norris F. Whittaker, Woodsfield, Ohio, who kindly loaned papers and photographs to the author.

[97]Luten, deposition (25 October 1913), *Luten v Dover: CR*, 391; Luten, *Luten Design Bridges* (1915), passim.

[98]Luten, deposition (4 September 1913), *Luten v Dover: CR*, 190; Daniel B. Luten, *Luten Patents and Litigation* (Indianapolis, 1914), 30.

[99]Luten, deposition (12 November 1913), *Luten v Dover: CR*, 423.

[100]Howard Newlon, "Concrete Bridge Patents: Useful or Otherwise," *Proceedings of the Third Historic Bridge Conference* (Ohio State University & Ohio Historical Society, 1990); Daniel B. Luten, *Do Patents Prevent Progress in Concrete Bridge Design?* (Indianapolis, 1915), 2, 11-15.

[101]For additional information on each structure listed, consult the inventory lists by county in Part Three.

III

Engineers and Artistry in Concrete Bridge Design

"Eminent engineers but professional vandals."
– *[H. G. Tyrrell]*

" ... To build without decoration is to build like an engineer, and thus in a manner proper to a Machine Age."
– *Rayner Banham*[1]

Henry Grattan Tyrrell, a bridge and structural engineer, ranked the United States near the top of the list of nations with the highest proportion of least attractive bridges, many but not all of which were made of metal. "Professional vandals," he reported in 1912, had been sacking the nation's natural beauty and littering the constructed landscape. Indifferent because of ignorance, engineers contributed substantially to the growing ugliness.[2]

Unlike France where engineers studied architecture, American students received "no instruction" concerning artistic bridge design in their engineering schools. Like others, most Hoosiers who selected engineering as a vocation appear neither to have come from homes where high culture flourished nor to have been exposed to much of it in their professional education. Engineering programs at Purdue University as well as at Rose Polytechnic Institute – the two institutions training the largest number of engineers within Indiana during the decades around the turn of the century – required little, if any, student attention to the arts and architecture.

"Engineers in America," Tyrrell concluded, "were therefore ignorant of the principles of aesthetics, or had given no time or thought to the cultivation of their taste in this direction. Where there is no desire for artistic production, it is certain that none will result." "Some engineers exhibited a willful, and most engineers a careless, indifference for design," another critic declared;" they exulted "over the disfigured city or the insulted landscape like a conquering savage."

III.A Professional Priorities Partly Deferred

In statements published a few years earlier, the editors of the *Engineering News-Record* [ENR] had implicitly endorsed some of H. G. Tyrrell's criticism of engineers and had also conceded considerable professional turf to the architect. In Europe prior to the time when persons "thought of figuring out the stresses in a structure before building it, the architect and the engineer were combined in the same person. As the constructive arts developed, two separate sets of builders grew up; the engineer, who built for strength and utility; the architect, who built for beauty." Each became a specialist in his own work but codependent "for the perfection of that work." "The architect, by virtue of his skill and training is the only professional judge of the beautiful in construction." The absence of "broad sociological or humanitarian aspects" to the engineer's training, the *ENR*'s editors noted wistfully a few years later, had also limited his access to the new and growing area of city planning.[3]

Neither the *ENR* editors' deference to architects nor Tyrrell's indictment of engineers, however, resulted in engineering's obeisance to architecture in concrete-bridge design. Indeed, the editors roundly criticized architects for preferring imitation over creative use of the new material. "The designers of concrete structures at the present day should abandon shams and deceptions and make the material stand forth as what it is. Imitation rock faces made with molded forms should be discarded; joints where no joints naturally come should be left out; intricate cornices should be done away with." In short, "beauty is not necessarily a matter of adornment" nor of "strict adherence to ancient forms of architecture."[4]

The beauty of concrete would be realized, the editors concluded, "when the engineer has succeeded in persuading the architect or at least the architectural mind in his own profession that there is at hand a material strong and cheap, but at the same time capable of an independent treatment which will give it an individual style and a beauty different from that of any material heretofore at the architect's command." The editors recommended, for example, the study of surface treatments of concrete and of the use of color.

III.B Engineering Principles of Artistic Design

H. G. Tyrrell's "general principles of artistic design" coordinated engineering values with architectural ones and gave priority to engineering: "The artistic motive is not the prevailing one and must be subservient to construction, but it must not be neglected." Above all else, he called for selecting "the most artistic form consistent with economy." He insisted on "the vital principle" that "`there is beauty in any useful structure designed on lines of true economy with the utmost simplicity and fewest parts.'"[5]

Tyrrell shared some of the *ENR* editors' disdain for "aberrations of aesthetic prudery – the desire to veil and drape the engineer's naked works." He, too, thought "a bridge must be a truthful creation and its appearance should show its purpose." Its strength and structure needed to be expressed, not hidden.

Among his other principles, Tyrrell included symmetry (presenting the bridge's general outline as essentially the same on each side of the center), harmony (maintaining where possible a single form of construction, varying span-lengths in a patterned way, or explicitly orchestrating contrast between changing elements or colors), and conformity to the environment. "The rule, generally, is to make the bridge more striking than its surroundings, so the eye will be naturally attracted to it," while also integrating the structure with its surroundings.

Finally, Tyrrell called for "a judicious use of applied ornament." The design "should please without apparent effect," by which Tyrrell meant "limited" and "small," not "excessive" and "large," ornament. In no case should ornament be allow to "add greatly to the imposed loads" on a bridge, for such violated the engineer's "fundamental principle of economy." Indeed, the bulk of Tyrrell's discussion about ornament concerned constraints on its use. As the *ENR*'s editors noted, "American constructive practice, fortunately, has never been much hampered by the incubus of `beautification.'"

III.1 PROFESSIONAL SKIRMISHING IN INDIANAPOLIS

While editors and college professors like H. G. Tyrrell might have hedged their assertion of engineering values and priorities and suggested some deference to architects, Daniel Luten, an entrepreneurial design and consulting engineer, never doffed his hat to members of that other profession. If there was one group of persons Luten lumped together as a class, it was architects, and he found little to value in that profession. Not surprisingly, Luten crossed swords with City-Beautiful advocates.

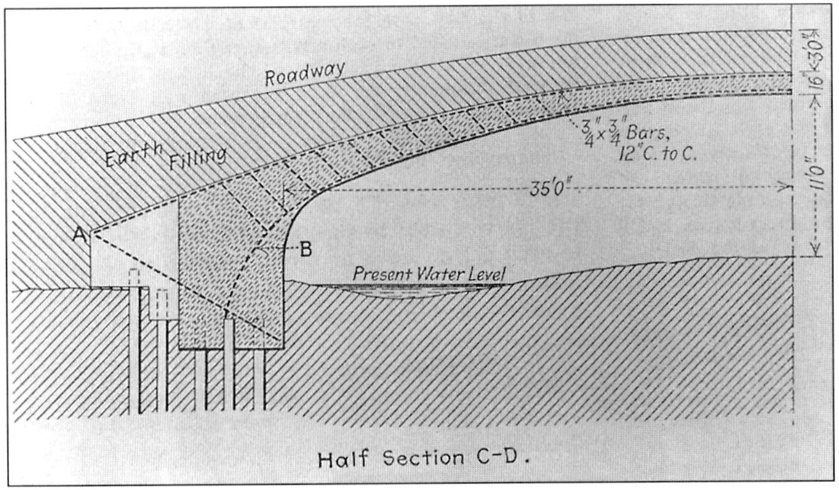

Judging the ribbed abutment (left) "entirely too light" to handle the lateral thrust of the essentially segmental ring, Luten predicted the abutment of the park bridge would crack at "A." Note the two-tier ring reinforcing with stirrups. The proposed "imitation of stone work" did not quite fulfill its promise (compare lower with upper right). Worse yet, the decorative rails with open-circle panels have been replaced with less harmonious filled ones.

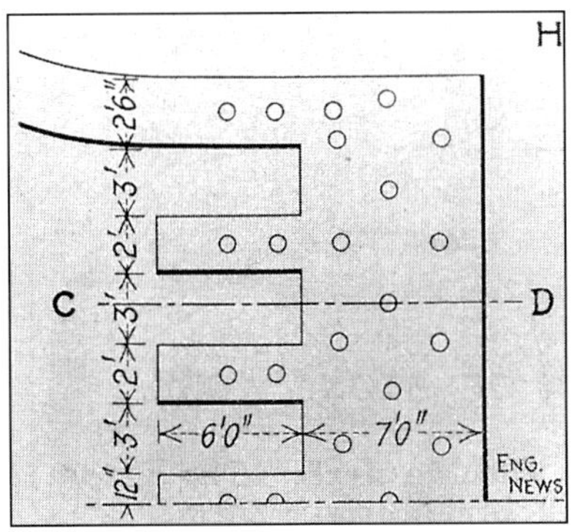

III.1A Attacking City-Beautiful Design

Just after J. Clyde Power retired as superintendent of the Indianapolis Department of Public Parks, Luten sent a letter to the editor of the *ENR* criticizing Power's bridge design.[6] As a long-time and very vocal advocate for City-Beautiful design, who seemed to find beauty only in cut-stone bridges, Power had turned to concrete design reluctantly at the very end of his career. Luten added insult to Power's injury from having to resort to the new material.

(Marion #3215)

Luten bluntly attacked the superintendent's Garfield Park Bridge (Marion #3215) for violating the engineer's first principles of utility and efficiency. The shape and size of the arch ring, the placement of the reinforcement, and the size of the abutments neither worked well separately nor complemented each other. Luten found the ring too heavy at the crown, its curve too full at the springing, and its reinforcing not continued into the substructure to take advantage of the metal's elastically. He thought the abutments were too light for the unnecessarily large amount of lateral pressure or thrust which the inefficiently-designed ring inevitably delivered to them.

The bill of indictment went further. Power, Luten claimed, was guilty of not treating reinforced concrete in its own right – the same critique the *ENR* editors and Tyrrell had made rather more generally of architect-dominated, City-Beautiful design. Power "failed utterly to appreciate the fact that he was working in reinforced concrete and has attempted to give the effect of a stone arch instead." Intent on imitation, the park superintendent decorated the ring and spandrels as though they were of cut stone and, to sustain the decor, filled the crown so heavily that the roadway humped over the span. Such pretense obscured the light and graceful lines to which reinforced concrete naturally lent itself.

III.1B The Counterattack on the Washington Street Bridge

The 1913 flood, which provided Luten a cornucopia of contracts along the Wabash River, also washed out a number of Indianapolis bridges including the steel structure on Washington Street over the White River. City and county officials had contended over this crossing in 1902, and the City-Beautiful advocates had narrowly lost.[9]

In 1913, however, the county surveyor sought to avoid yet another battle over the West Washington Street Bridge by developing a plan in consultation with the Indianapolis City Engineer and the engineer for the park department. The City Beautiful supporters came to an accommodation with the county engineer in advance. But all reckoned without the Marion County commissioners and Daniel B. Luten. When the commissioners opened bids in July 1914, they found that the lowest proposal came from the National Concrete Company based on a Luten design.[10] Much to the consternation of the city's officers, the county commissioners decided the lowest bid was the best and voted to award a contract to the National Concrete Company.

Luten planned for an eight hundred and forty-four foot structure for Washington Street with the seven arch spans ranging from one hundred and twenty feet at center to one hundred and six feet at the ends and all on a sixty-degree skew (Marion #4396).[11] The new bridge allowed seventy-five percent more waterway clearance than the plate-girder spans that preceded it.

The design was vintage Luten with linear or semi-elliptical rings, with a single line of reinforcing moving in stages from near the bottom at the crown to near the top over the haunches and then continuing down into the substructure, making as usual for slim piers with rounded cutwaters and somewhat modest abutments.[12] Luten used his spandrel walls to stiffen the rings, thereby allowing for some reduction in the size of both the walls and the rings. He put the walls alongside rather than on top of the rings. To integrate the walls with the rings, Luten ran transverse rods through and then out of the ring where they were bent upwards into the walls. The bridge carried a fifty-eight-foot roadway with interurban tracks and a pair of nine-foot sidewalks.

Wherever Luten could control it, as he could with the National Concrete Company, he insisted on a high level of craftsmanship in construction. He took pride in professional appearance, and he thought good craftsmanship contributed to public notice and approval. The Luten Bridge Company also instructed its superintendents of construction to remember "that the completed structure will in future years be measured solely by eye, and it should be so built as to present a pleasing appearance. Defects of all kinds are to be avoided, of course, but defects that will be apparent to the eye of the casual observer should by all means be eliminated."[13] For Luten, experienced engineering and skilled construction generated a form of

(Marion #4396)

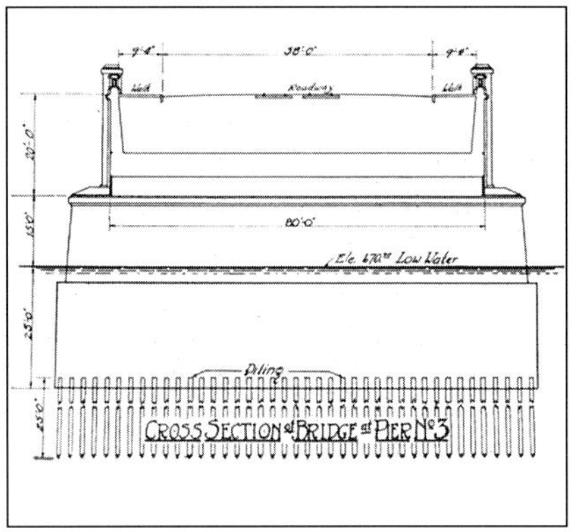

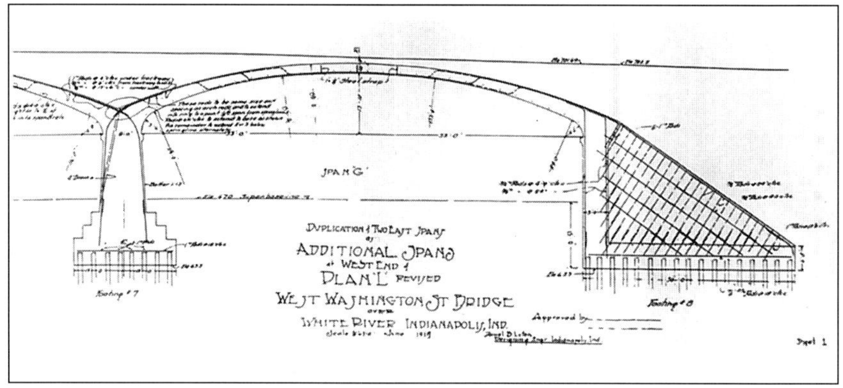

West Washington Street Bridge in construction (right, upper) and as completed (left, upper). Note how the bush-hammering of the spandrel walls and pier pilasters highlighted the rings and the rails. Timber piling anchored Luten's piers deep in the bed of the White River (left, lower). With Luten-system reinforcing extended across the arch rings and well into the piers and abutments (right, lower), the structural integration allowed for lighter bridge elements than when not harnessed together.

utility both artful and profitable: professional gentility merged with craftsmanship. Less efficient and poorly executed design, he believed, identified the inexperienced, the unlettered, and the unskilled.[14]

Luten's design conceded almost nothing, however, to the City Beautiful aesthetic as such. There would be no stone-facing on the West Washington Street Bridge, no refuge bays, no cartouches. The decor would all be appropriate for the material at hand and not imitative of another: the rings would be lined into the spandrel walls and polished, the pier pilasters raised out from the walls a little and panelled, and the concrete in the pilaster panels and the walls bush-hammered to give it a grainier texture and darker appearance. His one concession to the neoclassical came in the parapet rails where he inserted Pettyjohn balusters made with a special mixture of cement that turned them darker than the copings which held them.

Leave it to county commissioners to find a less costly deal better than a fixed one; mix in Indianapolis' City Beautiful officers who felt doubly betrayed by the award of a contract to a man who had ridiculed some of their bridge design in the national press; and the jilted party headed for courtly battle.[15] The city's civil engineer along with the park department's engineer and its consulting landscape architect, George Kessler, collected ammunition for a surprise attack not on decor but on structure. The engineers studied and photographed Luten arches constructed in Indiana. Through Kessler's extensive contacts, they also queried state-highway engineers in Missouri, Kansas, and surrounding states as well as some consulting engineers in the Plains states friendly with landscape architects. Finally, they turned to Dr. W. K. Hatt, a Purdue professor of engineering with whom Luten had occasionally scrapped in print since Luten, too, had taught at the university.

The City Civil Engineer's opened his formal public blast against Luten design with the claim that it was overpriced. Since Luten used considerably less steel and concrete than the county surveyor had prescribed, the ground on which the bidding had been conducted was not level. The total number of dollars bid to construct the Luten design might be lower, but the cost per unit of bridge structure was higher.

Because Luten apparently used almost one million pounds less steel and seven thousand cubic yards less concrete, the city engineer declared further that Luten's design was "much inferior in strength and stability" to the county plan. The collective critic engineers found Luten's rings and piers too thin and his reinforcing, because single-line rather than two-tier, inappropriately placed and too sparse. Indeed, the engineers predicted that Luten's arch rings would crack and his reinforcing rupture.

After the city's staff engineers had opened the battle, Indianapolis' Mayor Bell weighed in with a public statement of concern about the Luten design, and Henry Harmon, then state senator, tried to secure an injunction from the Marion County Circuit Court to check the county commissioners' contract award. The court refused to issue an injunction, and the city's offensive fizzled like a punctured balloon.

Luten had regularly refused to make any public comment on the charges, some of which he must have regarded as the expression of antiquated engineering while others screened matters of decor in smoke. The fact that in 1915 the city itself contracted with the National Concrete Company to extend the Washington Street Bridge with two spans built to the same Luten-design specifications that its staff engineers and their advisers had so roundly declared as unsafe just a year earlier suggests that the structural issues were mostly tactical parries in a fight largely over appearance.

THE WASHINGTON STREET BRIDGE

Two engineering firms evaluated the structural integrity of the contested Washington Street Bridge over the White River in 1989, some seventy years after its construction. The engineers "found no signs of distress in the concrete arch rings or spandrel walls that can be attributed to lack of enough reinforcing steel or concrete in the structure."

"In summary, I would say that none of the distress found in this bridge was due to structural underdesign of the bridge elements."
– *Stephen F. Weintraut, Butler, Fairman and Seufert, Inc., Indianapolis*[16]

"WHAT IS THE USE OF BUILDING UGLY BRIDGES, WHEN THEY CAN BE MADE BEAUTIFUL FOR THE SAME COST?"[7]

Selecting an appropriate bridge design for a particular site could involve, as Luten knew, more than structural matters. While, for example, park bridges ought to be as structurally efficient as any, their special context needed to be acknowledged in the form and the decoration used. In 1902, Luten designed the Columbian Park foot-bridges in Lafayette with flat and slim arch rings and decorative cast and wrought iron railings (Tippecanoe: West Br).[8]

For Indianapolis' Brookside Park in 1903, Luten used a more conventionally shaped tied-arch with wingwalls curved into the spandrels, all topped with a concrete railing of his own design. Following the curve of the wings into the spandrel walls, the railings carried open-rectangle panels under arced copings bordered by indented endposts capped with hemispheres.

To his arch, Luten suggested a way to enhance the vista at Lake Lure in North Carolina. He proposed a large refuge bay on each side at midspan, one of which would overlook Chimney Rock Mountain. For Shakilt Bridge, Luten designed a pair of concrete arch ribs topped by a timber deck and railing to harmonize with the rustic decor which parks professionals often preferred. While the less obvious ribs and abutments were made of "permanent" material, the more obvious and more bio-degradable deck and railings could be replaced relatively easily and inexpensively.

(Marion #2516)

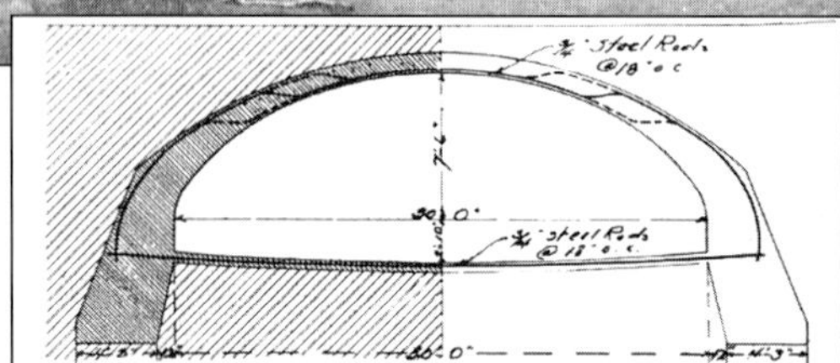

NOWLAND AVENUE BRIDGE, INDIANAPOLIS, INDIANA.

Proposed Island Bridge, Rutherford County, North Carolina (top). Shakilt Bridge (bottom).

III.2 "THE ARTISTIC DESIGN OF CONCRETE BRIDGES"

Luten never responded to City-Beautiful critics with rebuttals or extended explanations of his own principles of artistry in design. He regarded the writing of books and treatises on aesthetics or on theory as the work of academic engineers whose returns came more in increments of status and reputation than of dollars. When, however, a young and aspiring associate of Luten's found an opportunity to advance his career by writing on artistic design, Luten helped him provide the fullest statement of their largely shared views.

Born near Kirklin in Boone County in 1884, William J. Titus graduated from Indiana University with a major in mathematics and subsequently enrolled in civil engineering at Purdue University.[17] He worked temporarily for Luten before entering and ultimately completing a graduate program in civil engineering at the University of Wisconsin. Thereafter, Titus served in the engineering department of the Chicago, Milwaukee & St. Paul Railroad in Chicago before returning to Luten's National Bridge Company in Indianapolis.

At Wisconsin, Titus studied with George Hool, who was beginning an important and influential publishing career disseminating design instruction on reinforced concrete building. Titus' prior association with a leading national practitioner in concrete bridge design may at least partly explain why Hool asked Titus to draft an essay on artistic design of concrete bridges for the third volume of *Reinforced Concrete Construction*. Enormously successful, the three-volume work ultimately went through two editions and numerous printings before the Great Depression.

Titus must have been delighted to accept Hool's invitation, especially after the student had won his former and future employer's help. The photographs included in the thirty-page section reveal Luten's unacknowledged participation.[18] Of Titus' fifteen prints of artistically designed bridges, nine are easily identifiable as Luten's. Only the lampposts on one Luten design came in for secondary criticism from among the more than thirty prints of non-artistic work. Moreover, a number of the photographs of bad or, in Luten's own terms, inexpert design came directly from the files of the National Bridge Company.

III.2A The City "Ginger-Bread"

Nothing more distinguished the Titus-Luten position on artistry from that of H. G. Tyrrell or the *ENR* editors than their respective views on City-Beautiful bridge design. While both the academic engineer and the editors argued for cooperation between architects and engineers based largely on engineering values, Titus and Luten would not seek victory over the architect through some process of subversion or in nominal eclecticism. They declared war on the architect and his City-Beautiful structures.

"It is better that we should be content with plain, honest, solid, useful bridges than that they should be adorned according to the standards that have governed the ornamentation of dwellings and public buildings," Titus opened. "We have tended to load down many of our supposedly beautiful city bridges with ginger-bread ornamentation and every imaginable kind of inappropriate detail."[19]

Titus relegated his detailed discussion of ornamentation to the end of his essay, "because if it [ornamentation] should be forgotten entirely the result would more often be fortunate than otherwise."[20] Decoration "of any kind ... is likely to dwarf the bridge and destroy the harmonious effect." Titus and Luten, for example, would ban statuary and "totems" from the deck and omit refuge bays except at the ends of a bridge where they claimed pedestrians would find the best views of the scenery. They considered keystones and voussoirs on the rings to be "inappropriate and untrue." Figures in relief on the spandrel walls "always give the impression that the would-be artist has taken advantage of his subject to display his lack of taste." Without doubt "the spandrel of an arch is no place for a buffalo," Titus quipped sarcastically under one photograph. Both Titus and Luten objected to stone-facings which "disguise" concrete, dismissed other forms of decorative imitation, and criticized merely decorative, structural extensions like raised panels.[21]

Titus cited many examples of poor City-Beautiful design, but he placed George Kessler's locally-celebrated Capitol Avenue Bridge in Indianapolis (Marion #2501) at the head of his list. Under a photograph of the bridge, Titus inscribed the following indictment of the nationally recognized landscape architect's work: "Excess of details, too elaborate panels; lack of balance in arch spans, resulting in ginger-bread effect."[22] Luten may have inspired the decision to single out Kessler's design for criticism. Even if not the instigator, he would have considered it fair return for the City-Beautiful cadre's assault on him over the Washington Street Bridge.

The more out-of-sight, the less artistic H. G. Tyrrell thought a bridge needed to be: "less ornament is generally sufficient in smaller towns, and only in remote districts, seldom or never seen, should aesthetic design be neglected." Titus and Luten adhered to a different standard. "The truth of the matter is," Titus concluded, "bridges in the country districts are more often designed to suit their surroundings than bridges in the city." They can be, and many indeed are, "dignified and harmonious structure[s] ... with little ornamentation."[23]

III.2B Main Features of Artistic Design

Titus' preferred principles of artistic design – clarity of purpose, symmetry, style, harmony – underscored the general engineering values that the *ENR*'s editors had expressed in a piece-meal fashion and that H. G. Tyrrell had somewhat more systematically articulated. At most, Titus emphasized and shaded his description of those principles somewhat differently.[24]

Above all else, Titus wanted the designer to express clearly the structure at work, to display unmistakably its functioning strength while taking account of its setting. The bridge's style should conform to the environment and yet dominate its surroundings. Rugged terrain suggested bold and simple bridge lines; low country favored sweeping structure lines; parks called for more grace, detailing, and even quaint appearance; and commercial streets required spans which complemented surrounding buildings.

Like Luten, Titus endorsed a single form of concrete bridge, the arch, "one of the most beautiful of engineering forms." Titus thought arches "possess a boldness and an *inherent* beauty of line" that strictly horizontal bridge forms could not approach.[25] If for some reason the designer must mix forms in the same structure, the forms should be visually segregated from each other. Where, for example, beam approaches flanked arch spans, piers at the borders of the different forms should be enlarged to separate the one form from the other. Titus admired the parabolic shape of the open-spandrel arch, although he found the semi-ellipse from which Luten derived his preferred "linear" shape for filled-spandrels "probably the most graceful form."

The arched spans of a structure had to work and to appear to work as an ensemble. A well-designed bridge was centered around the middle of a span and not between spans. From this followed the rule that arch bridges carry an uneven number of spans, that the center span should be the widest, and that the other spans file away from each side of center by size with the shortest at the abutments. Since the rise of each of a bridge's spans must be related to its length to keep the rings' arcs alike, the top line of a lengthy structure might artistically follow "an easy vertical curve." A coped railing strong enough to guide and guard traffic needed to bound the roadway and sidewalks. The rails' dado might appropriately contain balusters or filled panels.

When describing "the basis" of his "aesthetic design," Luten periodically recited the aphorism that the "curve of duty is the line of beauty" that led Titus as well as Luten to emphasize the arch rings.[26] To accent them visually, the rings should be elevated, thickened towards the springing, and polished. Not surprisingly, Titus endorsed Luten's practice of roughening the spandrel walls to develop a contrast of color and texture with the ring.

Structurally integrated with the other elements of the bridge, piers and abutments must be included in the visual ensemble, too. Titus preferred the "tall and slender" piers that elastic treatment allowed and approved Luten's light and curved cutwaters. The abutments, in turn, had to harmonize with the piers, "but should seem enough heavier to carry the thrust coming on them." In all cases, the bridge should *seem* to be what it is.

III.2C Ungainly Beams

Not surprisingly, the bridge forms Luten widely condemned as inefficient, Titus found generally unartistic. Among these, the tied arch or concrete bowstring which carried arched chords alongside and above the roadway had a number of vocal admirers, especially in the region from Cincinnati, Ohio, to Portland, Indiana, and across the upper Midwest and Plains states where the form was widely known and appreciated as the Marsh arch. Titus dismissed bowstrings in concrete as "scarcely more than a novelty." But those strictly horizontal common carriers nestled here and there across the countryside and on secondary city streets fell prey to the scorn of Titus as they had to the disdain of Luten and H. G. Tyrrell.

As far as Titus could see, structural efficiency, the engineer's *sine qua non*, negated artistry in concrete beams and girders. Such spans did not even "approach" the "boldness or inherent beauty of line" found in arches. Because of their limited span-length, the predominant horizontal line of a long beam or girder structure was inevitably interrupted by many vertical piers often designed thin in the name of economy. If kept fairly low, as Titus thought best, they looked in Tyrrell's words like "perforated dams." If raised on tall columns, the vertical lines of the substructure overwhelmed the horizontal line of the superstructure, thereby precluding a harmonious whole. Titus though it "scarcely possible to hide an appearance of weakness and obtain a truly artistic effect" in such structures.[27]

Titus summed up artistic design for himself, for Luten, and for engineers generally as a kind of implicit moderate structuralism. "Expressiveness to most people is the chief source of beauty," he concluded. When the characteristics of the material and the functioning of the design "are destroyed rather than emphasized, the effect is not only flat and characterless, but may even be untruthful and insincere." Following "the curve of duty" as "the line of beauty" implied structuralism. Dressing those curves adequately "emphasized" or tastefully accentuated the structure without the distraction of ornate, pasty ornament.

"Girder bridge, well proportioned structurally, but inherently unsightly." – Titus

"A well-designed girder bridge showing clearly the limitations of the girder." – Titus

"Pile trestle in concrete. Supports appear to be inadequate, and the entire effect is one of cheapness accentuated by the gas pipe railing which is not even panelled in harmony with the spans. This structure is a wooden trestle molded in concrete." – Titus

(ENR, 1912)

Others before Luten planned bridges whose functions were not limited to carrying traffic safely from one river bank to another. In 1770, for example, Robert Adam designed the Pulteney Bridge over the River Avon in Bath, England, to support buildings (above). Adam extended the decor of the buildings to the spans.

Two major floods raged through San Luisito shortly after the new bridge's completion. While "practically destroying a great part of San Luisito," the floods inflicted only minor damage to the bridge. The ENR *reported rather fully on the construction of the San Luisito Bridge and the floods that soon followed (8 April, 23 September 1909).*

In 1908, Luten received the opportunity to design a replacement for two steel spans of the San Luisito Bridge destroyed by fire. The spans had carried a market house over the Santa Catarina River in Monterey, Mexico. Luten designed a non-combustible, reinforced-concrete replacement with a two-hundred-foot long and fifty-two-foot wide market building housing twenty stalls on each side of central arcade. The roof and its supports, the inner and outer walls of the market, and the spans themselves were all of concrete. Only the shutters for the windows, the screens on the stall fronts, and the stalls' counters were made of other materials.

NOTES

[1]H. G. Tyrrell twice repeated the first quote without attributing it. Henry Grattan Tyrrell, *Artistic Bridge Design* (Chicago, 1912), 25. Banham was quoted in David P. Billington, *Robert Maillart's Bridges: The Art of Engineering* (Princeton, N. J., 1979), 28.

[2]Tyrrell, *Artistic Bridge Design*, 10, 16, 24-25.

[3]"The Architectural Treatment of Concrete," *Engineering News-Record* [*ENR*] (8 August 1907) 58: 150; editorial, *ENR* (5 January 1911) 65: 17. Tyrrell agreed that for "large or important bridges, unless the engineer himself is proficient in aesthetics, which is seldom the case, it is better to have an architect associated on the work from the first, not merely to decorate the bridge after the engineer's work is finished, but to assist in the design from its inception." Tyrrell, *Artistic Bridge Design*, 38-40; William B. Wilson, *The City Beautiful Movement* (Baltimore, 1989), 2-3, 278, 285-90.

[4]"The Architectural Treatment of Concrete," *ENR* (8 August 1907) 58: 150.

[5]Sources for this section include Tyrrell, *Artistic Bridge Design*, 41-49; "Ornamentation in Engineering Structures," *ENR* (19 August 1909) 62: 203. For the views of two contemporary Canadian engineers, see Steven A. Walton, "Canadian Aesthetics of Early Reinforced-Concrete Bridges," *IA* (1995) 21: 5-14.

[6]For a discussion of Power and his design, see I.2A. Daniel B. Luten, "The Elastic Theory and a Faulty Arch," *ENR* (9 January 1908) 59: 43-44.

[7]Daniel B. Luten, *National Bridge Company* (Indianapolis, 1905), 18.

[8]Daniel B. Luten, "The Flattest Concrete-Steel Arch in the World," *ENR* (13 November 1902) 48: 402. See II.3B.

[9]See I.2A.

[10]Marion County, "Commissioners Record," 45: 191, 208, 236, 364; 46: 8, 9; Construction News, *ENR* (30 April 1914) 71: 257; (23 July 1914) 72: 60.

[11]The original contract with the county involved five spans for $331,110. In July 1915, the City of Indianapolis awarded National Concrete a contract for $110,000 to add two spans on the west end of the bridge as part of a flood control plan. Luten anticipated the extension in his original planning. Construction News, *ENR* (8, 22, 29 July 1915) 74: 29, 57, 76.

[12]"Construction of Washington St. Arch Bridge at Indianapolis," *ENR* (9 December 1915) 74: 1142-44; *The National Concrete Company of Indianapolis: Contractors for West Washington St. Bridge over White River, Indianapolis, Indiana*; Daniel B. Luten, deposition (3, 4 August 1916), District Court of the U. S. for New Jersey, *Daniel B. Luten vs. The Dover Construction Company: Complainant's Record (Supplementary Testimony)*, 9, 20; *The National Concrete Company of Indianapolis: Builders of the West Washington Street Bridge over White River* (October 1916).

[13]*Instructions to Superintendents of the Luten Bridge Company for the Construction of Reinforced Concrete Bridges* (York, Pennsylvania, undated), 20.

[14]Daniel B. Luten, *How Good Concrete Bridges May Be Secured* (Indianapolis, c.1914); Daniel B. Luten, *Artistic Design in Reinforced Concrete Bridges* (Indianapolis, 1927).

[15]The fracas is most fully reported, although without analysis, in "City Continues Fight on Bridge" and "Report on Investigation of Bridges of Luten Design Made by Engineer," *Indianapolis Star* (16 September 1914), 6.

[16]Wiss, Janney, Elstner Associates of Northbrook, Illinois, performed an extensive structural evaluation which was included in Butler, Fairman and Seufert, Inc.'s *Design Study Report for the Old Washington Street Bridge in Indianapolis, Indiana* (Carmel, Indiana, 1989). The quotations are from a letter of Mr. Stephen F. Weintraut to James L. Cooper, 22 November 1993, in possession of the author. Mr. Weintraut's full and prompt response to the author's queries is much appreciated. Patrick K. O'Brien, "Capital City Landing: Subsurface Archaeological Testing at Sites 12 Ma 622 and 12 Ma 623 and Historic Summary of the Old Washington Street Bridge in the Area of McCormicks Rock, White River State Park, Indianapolis, Indiana," (Glenn A. Black Laboratory of Archaeology, Indiana University, Bloomington, 1996) 96-1.

[17]For Titus' biography, see William J. Titus, deposition (12 August 1916), District Court of the U. S., *Daniel B. Luten vs The Dover Construction Company: Complainant's Record (Supplemental)* (Indianapolis), 77-78; Purdue University, *Alumni Directory, 1875-1934* (West Lafayette, 1934), 187; *Indianapolis Star* (16 July 1950), 2nd sect: 5.

[18]William J. Titus, "The Artistic Design of Concrete Bridges," in George Hool, *Reinforced Concrete Construction* (2nd ed., 3 vols., New York, 1916), III: 509-42.

[19]Titus, "Artistic Design," in Hool, *R C Construction*, III: 509-10.

[20]Titus, "Artistic Design," in Hool, *R C Construction*, III: 523-42.

[21]Titus, "Artistic Design," in Hool, *R C Construction*, III: 511, 514-15. Although Luten disapproved of decorating concrete arches like stone, he did so when the specifications required it (e.g, in Marion County). For a cautionary tale told about a stone turned concrete arch builder, see Luten, *Artistic Design*, 5.

[22]Titus, "Artistic Design," in Hool, *R C Construction*, III: 509. For earlier discussion of the Capitol Avenue Bridge, see I.2B.

[23]Tyrrell, *Artistic Bridge Design*, 35; Titus, "Artistic Design," in Hool, *R C Construction*, III: 510.

[24]Titus, "Artistic Design," in Hool, *R C Construction*, III: 511-22.

[25]Emphasis supplied.

[26]Daniel B. Luten to the Editors, *ENR* (June 1902) 47: 484.

[27]Titus, "Artistic Design," in Hool, *R C Construction*, III: 512, 517, 520, 522; Tyrrell, *Artistic Bridge Design*, 41-42.

IV

Engineers as Public Administrators: Good Roads and Rational Bridge Design

... Roadbuilding has developed from a mere pathmaking task of local import to a public works enterprise of interstate and national importance. It stands on the threshold of a further development exceeding in importance and cost all other means of transport.
–*C. S. Hill, Associate Editor,* Engineering News-Record *(1924)*[1]

The World Columbian Exposition boosted not only the landscape architects' City Beautiful but also the Good-Roads movement which had been gaining momentum since the early 1880s. Advocates seeking improved highway transportation outside the central city gathered in one of the White City's congresses to form the National League for Good Roads. Delegates from state road improvement groups, agricultural organizations, land-grant universities, and the League of American Wheelmen, a largely urban association of biking enthusiasts and manufacturers, recommended national aid through a road department within the Agriculture Department.

By appropriating $10,000 for the Secretary of Agriculture to begin "road inquiries in regard to the systems of road management throughout the United States," Congress endorsed the request of the National League for Good Roads.[2] Few if any League leaders or congressmen realized that the federal-state highway program they were beginning would grow well beyond their farthest-ranging dreams. They started a process which soon affected twentieth-century U. S. highways and bridges more than any other single development, established new patterns for highway design and finance, and eventually altered how and where Americans worked and lived.

The highway system and the bureaucracies arising around it following 1894 opened employment opportunities for hundreds of engineers, and spokesmen articulated a new public role for their profession. By 1900 no longer needing to argue that engineers required formal training in scientific thinking and rational analysis, the editor of the *Engineering News-Record* [*ENR*] now found being "expert in their profession" inadequate. "It is just as important that the graduates of our technical schools shall be men of character, fit for social, economic and political leadership."[3]

Editor Charles W. Baker regarded the leadership of engineers as necessary, not just optional. The "silent industrial revolution" had wrought an unholy alliance between politicians and large corporate trusts that had to be controlled for the general welfare. While the graduates of "certain institutions because of aristocratic traditions or the influence of wealthy benefactors may range themselves on the side of the barons," engineers graduating from land-grant colleges and state universities "will be found fighting under the banner of democracy:"

> The people must trust someone. Experts must conduct our public business or colossal waste and destruction will be wrought through sheer ignorance. We must have experts – engineers – who can be trusted, who are incorruptible … .
>
> In the community where he lives, in the states, in the national government, engineers are coming to the front more and more as safe and sane leaders. I can show you to-day communities, created by the engineer, where he is the main influence upholding law and order, planting the school, rearing the church, surrounding the community with all the things that make life worth the living … .
>
> The engineer is the great civilizer.

While engineers neither were nor are now much given to political theory, Baker's claim for leadership based on their special professional expertise offered an updated version of the Founding Fathers' arguments for filtering and refining the popular will. To the extent that an individual can convince the larger community that his or her knowledge and skill represent considerably more than common knowledge and sense, such a person may be anointed to a special status which allows him to monitor, administer, and adjust public or electorally-determined opinion.

Professionally-trained engineers regularly distinguished themselves from politicians who used native wit as a guide and whose expertise centered around electoral politics. Engineers claimed to favor the general over the personal or the special interest as well as the long-run over the ephemeral and fleeting. Politicians, on the other hand, weighed the interests according to their electoral clout and political contributions, and they responded to the whims of sometimes uninformed majorities. When looking into the mirror, engineers saw images of themselves as ideal civil-citizen servants putting the common good before personal reward and as able scientific administrators providing the best possible public products.

The engineers who came to dominate the Good Roads movement were not alone in promoting administrative science.[4] Since the end of the Civil War, civil-service reformers contested the tradition of the victorious party garnering the spoils of patronage. Towards the century's end civil service, in turn, merged neatly into the vision of Progressive reformers who saw no Democratic or Republican way to build a highway, regulate an industry, or administer the state. There was, instead, a precise, expert, scientific approach which yielded correct solutions. Leaders of a group of new professions from the applied natural and social sciences – agronomists, conservationists, economists, engineers, foresters, geologists, hydrologists, time managers – sought independence from partisan politics and special-interest groups under the banner of public administration, a banner that promoted additional and more-autonomous government bureaucracy.

(Washington County, off SR #56)

Left – Natural pier.

After *ENR* editor Baker, no one promoted engineers as public administrators more often and more publicly in the first decades of the twentieth century than Herbert Hoover, a mining engineer turned administrator and, later, President.[5] Hoover thought the engineer – "with his precision of thought, his capacity to analyze conclusions and to obtain the proper perspective of fact, and his ability to weigh the forces with which he deals" – could address social, economic, and political problems as few others might:

> There is somewhere to be found a plan of individualism and associational activities that will preserve the initiative, the inventiveness, the individual, the character of man and yet will enable us to socially and economically synchronize this gigantic machine that we have built out of applied sciences. Now, there is no one who could make a better contribution to this than the engineer, but to make that contribution our engineers in the future have got to have a broader and stronger place in our world affairs than they have today.[6]

(Boone #306)

House creep on Royal Street in Lebanon.

The engineers in the Good Roads movement accepted Hoover's and Baker's challenges to perform "public service."[7] College-educated engineers who had learned from their professors the virtues of practicing science, of studying a body of data systematically enough to find and arrange it around its general operating principles, could promote the general welfare through the development and application of technical knowledge. Operating in the public realm, scientific engineers were "technocrats" (before the term was coined) whose authority arose from their professional expertise.

IV.1 A GOVERNMENTAL PYRAMID FOR GOOD ROADS

Each of the three men who headed the federal road office from its inception through the First World War – Roy Stone (1894-1899), Martin Dodge (1899-1905), and Logan W. Page (1905-1918) – built upon the foundations laid by his predecessors. For Roy Stone, a civil engineer from New York and a bicycle Wheelman selected to head the newly created Office of Road Inquiry [ORI] in 1894, government's agricultural technocrats provided the precedents.

The Secretary of Agriculture insisted that the ORI concentrate on the gathering and dissemination of technical information and avoid all appearance of partisan politicking. The Department had a solid reputation as an applied scientific institution that tried to address the farmer's practical problems. Like his departmental colleagues, Stone understood that the technical knowledge which the ORI gathered could generate political clout by a different route than the conventional politician typically followed.

Stone laced his preaching of the gospel of rural-road improvement with the latest and best data and sought cooperative ways to convert others. To make the ORI into a central clearinghouse for information on road conditions and construction, he appealed to state officials for data that could be compiled into circulars and bulletins. ORI also inaugurated technical tests of materials and in 1897 began an object-lesson construction program. With great fanfare ORI engineers built about a mile of demonstration road, much as agricultural extension experts had planted and cultivated sample rows of corn, beans, or wheat in strategic places for farmers to see.

The first two twentieth-century heads of the federal road office, Martin Dodge and Logan Page, each in his own way, erected a far more elaborate agency superstructure on the Stone foundations. Dodge, a lawyer, articulated a rationale for the road bureau at the turn of the century that his successor, Page, an engineer, later repeated with more figures appended. Logan Page also promoted the engineer's scientific image.

IV.1A Farm-to-Market Roads

To engage his rural and agricultural audiences, Dodge regularly proclaimed that railroads and steamships provided quick and inexpensive transportation for much of America's industrial product. But "wagon road haul" remained critical for getting agricultural products to market. In "the bottomless state of rural highways" he found mostly "the dominance of King Mud." Dodge calculated that "it cost more to move a bushel of wheat or a ton of hay ten miles over the average country roads of the United States than to transport the same burden five hundred miles by railway or two thousand miles by steamship."[8] He made the bureau's principal focus explicit: farm-to-market roads.

House-moving day across a Luten-design arch in Washington County.

None of the thousands of units of local government which had charge of the roads could plan for the whole. "What has been everybody's business has come to be nobody's business," Dodge concluded, "and it is very largely due to this that the United States has probably the worst system of public highways of any civilized nation of the first class." To remedy the defect required state and federal intervention. The states would have to take greater charge of their units of local government while the federal government had to become the "advisory or directing member in the partnership." The power to direct the partnership put the federal bureau at the top, the authority to plan placed state highway agencies next, and the shortage of expertise left local government on the bottom rung of the ORI's ladder.

Page embroidered Dodge's rationale for "centralized control of highway construction and maintenance in the United States."[9] When speaking to regional audiences, he at least sometimes offered figures for how much producers who shipped over the better centrally-controlled farm-to-market roads would benefit above the local costs incurred for their improvement. With national audiences, he reasoned more generally about how "centralization makes for economy and efficiency in the administration of public roads." In these cases he turned to the international scene for comparisons, again using figures to support his general proposition. Page found the least efficient and most costly highway programs in Great Britain where local governments maintained control (and empiricism incidentally reigned) and the most efficient and least costly in France where highway control was centralized (and rational science dominated).

Parts of the engineering community found the arguments and figures which Dodge and Page used in their campaigns for good roads exaggerated and too free-wheeling. "The fool propaganda of good road enthusiasts has widely disseminated the notion that good roads are cheap to build and once built cost little to maintain; and this notion has caused a vast amount of trouble for engineers," commented the *ENR*'s editor who was aghast, for example, at states in the alliance system borrowing on a fifty-year repayment schedule to construct roads that would likely last no more than twenty years before having to be rebuilt.[10]

The writer of one letter to the *ENR* specifically parsed "the erroneous and wrong deductions" that Page drew with his figures for "the benefits to follow from the building of costly roads" under central control.[11] Indeed, no economic or technological reason early in the twentieth century impelled either Dodge or Page to leap from local to state and national control of highway development. Highway-serviced markets operated largely as secondary ports or rail heads at which agricultural produce was assembled for rail or ship transport to the national market centers like Chicago, St. Louis, New Orleans, and New York City. Highways did not serve as branch lines, let alone trunk lines. Kinetic energy powered steam engines on trunk or branch lines; animal power moved cargo short distances over local roads to rail-heads and wharves.

Given that animal-drawn wagons carried freight on the highway, almost any county government that cared could provide adequate roads locally on the basis of the state of the art. The quality of the roads in an area was not, as Dodge alleged, "everybody's business," since poor-to-nonexistent transportation for the product of a single local area would affect only the local producers, not the national market. Surely the local producers could throw the county scoundrels out at the next election if crops rotted in the fields because of inadequate roads!

Not until the First World War placed unusual logistical demands upon the rail system did need provide substantial grounds for more centralized control of highway development. The reason for Dodge's leap in 1901 and Page's in 1909 was most likely rooted elsewhere – in their judgment that, largely unaided by professional public administrators, local officials could not deliver good governance. Transferring control of road and bridge programs from elected local officials into the federal-state alliance, Dodge and Page implicitly argued, bypassed inadequately informed public opinion and its elected representatives who also sometimes succumbed to bribery and serviced special interests in favor of disinterested and scientific policy-making by professional experts.

IV.1B Towards Rational Administration

Logan Page, the third director of the (renamed) Office of Public Roads [OPR], did far more than reiterate Dodge's general prescription for which level of government should have what kind of road responsibility. Page invested his considerable background in scientific engineering in his new position. Before serving as director of the OPR's testing laboratory, he had graduated from Virginia Polytechnic Institute as well as from Harvard University's highway-engineering program, headed Harvard's Lawrence Scientific School testing facility, and made a pilgrimage to the Mecca of engineering science, France, from whence he borrowed and adapted both method and equipment.

Page used the ever-growing appropriations Congress made to expand the staff of college-educated engineers in part to solve practical highway problems through research. As graduates of engineering programs in which the emphasis on mathematically demonstrated mechanical theory increasingly reigned, Page's recruits slowly consolidated the OPR staff's preference for scientifically centered or rational engineering and extended the bureau's reputation for expertise in highway matters. The OPR exhibited its expertise in part through additional laboratory and field study research projects as well as with ever more demonstration roads.[12]

Page's accent on the OPR staff's credentials for rational or scientific analysis of problems fit hand in glove with his active lobbying for the creation and strengthening of state-aid road programs. Such programs transferred some county control of roads to an engineer-led state highway commission and often encouraged additional local deference to state advice. The OPR's visibility for rational analysis in turn elevated its status above the state agencies, strengthening the case for why they, too, should follow the advice and specifications that the national experts provided.

IV.2 TOWARDS RATIONAL, GOVERNMENT-CONTROLLED BRIDGE DESIGN

... As the business of highway building has developed under state and federal aid there is little occasion for the independent practitioner. Practically all engineering of main roads is in the hands of federal, state and county highway department engineers.
– *C. S. Hill, Associate Editor,* Engineering News-Record *(1924)*[13]

Federal road officials valued hierarchical integration far more than autonomous diversity as they expanded their highway domain. Besides bringing more states into the federal-state highway alliance and incorporating an ever larger number of roadway miles, they wanted to apply national standards within the system as well as beyond it. A subsidiary part of highway development, bridge design and construction was too important in the transportation chain to escape the federal-state alliance's will to standardize. Bridges accounted for about a third of all highway expenditures.

The OPR established a Division of Bridge and Culvert Engineering in 1910 to collect data, publish circulars, and construct demonstration bridges. Within a few years, the OPR published standard bridge specifications and prepared standard plans of various structural types for state and local use. The Bridge Division soon grew to a staff of sixteen in Washington, D. C., and each of the OPR's dozen regional offices contained at least one engineer assigned to bridges.

Since far more expertise went into designing and constructing bridges than went into roadways, engineers almost intuitively agreed that they ought to take full charge of bridge design and set the terms for bridge construction. The question among engineers was not whether they should design bridges but who among them should and under what terms. The leaders of the federal-state alliance favored socialized design for within the system – that is, design by government-employed engineers – and as much hegemony over local bridge building as was practicable although they were rarely entirely candid about the scope of their desired empire.

The decision of the alliance to monopolize bridge design within the system raised little immediate complaint, even though the OPR's insistence on only rational or scientifically correct design inevitably introduced less innovative and less efficient as well as more secure and more expensive

(St. Joseph #210)

St. Joseph River via the LaSalle Street Bridge, South Bend.

structures.[14] To be scientific, knowledge must be studied systematically and placed within the context of its general principles. It must, in short, be understandable within an accepted body of theory and measurable through mathematical expression. Invention, on the other hand, arises from imagination, insight, or intuition and is not necessarily packaged within some accepted body of theory or accompanied by mathematical formulas. The invention must function or work, but it need not be theoretically or mathematically explained to be considered as empirical knowledge. Today's empirical knowledge may become tomorrow's science when it is fitted into theory and measured. In short, the empirically-based generally precedes science-centered design because of the latter's additional requirements for theoretical and mathematical demonstration.

To begin to bring local governments within the federal-state alliance, state highway commissions tried the carrot before using the stick. They offered to provide local governments with rational bridge plans without charge, sometimes stipulating that local officials promise not to consider any competing designs for a structure. But all too frequently the state's offer of "free" bridge design

(Lawrence #197)

Over a straight and narrow arch to church in Heltonville.

– that is, design paid for with state and federal but not local taxes – did not win over local officials. Calculations of the *total* cost of erecting a bridge often dominated local decisions not to adopt state design: the cost of building with "free" state design might exceed that using private or local design, even though the latter had to be paid for from local taxes.

While encompassing a good bit of variation in different places and times, bridge design and construction in the localities usually followed a relatively common set of practices. Many Midwestern counties and cities employed their own surveyor-engineers as did the larger bridge fabricators or construction companies. Left to their own devices, through the bidding process the counties and cities regularly combined their in-house expertise – ranging from those employees trained on the job to those formally-educated – with the expertise of the construction companies. The county engineer typically prepared specifications for each particular site that all bidders had to address, and he sometimes proposed a bridge design as well. More often than not, a company might submit a proposal based on a plan of its own and a separate bid based on what the local engineer had prepared. The local surveyor or engineer and the elected officials who awarded contracts could then choose among the packages of plans and construction prices. Local practice regularly admitted empirically based designs into the bidding process and often rewarded them when proposed by persons whom the local officials respected and where they promised the lowest total cost.

Although required specifications might vary from place to place within as well as between counties and cities, and the empirical plans submitted in the bidding process could cover a dramatic range, the federal-state alliance system sought rational regularity. In a 1908 report to the state commission, for example, Illinois State Highway Engineer A. N. Johnson laid out some of the assumptions behind the bridge plans for short spans that he offered to localities without charge. Johnson prepared a scientifically correct set of plans, a copy of which he proposed to give to each community. His standardized plans presumed that requirements for every bridge of a particular length anywhere in the state should be considered the same and should, therefore, ensure safe passage for the heaviest, fastest, and widest loads found anywhere in the state.

To ensure that local communities used his standard plans, Johnson also supported legislation to require "that all plans for this work should be approved by the state and the work carried on under State specifications. This would be necessary to safeguard and secure to the taxpayers the standard of design and construction that work of this character requires and public safety demands."[15] The Illinois Highway Commission was by no means alone in considering or securing laws requiring local governments to obtain state approval of bridge plans, although not all states attempted the level of standardization that Johnson proposed.

When state highway commissions or the OPR began to press local governments over bridge design, entrepreneurial engineers, who would lose some of their design business to the state, were the first to complain. Wherever state approval of design became mandatory, local officials often protested reduction of autonomy and higher prices of construction based on rational or scientific design. But nothing unified and galvanized opponents of mandated bridge design more effectively than the OPR itself, when in 1911 it followed the by-then standard gambit of offering "free" bridge design to local governments anywhere in the United States.

The editor of the *ENR* reprinted parts of the OPR bulletin and then attacked as monopolistic federal intrusion into local bridge design and construction.[16] The editor found no fault with the alliance's designing and constructing bridges within its own system, and he waffled over state intervention in local government. But he saw "no reason why bridge engineering should be turned over to the Government as a monopoly. The achievements of American engineers on bridge design and construction would never have been recorded under any central bureaucratic organization." If, as the editor insisted, those whose names appeared on the OPR bulletin "have any regard for professional ethics or desire to retain their standing with their professional brethren, they will be prompt to withdraw from their present position."

Logan Page tried to deflect the heat of the *ENR* editorial by sidestepping the issue of design and speaking almost exclusively of construction. "It is not, nor has it ever been, the policy of this Office to enter into competition with private engineers in road or bridge construction," he claimed. "Our work is almost entirely of an educational and experimental nature ... and follows the identical policy pursued for the past twelve or thirteen years in its object-lesson road work." Since the OPR generated ever more jobs for engineers, Page concluded, "for the *Engineering News* to take the stand that this Office is trying to destroy the engineering profession is preposterous."[17]

Sensing or at least claiming retreat as well as an absence of candor in Page's answer, the *ENR* editor advised private engineers to notify OPR of any case of apparent competition between the office and the engineer so the office might "withdraw its offer in the particular case." If, though, Page would not argue the issue with the *ENR*, the editor of a widely-used pocketbook for civil engineers did enter the fray. John C. Trautwine, Jr. broadly and lengthily defended the OPR's policy.[18]

Trautwine characterized the *ENR*'s defense of bridge design by individual engineers as "curiously archaic" and "a bit of reactionism." The individual design engineer, Trautwine argued, rightly "has no chance" against the OPR in the American system: "The stronger organization absorbs or crushes the weaker organization and the individual, with the result that the public gets better service." "Irresistibly advancing from individualism (savagery) to socialism (civilization)," Americans had "but a foretaste of the wealth in store for us all when the evolution shall have been consummated."

For the engineering profession, Trautwine offered "snug places at the public crib." "Within the next half-century the private engineer ... will have gone the way of the private pump and well, and with similarly happy results all around." Trautwine found no better educational model for engineers than the French one, which in the United States appeared in the military academies of West Point and Annapolis. "One of these days," he predicted, "we shall see governmental schools of civil (non-military) engineering, whose students will be not only prepared, but destined, for `employment in a Government Bureau,' and whose scholarships will be prized as are those in our present military and naval schools." The OPR and its adherents thought they already had the academic anchor they needed for the highway alliance with the government-funded colleges of engineering in the states.

The *ENR* commended Trautwine for showing "directly" where "the terms of the offer published by the Office of Public Roads ... would lead if it were taken at its face value." The editor then met "our correspondent on his own ground" by suggesting that "if the Roads Office proposed to demonstrate that it really is able by `survival of the fittest' to do better and cheaper bridge engineering than the private engineer can do, it should charge to the clients the actual cost of the work," not subsume those costs in national taxes. "When a Government Bureau offers its services gratis for the public good, somebody has to pay the cost of performing that service. There is no economy or advantage in robbing Peter to pay Paul, or in making a gift to a certain small part of the public for which the whole public have to pay in taxes." Whether corporate or government, the editor believed, large-scale organizational work so stifled "personality" and initiative that the unfettered private engineer would inevitably triumph in a fairly-run race.

(Marion #2501)

Fall Creek via the Capitol Avenue Bridge, Indianapolis.

Trautwine identified and exploited an apparent fault-line in the *ENR*'s reasoning. The editor argued that large-scale organizations are inherently less efficient than individual entrepreneurs and that the OPR's single competitive advantage in bridge designing came from its sleight-of-hand on cost. Trautwine wondered why the editor did not raise the same indignant protest against state highway departments that meddled with bridge design. Indeed, the *ENR* had equivocated about state government intervention that some of its correspondents criticized.[19]

The *ENR* endorsed much "state supervision of engineering work." The editor accepted state management of bridge design on state highways. He also approved "state inspection" of local design and construction as "of great benefit to the taxpayers in improving the character of bridge construction, putting a check on petty graft in bridge contracts, and fostering better types of structure." Monitoring of local bridge design needed to be limited, however: "state supervision should not kill honest private competition." "The State Engineer should not undertake work as a designer except on purely state work built with state funds. His duty should be to guard the interests of the various communities in matters of safety and economy."[20]

While the *ENR* had helped to check hegemonic moves by the federal-state highway alliance over local bridge design and construction, the editor had by no means checkmated them.

IV.3 PATENT PAWNS IN THE WAR TO CONTROL DESIGN

Once the *ENR* had forced the OPR into a strategic retreat over direct federal intervention in local bridge building, the battle between private and federal-state bridge design returned to the state level. Many state highway departments continued to offer local governments "free" bridge design, and some lobbied legislatures for mandated design. In their struggle against the alliance, a few private designers deployed a federally-approved weapon, the patent. Alliance leaders, in turn, saw patents as an obstacle to their desired control of local reinforced-concrete bridge design as well as to design within their own system.

The patent system provided legal protection to empirically based notions and their promoters. To secure a patent, one needed to propose a new idea or device that worked. One did not have to offer a theoretical explanation for an invention to explain or justify it scientifically. Just as a host of persons patented metal trusses from the 1850s through the 1880s, so another group secured patents for concrete bridge systems starting largely in the 1890s. Major holders of concrete bridge patents included several mentioned above – Joseph Melan (1893), Fritz von Emperger (1897), Edwin Thacher (1899), and Indiana's Daniel B. Luten. Luten secured the largest number and widest range of patents with over forty by 1916.[21]

Patents gave individuals or groups monopolies over their inventions for a stated period of time.[22] They involved a *quid pro quo*. Consumers benefited from a stimulus to invention that introduced some desirable new service or that made it possible to perform an old service sufficiently more efficiently to reduce cost and price. The inventor enjoyed a legally conferred monopoly which might generate profits in excess of a competitive return for the life of the monopoly.

Two explicit assumptions underlay American patent law. First, inventors would inevitably invest less time and money in developing new ideas should competitors be able to copy an invention and thus share the profits. Second, availability of a desirable new product or process would make the patentee's monopoly prices less burdensome – better to have the product at a premium price during the life of the patent than never to have it at all. Whether, in fact, patents stimulate invention to the degree that the gains of having patent laws outweigh the losses remains an unanswered question.

The OPR-state alliance had very different objections to the patent system than the possible ill effects of temporary monopoly. The alliance leaders were, indeed, less interested in substituting competition for monopoly than in replacing one monopolist (the patent holder) with another (the federal-state alliance).

IV.3A Daniel Luten's Skirmishes with the Federal-State Alliance

Daniel Luten first confronted alliance resistance to private designers rather close to home. In 1906, Illinois State Engineer, A. N. Johnson, had started designing as many structures as he could for local governments without charge provided that they entertained no other designs during the bidding process.[23] Learning that Johnson was consistently refusing to allow Luten designs into competition with the state's, Luten conferred with the state engineer to no avail. Johnson apparently reasserted without explanation his refusal to allow Luten designs to compete with his own.

As Luten reported under oath, "I know of no objections that Mr. Johnson may have had against our design except that he objected to competing with us. I have never heard any criticism by him of any of our designs specifically." Indeed, Luten did "not even know that Mr. Johnson is opposed to patented bridges, but I have considered that that must be one of his reasons for his attitude towards our designs."[24]

Not all state highway engineers opposed to private and empirical bridge design could get the public doors slammed before Luten had entered, and not all were willing to sidestep his patented claims. In Iowa a series of confrontations between Luten and other private designers and patentees with the federal-state alliance changed the national landscape for reinforced concrete bridge design.

Iowa provided a classic antagonist for private designers when in 1904 the General Assembly appointed the Iowa State College of Agriculture and Mechanic Arts [Iowa A&M] at Ames as the state's highway commission under the direction of the deans of engineering and of agriculture. The legislature expected the commission to devise plans, establish roadway construction and maintenance demonstrations, and advise the counties on highway problems. The General Assembly also established and funded an engineering experiment station under the Iowa dean of engineering to test and analyze materials scientifically, to develop new devices, and to disseminate technical information. In the commission and the experiment station, the interests of academic engineers and federal-state highway bureaucrats soon become inextricably intertwined and stood revealed unmediated and unshielded. Here rational, scientifically-centered engineering could ultimately exercise a decisive, paternal sway over empirically-driven, patent-wielding entrepreneurial design engineers.[25]

To make it easier for Iowa A&M's engineering professors to serve as highway commissioners, the state commission established its offices at the college in Ames rather than with most of the other state bureaus in the capital at Des Moines. In this environment agricultural and engineering programs lived together as amicably as they did in the federal Department of Agriculture and the OPR. Anson Marston, who administered the engineering programs as dean, preached rational engineering in and out of the classroom. As one of Iowa A&M's "pioneers in research," he practiced engineering in the laboratory.[26]

(Morgan #163)

Built in 1911, the Stotts Creek or Exchange Bridge still served the motorist in 1980, although the spandrel walls were already separating and pushing out from the ring especially over the pier.

In 1906 the commissioner professors appointed one of their own as their first full-time highway staff member. Commissioner Marston promoted the selection of Thomas H. MacDonald, an Iowa A&M graduate whom Professor Marston had taught and whose appointment as assistant professor of civil engineering Dean Marston had supported at the college two years earlier. In his multiple roles, Marston had "passed on to MacDonald" his inherited "emphasis on a scientific approach to engineering."[27] MacDonald's place in the state commission grew as the organization gained strength, and his influence soon extended beyond the state to the regional and then to the national level. Indeed, MacDonald served as national director of the federal-state highway alliance from 1919 to 1953.

Iowa had almost no concrete bridges at the time MacDonald was a student. When his engineering instructors addressed bridges, they focused more on steel trusses than on forms with the newer mate-

(Morgan #163)

By the early 1990s, key segments of the Stotts Creek bridge's wall and rail had fallen into the creek. The earth fill that the wall had once contained was now rapidly washing off the ring behind it.

rial, concrete. But, out of the classroom and into the community arena, the professors turned commissioners soon assigned culvert and small-span construction, where concrete forms worked especially well, a very high priority. Consequently the experiment station under Marston's direction built and tested concrete slabs, beams, and small arches. From these tests, the college/commission developed standard specifications and plans in 1904 and 1905 for culverts and concrete slab bridges. MacDonald and his colleagues apparently distrusted concrete arches because they thought temperature changes were "everlastingly at work tending to destroy" them. "As a general proposition," MacDonald reported in 1911, "we do not believe the arch to be the best type of bridge, especially for the shorter spans."[28]

Luten designed arches which began to appear in Iowa as early as 1906 through the efforts of N. M. Stark & Co., of Des Moines. They quickly started to compete directly with commission plans. When Stark with a Luten design beat out other bids on the commission's plans for a bridge near Ames, MacDonald allegedly filed a letter with the county supervisors asserting that "Luten bridges are not generally recommended by engineers."[29]

The more that state design lost out in open competition with Luten design in the coming years, the more MacDonald, the college, and the commission chafed at the defeat of rational state planning at the hands of the private and empirically-oriented Hoosier. MacDonald assigned an assistant – probably Conde B. McCullough, another Iowa A&M engineering graduate – to investigate and to report on the Luten-design structures being built across the state. MacDonald felt that the Iowa Highway Commission needed to address "two distinct principles" in its consideration of Luten designs: "first, their structural integrity from the engineering standpoint, and second, the payment of royalty charges on patent claims whose validity had not been determined by any court having jurisdiction."[30]

With reports in hand and a common position hammered out among them, the three Iowa engineering professors acting as highway commissioners summoned Luten to appear before them in December 1911. The meeting opened with the report of a field study indicting Luten-design bridges in Iowa and a list of changes required before Luten plans might receive the commission's blessing. These changes included, MacDonald reported, "the addition of both steel and concrete ... to carry the specification loadings without excessive stresses in the materials."

Luten provided a more detailed account of the changes demanded. The commissioners insisted that he deepen foundations by a foot, increase the thickness of the flood-proofing pavement of his tied-arch by a third as well as place the pavement two feet below the streambed, untie the spandrel walls from the arch rings, and reinforce the rings as in the Thacher system. In short, the commissioners asked Luten to concede a number of the empirically derived ideas he had patented in favor of older, more conventional, and rationally confirmed notions.

Not surprisingly, Luten rejected all these demands except for extending the foundations. Significant increases in the flood-proofing pavement eroded its economic advantage, and placing it below the streambed eliminated a key structural purpose – limiting scour of the abutments. Untying the spandrel walls from the arch rings cut out the spandrel's ability to stiffen the ring substantially. And moving from single to two-tier reinforcing in the ring doubled the amount of steel used.

Ignoring the highway commission's demands, N. M. Stark continued to sell Luten-design bridges in counties and cities that were ignoring the commission's pronounced opposition to the design. Luten's "Standard Specifications" carried a bonded contractor's guarantee of "the proposed work for one year from completion against failure on account of any defect of design, workmanship and materials." Luten regularly insisted on this contractual assurance to clients that his empirically based designs worked and that his bridges were properly erected.[31]

Luten also paraded his battalion of patents before the Iowa authorities. He warned the highway commission, county governments, and bridge builders through circulars not to transgress his patented claims in their plans. Later in the year, Luten took Marsh Engineering Company, an Iowa-based concrete design and construction firm, to court for allegedly violating his patents in a bridge they built in Minnesota.

Within the year, the Iowa highway commission opened a broad offensive against Luten and the other private-design patentees. In 1913, the Iowa General Assembly required counties and cities to secure state approval of all bridge plans and thus ended construction of Luten-designed structures throughout the state. The legislature also authorized the state executive in June 1913 to intervene in patent cases concerning road and bridge design and construction, and the governor quickly directed the attorney general to join the defense of Marsh Engineering against Luten's patent claims.[32]

IV.3B Iowa *et al* v Luten

The legal battle that evolved was not between equals. Besides state resources, Iowa marshalled the federal-state highway alliance, used the attorney-general staffs of surrounding states, hired nationally-recognized patent attorney Wallace R. Lane, and enlisted prominent political figures from the Prairie States against Luten, his personal attorney, and his patents. For four years, Lane, Assistant Attorney General Henry E. Sampson, and Conde B. McCullough canvassed the literature and collected testimony on the development of reinforced concrete and visited sites and persons across the country including Luten's City-Beautiful critics in Indianapolis. Indeed, Iowa intelligence was so good that the state apparently learned the contents of depositions being prepared for a Luten case to be tried in New Jersey.[33]

Iowa's Attorney General called a regional conference in early 1914 of those interested in "having [Luten's] patents knocked out by the courts." State highway officials and attorneys from Kansas, Wisconsin, and Minnesota huddled with their Hawkeye counterparts.[34] The OPR-state highway alliance increased the pressure on Luten still more when in March 1914 the U. S. Senate passed a resolution introduced by Senator Bristow of Kansas directing its Committee on Patents to investigate the patents issued to Luten and to report on their legality as soon as possible.[35]

MacDonald introduced a resolution at a meeting of the American Association of State Highway Officials in December 1916 to refer Luten's patents to the Federal Trade Commission, and the organization adopted it. The Iowa attorney general, in turn, prepared the complaint presented to the Trade Commission, claiming that Luten had "obtained innumerable minor patents on ordinary mechanical processes or devices" concerning reinforced concrete bridges. Luten, the attorney general asserted, had "consistently and generally" exaggerated the scope of his patents, had misrepresented consent decrees as legal judgments on the merits of his patents, and had bullied and intimidated local officials and contractors with his claims for royalties. In sum, Luten allegedly engaged in "unfair methods of competition."[36]

IV.3C Private Designers Lose Patent Protection: The Courts Find Professional Skill, Not Inventive Genius

State and federal forces arrayed against the private, patent-holding concrete bridge designers landed some serious blows well before the Marsh Engineering case could be fought to a conclusion. The first blow came from Judge R. E. Lewis of the federal District Court in Denver, Colorado, who voided six of Luten's patents in November 1915 on the ground that Luten in each case "made application of mere mechanical knowledge and skill ... in no sense a demonstration of inventive genius."[37]

As Judge Lewis saw it, the core invention rested in the principle that embedding metal in concrete strengthened it, a principle which had been discovered almost a half-century before Luten. The judge made central use of Luten's own testimony that he did not add to the principle but applied it: his patents did not embody "anything beyond or more than placing `the steel in a new way that produces better results in a more efficient form.'" Luten's applications, the judge held, illustrated mechanical or engineering skill belonging in the marketplace of professionals rather than invention around whose exchange the federal government should provide some protection. Thus "each and all" of Luten's "claims were nonpatentable at the time he made his several applications."

A second blow came when the editor of the prestigious *ENR* reprinted the full text of Lewis' decision and publicly applauded the judgment. The decision "is important," the editor argued, "because it lays down with commendable lucidity a principle long contended for by engineers but rarely, and then reluctantly, admitted by lawyers and the courts. This principle is that the arrangement of metal parts in a body of reinforced concrete to resist the stresses which are set up involves merely mechanical design and not invention." Indeed, "any structure which a competent engineer would naturally arrive at through the orderly development of the principles of design cannot be held as an invention."[38]

The *ENR* editor also directly indicted reinforced concrete designers for unprofessional conduct when they resorted to the patent system. "Engineers," the editor proclaimed, "have constantly resented" court deference to mechanical cleverness and have "consistently regarded as impositions" the demands for royalties, "not because of a wish to deprive the so-called inventor of the profit of his thought, but because of a feeling that the inventor was capitalizing for his own benefit the accrued knowledge of the profession." In a later issue, the editor quoted a federal Supreme Court judgment that "to grant a single party a monopoly of every slight advance made, except where the exercise of invention somewhat above the ordinary mechanical or engineering skill is distinctly shown, is unjust in principle and injurious in its consequences." The editor endorsed the court decision and called the Patent Office, lawyers, and lower courts to task for not applying it.[39]

Following some letters attacking the Lewis opinion and the *ENR*'s endorsement of it, the editor elaborated his position further. Improvements similar to those which the editor believed to be properly patentable in the nineteenth century were not appropriately patentable in the twentieth, in part because engineering had in the meantime become a profession. When, for example, Albert Fink, Wendel Bollman, and S. S. Post were issued patents for their metal truss forms,

> little was known by practical bridge builders of the stresses in structures, and the forms of truss referred to were real inventions – valuable contributions to the art. But with the advanced knowledge of bridge design current today, such particular arrangements of trussing could be made by any competent designer to suit the local conditions, and any monopoly of any such special arrangement would be justly opposed on the ground that no invention was involved in its creation.
>
> The position of the engineering profession almost without exception is that the designing engineer should not be obstructed in his regular routine work by patents issued to some other designer merely because the latter has devised some particular arrangement of parts, without making any real advancement in the art such as justifies the Government in its grant of a monopoly.[40]

The Iowa Highway Commission – self-proclaimed "leader in the fight against the bridge patent royalty" – secured two more victories in 1916, this time over Edwin Thacher's concrete bridge patent. Once again, "considerable cooperative work" by the state highway departments of Kansas, Wisconsin, and Illinois prepared the way for the court tests. Judge Martin Wade found the Thacher patent invalid in both cases. The Iowa Highway Commission crowed: "Iowa with decisions by an Iowa judge in an Iowa district of the federal court, has come to the front with the two most recent and most sweeping of these decisions."[41]

Luten could find no comfort in the judgments or in the fact that his case against Marsh Engineering was scheduled to be heard by that same "Iowa judge in an Iowa district of the federal court." Indeed, when Judge Wade finished listening to the oral arguments of Luten v Marsh in January 1918, he did not even adjourn "to study the various items of evidence presented here upon both sides."[42] Having already made up his mind, the judge issued an oral opinion from the bench voiding three of Luten's patents "for lack of invention" and finding a fourth "not infringed." Having given the subject of concrete reinforcing "considerable study at the time of the trial of the Thacher Case," Wade expressed

> very grave doubts whether any of this method of reinforcing is patentable invention, or has been for twenty years, or has been at all after some man first found that reinforcing could be put into concrete and maintained, and that it would add to its strength .... All the rest is a matter of engineering knowledge, pure and simple.
>
> ... Long before this patent was granted, Monier and Von Emperger had given it to the world – not, perhaps, the particular piece of wire or iron which the Luten patent contemplated, but the method of doing that very thing. The art is old.

Wade also castigated Luten for one of the ways in which he had tried to protect his patented design ideas. Luten came into court without "clean hands," the judge concluded, because the patentee had not indicated in his warning circulars to officials and builders that the court decrees listed therein had been reached by consent of the parties rather than by trial and judicial pronouncement. When "a man recites a list of cases as having been tried, or in which decisions have been rendered, without disclosing that nearly all of them have been consent decrees, it is not the truth. It is only half the truth." Wade not only invalidated three of Luten's patents; he also impugned the patentee's integrity.

Judge Wade did allow "that Mr. Luten may have added very materially to the usefulness of the art of building bridges" by placing reinforcing or constructing more efficiently; "but this mere reduction of the amount of material, after all, is purely an engineering proposition, and, while to the layman it may seem to involve invention, it is simply the application of knowledge which he [the engineer] possesses. In judging whether these things involve inventive genius, we have got to consider the field of scientific knowledge in which the party undertook to work. Things that seem very dense to us, who are not educated in that field, would be matters of common knowledge there." In short, the professionalization of engineering had changed the rules of the patenting game. Patentable invention in bridge design required an uncommon idea among a community of certified experts – specifically engineers – not just among the populace at large.

Judges, professional peers, and opponents in the emerging federal-state highway bureaucracies had collectively and effectively undermined the claims of Luten and others to inventive genius and assured the dominance of rational engineering over more empirical bridge design. The patents of Albert Bone, Edwin Thacher, and Daniel Luten had been invalidated, and the campaign for central control over local bridge building was being waged with some success in state after state.

IV.4 FEDERAL AID: THE TIE THAT BOUND

Even before MacDonald and his Prairie States collaborators defeated the last reinforced concrete patent holder in court, the federal road office had consolidated its ascendancy over the nation's highway development though legislation. MacDonald again emerged at the front of the highway alliance's effort, this time in a successful drive for congressional passage in 1916 of the Federal-Aid Road Act which inaugurated a systematic program of federal matching grants to the states for highway design and construction.[43]

The act's aid provisions reflected vintage federal road office strategies. First, they removed fund allocations from the more obvious influences of partisan politics by determining each state's share of the federal fund through a formula based on population, area, and road mileage. Second, grants acknowledged the road office's home in the Department of Agriculture and its ties to agricultural and mechanical colleges by restricting matches to no more than ten thousand dollars per mile. The low-ceilinged matches addressed rural dirt, gravel, and stone roads quite well but provided less than half of the cost of building the hard surfaces used in and near towns and cities.

Finally, the act shaped and formalized key elements of the federal-state highway alliance system itself. The (again renamed) Bureau of Public Roads [BPR] administered the federal aid. The BPR made grants to state highway departments whose authority and organization, while locally created, now had to be federally approved and which had to follow federal standards and guidelines. By 1917 every state in the union save one – Indiana – had a federally certified highway department.

NOTES

1C. S. Hill, "Half a Century of Road Development," *Engineering News-Record* [*ENR*] (17 April 1924) 92: 685-86.

2W. Stull Holt, *Bureau of Public Roads: Its History, Activities and Organization* (Baltimore, 1923), 5. Holt provides a handy and official description of the early legislation, administrative organization, and federal activities. For a more complete and analytical study of the federal dimension in the evolving federal-state highway system, much is owed to the impressive work of Bruce E. Seely, *Building the American Highway System: Engineers as Policy Makers* (Philadelphia, 1987).

3Charles Whiting Baker, "The Engineer as a Man Among Men," *Engineering News-Record* [*ENR*] (20 February 1908) 59: 197-198.

4James A. Morone, *Democratic Wish: Popular Participation and the Limits of American Government* (New York, 1990), 97-119.

5Stanley Schultz & Clay McShane, "To Engineer the Metropolis: Sewers, Sanitation, and City Planning in Late-Nineteenth-Century America," *Journal of American History* (September 1978) 65: 389-411.

6Herbert Hoover, "The Engineer's Place in the World," *ENR* (24 January 1924) 92: 160. Hoover had been writing and reporting essentially the same sentiments since at least 1905.

7Charles Whiting Baker, "A Broader Training for the Engineer," *ENR* (20 May 1909) 61: 536-38. As an example of others who found the formal training of engineers too narrow for the mission of leadership, see "`Why Should the Profession of Civil Engineering Appeal to a College Man?'" *ENR* (23 April 1908) 59: 463-64.

8Martin Dodge, "The Government and Good Roads," *Forum* (1901) 32: 292-97.

9Logan Waller Page, "Problems of Road Administration: A Plea for Centralized Control of Highway Construction and Maintenance in the United States," *Engineering Magazine* (November 1909) 38: 265-67; "A Critique of the `Good Roads' Propaganda," *ENR* (10 August 1911) 66: 183.

10"Some Engineering and Economic Results of the Commercial Motor Vehicle," *ENR* (22 June 1911) 65: 762-64; "Some Good Roads Problems and Fallacies," *ENR* (21 September 1911) 66: 345.

11"A Critique of the 'Good Roads' Craze," "The Good Roads Craze," *ENR* (10, 31 August 1911) 66: 183, 264.

12Bruce E. Seely, "The Scientific Mystique in Engineering: Highway Research at the Bureau of Public Roads, 1918-1940," *Engineer in America*, ed. Terry S. Reynolds (Chicago, 1991), 311-13.

13Hill, "Half a Century," *ENR* (17 April 1924) 92: 686.

14For a more general and more extensive historical analysis in support of the propositions presented here of how empiricism and science relate to invention, see Abbott P. Usher, *History of Mechanical Inventions* (2nd ed., Cambridge, Mass., 1954), 56-115.

[15]A. N. Johnson, "A Recommendation for Concrete, Plain or Reinforced, for all Bridges of 50-ft. Span or Less in Illinois," *ENR* (7 January 1909) 61: 10. Wisconsin, Iowa, and Minnesota, for example, also provided "free" bridge designing for local communities with strings attached. "Competition of State Highway Departments with Independent Bridge Engineers," *ENR* (28 December 1911) 66: 777-78; "Annual Meeting of the Iowa Engineering Society," "State Competition with Private Engineers," *ENR* (25 February, 4 March 1915) 73: 411, 450.

[16]"An Offer of Free Highway Bridge Engineering Work by a Federal Bureau," *ENR* (16 November 1911) 66: 600-1.

[17]"The Offer of Free Highway Bridge Engineering by the Office of Public Roads," *ENR* (7 December 1911) 66: 687-88.

[18]For the exchanges between the *ENR* editors and Trautwine, see "The Offer of Free Highway Bridge Engineering by the Office of Public Roads," "More Concerning Free Engineering Work by Government Bureaus," *ENR* (7, 28 December 1911) 66: 688-89, 778; "Can the Independent Compete with the Trusts," "The Doom of the Independent Producer," *ENR* (11 January, 1 February 1912) 67: 71, 213.

[19]"Competition of State Highway Departments with Independent Bridge Engineers," *ENR* (28 December 1911) 66: 777-78.

[20]"State Supervision of Engineering Work," *ENR* (12 October 1911) 66: 444; "State Control of Highway Bridge Construction," *ENR* (13 June 1912) 67: 1137-38.

[21]See I.1A and II.2; "The Luten Patents," U. S. Senate (19 July 1916), *Congressional Record*, 11273-74.

[22]To extend the following summary of economic thought on the patent issue, see John M. Blair, *Economic Concentration: Structure, Behavior and Public Policy* (New York, 1972), 386-88.

[23]Daniel B. Luten, deposition (30 September 1913), District Court of the U. S. for New Jersey, *Daniel B. Luten vs. The Dover Construction Co.: Complainant's Record* (Trenton, N.J., 1913), 338.

[24]Luten, deposition (24 October 1913), *Luten v Dover: CR*, 378-79.

[25]For this section in general, see Seely, *Building the Highway System*, 32, 53-54; Thomas H. McDonald, "Bridge Patent Litigation in Iowa," *Iowa Engineer* (January 1918) 18: 117-27; Daniel B. Luten, "What's Wrong With Iowa?" *Engineering and Cement World* (15 May 1918): 24-30. The author is especially indebted to Eric DeLony, Chief of the Historic American Engineering Record, for locating materials on the state of Iowa's activities concerning private bridge designers and patentees in this period and for providing copies of articles published by the Iowa Highway Commission staff.

[26]On the college environment and the commission, see Earle D. Ross, *History of the Iowa State College of Agriculture and Mechanic Arts* (Ames, Iowa, 1942), 128-29, 158-61, 172-73, 254-57. Marston had worked with Robert Thurston of Cornell University.

[27]"New Roads Official Named," *Public Roads: A Journal of Highway Research* (March 1919) 1: 3; Seely, "Highway Research at the BPR," *Engineer in America*, 321, 338-39.

[28]T. H. MacDonald, "Concrete Bridges and Culverts," *Wallace's Farmer* (24 March 1911): 558, as quoted in Luten, "What's Wrong With Iowa?" *Engineering and Cement World* (15 May 1918): 27.

[29]Luten, "What's Wrong With Iowa?" *Engineering and Cement World* (15 May 1918): 26.

[30]There are two reports of this meeting. MacDonald offered his in "Bridge Patent Litigation," *Iowa Engineer* (January 1918) 18: 120-22. For Luten's recount under oath, see his deposition (24 October, 11 November 1913), *Luten v Dover: CR*, 380-90, 402-4.

[31]"Guarantee," *Luten's Standard Specifications for Reinforced Concrete Highway Bridges* (Indianapolis, 1910), 9.

[32]Luten, deposition (11 November 1913), *Luten v Dover: CR*, 405-6.

[33]Luten, deposition (13 November 1913), *Luten v Dover: CR*, 441. D. C. Hayne, Assistant City Engineer for Indianapolis, provided Henry Sampson, Iowa Assistant Attorney General, with access to photographs that Hayne took "in connection with the investigation of the Washington Street Bridge to be built in" the Indiana capital city. Hayne was apparently quite willing to share the city's photos "of Luten bridges showing defects." Henry Sampson, "Notes," Attorney General Cases S.II (State Historical Society of Iowa Library and Archives, Des Moines).

[34]"The Luten Patents on Concrete Construction," *ENR* (5 February 1914) 71: 329.

[35]Bristow introduced the resolution on 8 January 1914. It passed on 1 March. "An Investigation of Patents on Concrete Construction," *ENR* (11 March 1914) 73: 509. J. T. Newton, First Assistant Commissioner of the U. S. Patent Office, reported "that the records of the office have been carefully scrutinized without finding any indication of fraud therein" for forty-one Luten patents. "The Luten Patents," U. S. Senate (19 July 1916), *Congressional Record*, 11273.

[36]T. H. MacDonald, transcript of testimony (2 January 1918), "Luten v Marsh *et al.*," Attorney General Cases S.II (State Historical Society of Iowa Library and Archives, Des Moines), 21-22; "Before the Federal Trade Commission – Complaint: State of Iowa, through Its Attorney General v Daniel B. Luten and the National Bridge Company," Attorney General Cases S.II (State Historical Society of Iowa Library and Archives, Des Moines).

[37]"Six Concrete-Bridge Patents Void on Broad Grounds," *ENR* (2 December 1915) 74: 1094-95.

[38]"Broad Patent Principles and Reinforced Concrete," *ENR* (2 December 1915) 74: 1089.

[39]"United States Supreme Court on Invention and Design," *ENR* (9 December 1915) 74: 1138-39.

[40]"Patents on Reinforced Concrete," *ENR* (23 December 1915) 74: 1236-38.

[41]"End of the Bridge Patent Royalty in Sight for Iowa," *Iowa Highway Commission Service Bulletin* #10 (October 1916): 8-11.

[42]Judge Martin J. Wade, Southern District of Iowa, "Luten v. Marsh *et al.*" (7 January 1918), *Federal Reporter* 254: 701-7.

[43]Seely, *Building the Highway System*, 40-47; "Good Progress Made in Operation of Federal-Aid Road Act," *ENR* (20 December 1917) 79: 1148-50.

LUTEN DESIGN

WABASH RIVER BRIDGE
GEORGETOWN, INDIANA

LUTEN ENGINEERING COMPANY
INDIANAPOLIS, INDIANA

SUBSTANTIAL
ARTISTIC
ECONOMICAL

V

Resolved: That State Highway Commission Control of Public Bridge Design Best Promotes the General Welfare

While the words not enclosed in quotes are paraphrases or sometimes imagined, the issues raised and the opinions expressed are intended to reflect those of each speaker accurately.

An Imagined Debate between Thomas H. MacDonald, Affirmative, and Daniel B. Luten, Negative, in 1918

MacDONALD: J. A. L. Waddell was right in his recent book, *Bridge Engineering*, when he stated that "for nearly half a century the designing, letting and construction of highway bridges have been synonymous with ignorance, cupidity, and graft; and it is only lately that there has been any genuine improvement in the highway bridge business." If it weren't for the Good Roads movement, we'd still be living in those dark days. Fortunately, Progressive leaders organized the federal Office of Public Roads and state highway commissions to reform our highway practices.

Thomas H. MacDonald.

In the last century, engineers learned largely through apprenticeship, behaved like mechanics and inventors, and lacked "the rewards, not necessarily material, which generally accrue to recognized authority." "To master and thus lead in a new and major translation in our national life," engineers today must practice scientific methods and operate within scientific precepts.[1] Our colleges are educating more and more men who practice engineering as a science and not just as a craft. And increasingly the public is willing to allow professional experts to make public policy. Today we are less at the mercy of ward bosses and uninformed county commissioners, on the one hand, and self-promoting and sometimes grasping businessmen in cahoots with would-be engineers, on the other.

Daniel Luten represents the world we need to put behind us. The Iowa Highway Commission had barely begun to operate when "agents appointed by Mr. Luten" successfully began to promote bridges built on his patented designs. "The impression has gone forth that these designs were used because of their popularity with the public generally. Such is far from the case. These designs were used because Iowa, until 1913, had no bridge laws, no county engineers, and few public bridge lettings were held.

Each bridge company controlled, or at least attempted to control, the work in certain districts. The agents handling the Luten plans used these plans for the most part in the counties in which they controlled the bridge work and supplied their own plans. Probably ninety per cent of the bridges built on the Luten patented designs in Iowa were privately let without other plans or other competition being considered. In fact, in some cases, there was not even a written contract covering the transaction."[2] Here we see a number of the practices Mr. Waddell deplored and that our state highway commissions are trying to eliminate.

Daniel B. Luten.

***LUTEN*:** Like Mr. MacDonald, I applaud the Good Roads movement and welcome the number of engineers our colleges are turning out. The central-government road crowd are not the only members of the profession, though, and their infatuation with academic science does not produce the most practical and useful engineering. I'll match my engineering degree from the University of Michigan and my Purdue teaching of civil engineering against the credentials of those working in central-government road bureaus. But I'm not interested in spending a lot of time theorizing about why something functions or in doing figures to demonstrate numerically what I already know will work. A professional engineer is parsimonious: he uses nature to meet human needs at the lowest expenditure of effort and does not spend public resources unnecessarily on mathematical and theoretical rituals. A professional engineer should keep explanation as simple as possible.

I'm not the only one you road-bureau fellows denigrate for experimenting with sound structures that give good value instead of copying pedigreed designs. If you listen to MacDonald, you'd think that only the ignorant and the greedy run local government in Iowa. At least some of MacDonald's Iowa counties and cities doubtless follow the widespread practice of local governments in eastern and west-coast states: they hire and use trained engineers. I'm also not as quick as the road-bureau men to dismiss out-of-hand the native shrewdness of many county commissioners and city councilmen. The country would be better served if counties and cities hired more and better trained engineers to work with our elected local officials rather than turn all road control over to the road bureaus.

In singling me out for criticism, Mr. MacDonald spoke about bridge construction. As he knows, I am not a bridge contractor. I'm a designer, and my plans are available to all honest builders, including those honest and competent contractors in Iowa who have built bridges to my design. Although I am not involved in or fully informed about their contracting practices, I do insist that a contractor who uses one of my designs must give every client a year-long warranty on both design and construction. MacDonald is the type who would blame Henry Ford for every horse struck down by a Model T.

Mr. MacDonald has attacked my professional reputation. To discredit my design, he has had one of his assistants paint a bridge with mud to make it appear cracked, and then the assistant photographed the made-up structure. Thereafter prints of this libelous indictment circulated to newspapers around the state.

The willingness of these public servants to fabricate evidence didn't stop with this instance of photographic forgery. MacDonald's highway assistant, Conde McCullough, is such an artistic fellow that for the recent court case on my patents, he "actually manufactured evidence to show that one of the inventions involved had been used in 1890; he molded the device in clay as a part of the bridge and then photographed it and secured the testimony of a layman that the photograph and drawings were accurate representations of masonry built sixteen years ago, when the actual masonry was still in existence beneath the clay and of far different construction."[4]

***MacDONALD*:** I will not comment on Mr. Luten's unfortunate attacks on Iowa Highway Commission staff members. But I will talk about bridge design and be particular about Luten's. Indeed, I'm proud that we have put a virtual end to the use of Luten design in

NOT GUILTY, BUT GUILTY OF PHOTOGRAPHIC FORGERY: TESTIMONY UNDER OATH[3]

Lawyer: Have you a copy here with you of a photograph you exhibited at that meeting of one of Mr. Luten's bridges on which you had outlined alleged cracks by painting mud on the face of the bridge?

MacDonald: No; no; and there are no photographs in existence of that character that I took.

Lawyer: They have been destroyed?

MacDonald: No, sir. The photographs to which you refer were taken by one of my assistants. I didn't know anything of this circumstance until sometime afterwards - after the photographs were taken. I think, however, if you wish that in the record, that has been attributed to me in the cross examination, but I didn't take those photographs.

Iowa. In 1913, our legislature decided that the plans for all bridges which counties proposed to erect would have to meet the approval of the State Highway Commission. The commission has since adopted detailed, mandatory specifications for new county and city bridges. "State control … replaced county control" and put a stop to private operators like Luten preying on unsuspecting local officials with inferior, cut-rate bridge plans.[5]

In the case of Luten, the commission put his designs to a "theoretical analysis." When we considered them "in accordance with the method of design commonly employed by bridge engineers," we found his designs inadequate in "both steel and concrete." We also "carried on very thorough field examinations of all structures built under the Luten plans in a number of the counties." His bridges "were not correct or adequate, based on what has been demonstrated to be the best engineering practice in the construction of reinforced concrete bridges."[6] Luten design involves too much rule-of-thumb and too little compatibility with established theory to be considered scientifically sound.

LUTEN: From the lengthy conference I had with the State Highway Commission's staff in Ames, Iowa, several years ago, I know quite well how Mr. MacDonald thinks about design. Indeed, as I have stated on many occasions and as my patents show, I have found ways to build concrete bridges less expensively. I have tried to eliminate unnecessary steel and concrete, in part by uniting elements of an arch bridge so they work in harness, and I have sought and found ways to erect bridges with less labor. Whether you call this invention, engineering, or mechanical skill is not the point. I serve the public by helping it get good bridges for fewer dollars, including the fee for my license.

As your assistant, Mr. Conde B. McCullough, has acknowledged, road-bureau control of bridge design in Iowa "has not been instituted without criticism. The chief criticism has been that bridges and culverts built under state control are too massive and too expensive."[7] Indeed, in your crowd's insistence on being scientifically correct and safe on almost everything, you become guilty on both counts of the criticism. Designing conventionally by the book seems to suit the academic engineer and the government bureaucrat by taking all the risk, innovation, and efficiency out of bridge design.

You insist, for example, on two-tier reinforcing of the ring of every arch you design. The double line of reinforcing runs steel through regions where it isn't needed – where virtually always under compression. To keep the redundant steel from buckling out of the concrete, you strap the rods of the two tiers together with stirrups, thereby fixing the problem you unnecessarily created by using still more steel. Two-tier reinforcing with stirrups makes structural sense in a few special cases, but you don't need to double the reinforcing in every arch as if each were exceptional. In your hands, correct and safe design becomes synonymous with public waste.

The road-bureau crowd compounds its inefficiencies by insisting on standardization. Instead of designing each bridge to fit the specific needs of its particular site, the engineering bureaucrats offer a cookie-cutter model designed to address the most strain to which any such span might be subjected. Again, it provides an overdesigned structure for all but the most extreme cases.

The state engineers' search for standard design introduces other kinds of inefficiencies, too. Most state highway designs for shorter spans are for concrete slabs and beams, largely because their design is most easily standardized. The arch, however, might be a more efficient form for many of these short spans. But the arch gets used less by road-bureau designers since it is not so readily standardized. If, Mr. MacDonald, you and your friends were a little more practical and flexible and a little less theoretical, you'd serve the taxpayers a lot better.

***MacDONALD*:** Had you read beyond the headline of Mr. McCullough's report, Mr. Luten, you would have found the Commission's quite thorough refutation of the public criticism. As I pointed out earlier, a considerable number of the concrete bridges built when county officials were still in charge – yours included – were too light. Leaving bridge specifications and design completely to the discretion of local governments tempts bridge companies to propose and county officials to purchase the cheapest, not necessarily the best, structures. One way to reduce price is to offer less carrying capacity. "I know of bridges built not over three years ago, with concrete floors and supposed to be modern in every respect, which have failed under the load of a traction engine. Is there any excuse for such failures? Would any man design a bridge which would not carry a traction engine if he wanted to give value received?" "Everything is left to the honesty of the bridge company. Bridge companies are looking out for number one. Who is looking out for the county and the taxpayers?"[8]

"The excess material used under the state specification has been a direct attempt to overcome the defects found in the lighter type of construction." It "is not only warranted but absolutely essential" to make our bridges heavier if we want them to serve us permanently.[9] A bridge designed to inadequate specifications and with a plan shaved to the limit that fails – as it must over time – is no bargain to a community that has to pay for major periodic repairs and then to replace the whole in an inordinately short time.

Luten's arguments for tailoring bridge specifications and designs to each particular site ignores the situation of today's highways. In the United States, "a road is a public highway free to the traffic of all persons and all vehicles. A bridge is part of a public road. Any and all road vehicles may therefore pass over any bridge" and should be able to do so safely. Variation in the surface of a road can limit travel over it; bridge breakdowns threaten life and block travel altogether. Today's "state-road system is in no way physically separate from the network of town and county roads. The same kinds of traffic move over roads whether under state or under local jurisdiction." So, "the same load limitations should apply" to the bridges on both. There is no inefficiency in securing a common denominator of safety for all.[10]

Let's turn to the second criticism: cost. Again, we need to use a more sophisticated comparative measure than just contract price since the state is now asking for heavier structures than the counties had earlier required. Nor can we simply compare cost of materials by the yard or pound for a particular crossing. "As one contractor put it 'The county isn't buying a bunch of concrete. It wants a bridge.'" What we need is a "measure of service" which can best be judged by the structure's "waterway carrying capacity." The cost per square foot of waterway clearance for bridges built in Iowa under county control came to $5.38, while state-controlled spans cost $3.93. The taxpayer has saved almost twenty-seven percent from state-controlled planning, and the public has been assured safer, more permanent bridges.

The American people will be much better served with a highway system that professionally-trained and government-employed experts manage. These public servants have no political axes to grind or pockets to fill, and they can see to it that businessmen serve the community's interest more than they do their own self-interest. We are working hard to give the people the best possible roadways in part by keeping unprincipled bridge design and construction hucksters at bay.

***LUTEN*:** You have to watch Mr. MacDonald as he wiggles around on the questions of the road bureau's role and its costs. Notice how he condemns private design because a number of relatively new, county-built bridges gave way under traction engines. His examples are of steel-truss bridges, and he ought to know a lot about them, since he and his colleagues still regularly favor steel trusses – the last generation of bridge design – over reinforced concrete arches – the preferred form of the present and future. Now, indeed, had the local governments used concrete arches instead of metal trusses, traction engines would not have broken them down, since concrete arches can carry considerably more load. Please note that not even MacDonald has claimed that the first of the seven hundred Luten arches built in Iowa has yet failed because of load. But the Iowa state commission keeps building lots more temporary steel-trusses and concrete slabs and beam spans, all of which carry loads less efficiently than do reinforced concrete arches.

On the matter of costs, MacDonald doesn't dare compare contract prices, since his scientifically correct and standardized design requires heavier bridges. Still, he can't dispute the common knowledge that the more materials you use, the higher your costs go. Since he has already argued that the state requires the use of more concrete and steel placed more deeply into the earth than the counties have done, common sense suggests that a state-controlled bridge would cost more than a "lighter" county one. Indeed, he and his colleagues have admitted that the state requires "extra expenditure." He also knows he can't get away with a simple comparison of prices per foot or yard of steel and concrete. And so he and his colleagues have come up with a "waterway carrying capacity" unit.[11]

You have to hand it to the bureaucrats for making simple things sound complicated! They are magicians with figures. Watch their sleight-of-hand when they compare county-controlled bids and contracts, which include design and construction costs, with state-mandated bids, which include little more than construction. I agree that we need to evaluate bridges on the basis of cost incurred to attain a given standard of performance, but uniform accounting standards ought to be applied in preparing the comparisons. To make an honest comparison, for example, the bureaucrats should add to the contract figures for state-controlled structures the planning and design costs they incur and then take out of a different public pocket. They need to include the cost of their government offices just as I need to account for the cost of my staff in the price I charge for the use of my designs.

As I see it, three methods of bridge design and building are now operating in this country. I call these methods "English," "German," and "American," and I bet you can tell from the labels which I prefer![12]

The road-bureau crowd practices the "English" method in one form or another. Their engineers draft detailed and standard specifications and often draw standard plans. Mr. MacDonald's specifications are so particular that almost no room is left for the designer, and that's on purpose. State control means what it says, and it reduces the private designing engineer's office into a job shop of draftsmen. Virtually all room for innovation has been fenced off.

The engineers who prepare these standard specifications and plans may know a lot of mathematics and theory, but unfortunately they often have very "little or no experience in concrete bridges" outside the classroom and office. The government engineer inevitably "designs an excessively heavy, costly structure in part to protect himself against the inexperience of the probably unskilled low bidder and possibly against his own inexperience." The "English" method is "the easy way" for science-centered, engineer-bureaucrats because the lowest bidder is determined without difficulty and the building contractor tied down within a web of specific requirements. The public, though, pays a high premium in both design and construction.

Above – The ENR *listed the Squaw Creek Bridge near Ames, Iowa, among the casualties of a June 1918 flood. The flooding undercut one of the piers of this three-span, Luten-design arch. Below – Undermined by five-to-seven feet, the pier settled into the abyss, first cracking the ring and spandrel of an adjacent span and then pulling down all the arches within a day.*

The hurricane which swept over the Miami area of Florida in September 1926 with one-hundred-and-thirty-mile-an-hour winds rendered fifty thousand persons homeless and destroyed an estimated one hundred million dollars of property. Baker's Haulover Bridge, (above) a two-hundred-and-fifty-foot, three-span arch structure which the Luten Bridge Company erected two years before, stood in the direct path of the hurricane over a man-made inlet just off Biscayne Bay. Twenty-to-thirty-foot waves crashed over the bridge in the storm, completely washing away both approaches (below) while, in the judgment of the county engineer, leaving the arches "standing just as good as when finished."

(*Luten Bridge Company,* Luten Bridge in Tropical Hurricane)

The "German" method is the opposite extreme. It was widely used earlier – often against the public interest – by county and city governments across the country. Under the "German" method, each builder prepares and submits his own plan. This forces builders either to seek competent designers or to cobble a plan together on their own.

At its best, an award under the "German" method would go to the bidder who most effectively combines good planning and building. But the best bidder may not necessarily be the lowest. The difficulties with this method – and they are considerable – come in identifying in a prompt and acceptable manner within a democratic society the proposal offering the public the best value. These difficulties have defeated the public interest when left – as it unfortunately all too often has been – in the hands of incompetent or corrupt local government officials. Uninformed officials might not know how to determine which bid is best. Corrupt ones have little difficulty deciding; they can easily size up the bribes offered.

I favor a third way, the "American" method, under which the local or state engineer prepares a mandatory general plan and set of specifications for all bidders. The plan outlines only the principal architectural features and external dimensions of the proposed bridge. The specifications set forth the loadings and capacities required plus any special requirements of the work. In other words, the plan and specifications ought to emphasize how the bridge should perform, not the way in which the performance is achieved. Each bidder needs to submit detailed sections covering thickness of parts, arrangement of reinforcement, and guarantees. The prospective builder should also provide a secured guarantee of his work for a year against defects of design, workmanship, and materials.

With the aid of their local or state engineer, the elected or appointed officials can select the best and lowest bid with relative ease from among competitive designs and builders. This "American" method turns the local or state engineer into a Consulting rather than a Design Engineer, giving him control over the whole process for which he is trained and distancing him from matters in which he is probably less experienced. The method encourages the experienced builder to team up with an "expert engineer who can design the most economical plan," since the contractor must both submit a winning bid and guarantee the completed structure. The "American" method puts a premium on ingenuity in design and construction, giving the public a better and less expensive bridge while profiting the most efficient design and construction firms.

I'd like to see the "American" method prevail. But with Mr. MacDonald becoming the new Director of the federal road office, I can't be too optimistic. In Washington, he may adopt the latest lingo about "technocracy," that supposedly "new system and philosophy of government in which the nation's industrial resources should be organized and managed by technically competent persons for the good of everyone instead of being left to the management of private interests for their own advantage."[13] Ironically, of "the things complained about by the `crats,' all have been made possible by private ownership" and initiative. In whatever guise, Mr. MacDonald and his colleagues seem to want to follow Eddie Cantor in proposing the production of self-turning pancakes by mixing Mexican jumping beetles with batter. What we need, instead, is to let the marketplace rather than a bureau of technocrats discipline professional bridge design!

NOTES

[1]Thomas H. MacDonald, "Objectives of Highway Research," (1923), 2-3, as quoted in Bruce E. Seely "The Scientific Mystique in Engineering: Highway Research at the Bureau of Public Roads," *Engineer in America*, ed. Terry S. Reynolds (Chicago, 1991), 339.

[2]Thomas H. MacDonald, "Bridge Patent Litigation in Iowa," *Iowa Engineer* (January 1918) 18: 118, 120.

[3]T. H. MacDonald, transcript of testimony (2 January 1918), "Luten v Marsh, *et al.*," Attorney General Cases S.II (State Historical Society of Iowa Library and Archives, Des Moines), 20.

[4]Daniel B. Luten, "What's Wrong With Iowa?" *Engineering and Cement World* (15 May 1918): 30.

[5]C. B. McCullough, "Are Highway Commission Bridges Too Heavy?" *Iowa Highway Commission Service Bulletin* #12 (December 1914): 3-6. Since MacDonald sometimes referred to McCullough as "Bridge Engineer for the Commission" and assigned him to build the case against Luten and since the commission published the article noted above, some of McCullough's words have been put into MacDonald's mouth in this imagined dialogue.

[6]MacDonald, "Bridge Patent Litigation," *Iowa Engineer* (January 1918) 18: 120-1.

[7]McCullough, "Are Highway Commission Bridges Too Heavy?" *Iowa Highway Commission Bulletin* #12 (December 1914): 3-6.

[8]The words quoted above are taken from a speech of a Professor of Mechanics at the University of Kansas at a meeting of the Kansas Engineering Society. The Iowa and Kansas academic and government engineers faced many similar circumstances, thought a lot alike, and worked together rather closely. H. A. Rice, "Needed Reforms in Highway Bridge Construction," *ENR* (12 February 1914) 71: 349.

[9]McCullough, "Are Highway Commission Bridges Too Heavy?" *Iowa Highway Commission Bulletin* #12 (December 1914): 4-6.

[10]The lines quoted above come from the editor, "The Highway Bridge as a Public Engineering Problem," *ENR* (21 January 1915) 73: 130-31. The sentiment ran so widely through the federal-state highway alliance that it could be considered a commonplace one there.

[11]McCullough, "Are Highway Commission Bridges Too Heavy?" *Iowa Highway Commission Bulletin* #12 (December 1914): 5-6.

[12]Daniel B. Luten, *How Good Concrete Bridges May Be Secured With Contract Awarded to the Lowest Bidder at Public Letting* (Indianapolis, c1914): 3-15. Luten explored the application of these same principles within the Indiana context in a talk he delivered to the Indiana Engineering Society in January 1914. He published the same as *Why Indiana Leads the World in Concrete Bridges: The Indiana Bridge Law* (Indianapolis, c1914).

[13]The words quoted herein are taken from the report of a speech Luten made on "Technocracy," SCIENTECH Club of Indianapolis, *News Letter* (23 January 1933) 14: #4, (in Indiana State Library).

The Luten Bridge Company of York, Pennsylvania, erected this Luten-design footbridge at Washington and Lee College, Lexington, Virginia in 1927.

PART TWO

Consolidation and Standardization, 1917-1942

(Elkhart #387)

Bridge Street Bridge over the St. Joseph River, Elkhart.

VI

THROUGH THE REAR-VIEW MIRROR: INTO THE AGE OF AUTOMOBILITY ON STATE HIGHWAYS

(INDOT)
Hoosier trenches.

Hoosier skepticism of federal-government programs has been longstanding and unabashed. Almost never advocating new federal initiatives, Indiana residents did not rush to demand the creation of a state highway commission, nor did most local officials with special responsibility for roads press the state government to redistribute authority. The very considerable Hoosier automotive and concrete interests behaved differently, however, since good roads increased their business prospects. They actively lobbied state and national leaders for improved highways. With the passage of the Federal-Aid Road Act of 1916, the political balance in Indiana finally tipped in favor of state-administered roadways.

Although the promise of federal aid corralled Hoosiers into the evolving federal-state highway alliance, the First World War postponed Indiana's full participation and allowed for an effective start in accord with automobility. "When Johnny came marching home" to Indianapolis, Popcorn, Fickle, and other Indiana communities at the Great War's end, both he and many of the Hoosiers whom he rejoined thought differently about vehicles and roads than when he had left. The veteran did not welcome memories of trenches, which Hoosier earthen roads seemed to emulate in springtime, nor did he fancy going to town by wagon. The war had demonstrated that he lived in a motorized age, and, depending on the number of dollars in his pocket, he liked what he heard about Ford, Apperson, Cord, Duesenberg, Haynes, Studebaker and other promoters of the automobile. An increasing number of his friends and relatives had also come to accept, if not always to welcome, the usefulness of motor vehicles.

Late entry into the federal-state highway system allowed Hoosiers to sidestep some of the first effects of automobility, a respite not available to the established alliance leadership. Indeed, war-related motorized transport shook the alliance to its foundations and forced the Bureau of Public Roads [BPR] to scramble to retain its leadership of roadway development. By consolidating an alliance which masked national control in a sea of seemingly autonomous, state departments, the BPR under Logan Page had achieved a major political triumph, one that still dominates highway development. But, by dedicating federal aid almost exclusively to wagon farm-to-market roads – in part to secure the political victory and in part out of loyalty to its government-collegiate partnership with rural and agricultural interests – the BPR had blundered. Its short-sightedness over automobility threatened to undermine its major political triumph.

VI.1 THE BUREAU OF PUBLIC ROADS BACKS INTO THE FUTURE

As, in part, a counterweight to the BPR's rural focus and intrastate orientation, the leaders of several east-coast state highway commissions had organized the American Association of State Highway Officials [AASHO] in 1914.[1] AASHO members lobbied for auto-friendly policies within the BPR network. Organizations like the American Automobile Association [AAA] and the National Association of Manufacturers [NAM] worked to shape road policy from without. These organizations had favored passage of the 1916 highway bill on the principle that part of a loaf was much better than none.

If the truck and auto interests and their allies could have had their way, they would have preferred a determined move towards what the industry's more adventurous publicists had been actively promoting for years – a nationally-routed, hard-surfaced highway system. Because motor vehicles could move at higher speeds, carry heavier loads, and travel greater distances in a given period of time than animal-drawn wagons, the motor vehicle interests did not view trucks and cars as merely local carriers – from farm and home to town, railroad station, or water terminal. Motor vehicles could substitute for wagons and carriages locally and for the trains and boats regionally, if not nationally. Thus automobility called for an interregional network of roads articulated in a continuous national pattern.

Besides establishing neither a national nor an interregional system, most of the federal-state highway alliance's Good Roads built between 1895 and 1916 were poorly designed for motorized use. The speed, weight, and powered wheels of automobiles ripped apart the surfaces of dirt, gravel, and stone roads and covered passers-by and roadside residents in a cloud of dust.

Engineers understood well before 1916 that motor vehicles required hard-surfaced roads. In a talk before the Indiana Engineering Society in early 1909, for example, C. A. Kenyon of Indianapolis concluded that "the great increase in automobile traffic and the very destructive influences of such traffic have. . .put an entirely new aspect to the road maintenance question. Macadam [stone] construction seems quite unable to withstand these influences, and the plan of sprinkling tar or other compositions upon a finished road is little more than a dust preventive, as it cannot strengthen a macadamized road."[2]

Later in 1909, the *Engineering News-Record* [*ENR*] underlined Kenyon's conclusion when it reported on the impact of large busses upon a heavy stone-surfaced road in Germany. The busses rutted the stone up to half a foot deep and a foot wide within a few months. After repairs with coarse broken stone, bound with courses of cinders and sand and pressed in place with a steam roller, the road became badly cut up once again following two additional weeks of bus use. This "accelerated test of road destruction by motor-car traffic" helped convince the ENR that "we are only at the threshold of our troubles with highways due to automobile wear. For two years automobile factories have been working overtime, and it is said that two hundred thousand automobiles are to be built next year."[3]

(INDOT)

Choosing one's ruts carefully.

Federal-state highway alliance engineers were also in 1909 discussing road surfaces appropriate for motor vehicles. Logan Page of the federal road office, acknowledged that "the introduction of motor traffic ... has completely upset" the "principles involved in the proper construction of a modern broken stone road." In a move to restore balance to the old principles with the new vehicles, Page advocated addition of a bituminous binder atop the stone road. A. N. Johnson, the State Highway Engineer of Illinois, however, disagreed with Page, his federal superior. Heavy motor vehicle traffic, Johnson predicted, would likely require "brick or concrete construction, rather than in the use of a plastic binder for broken stone." Page took another short step forward later that year in a speech to Portland Cement Manufacturers in which he allowed "there could be no better form of road construction, at so low a cost, to meet only automobile traffic conditions than a cement macadam road."[4] Yet Page still hesitated, claiming he had not seen a cement macadam road of which he fully approved.

Over the years, federal engineers did build experimental or demonstration roads with differing surfaces, and they did try a variety of road treatments to contain dust.[5] But federal officers kept sidestepping the increasingly obvious reality of automobility and its implications for road-building.

VI.A "See America First"

Was it when perched atop a signpost in the dark some nine miles outside Indianapolis after a run in a car that Carl G. Fisher decided it was time to move from speculation to demonstration of an automotive roadway? The headlight manufacturer had come to a three-way fork in the road on his return. Spotting what appeared as a signpost, he shimmied up it and lit matches only to find "Chew Battle-Ax Plug" advertised and no directions homeward offered.[6]

In September 1912, Fisher announced a campaign to construct a coast-to-coast, hard-surfaced highway suitable for automotive travel to serve as an object lesson for the American public. Fisher secured immediate pledges of over three hundred thousand dollars from Indianapolis car manufacturers towards a goal of ten million dollars. Within six months, the president of the Lehigh Portland Cement Company reported that he was talking to industry leaders about a donation of at least a million and a half barrels of concrete for the highway. The Indiana Automobile Manufacturers Association further promoted what would become the Lincoln Highway with a caravan of twenty-two Indiana-built cars and two trucks to travel from Indianapolis to San Francisco and Los Angeles in the summer of 1913. Elwood Haynes and Carl Fisher led the Trailblazers on their widely publicized six weeks' overland jaunt which ultimately brought all the Hoosier autos and trucks to their advertised destinations.[7]

In 1914, the Lincoln Highway Association began to distribute some of its donated cement for the construction of a demonstration or "Seedling Mile" of hard-surfaced road in each county through which the route ran from New York to San Francisco. With a seedling-mile donation promised from the Association, for example, St. Joseph County, Indiana, issued bonds for the construction of sixteen miles of concrete pavement on the Lincoln Highway in 1915, and the Association claimed credit for inspiring the financing of fifty additional miles of concrete roads in northern Indiana to be built within a year.

The governor of a state at the terminus of the Lincoln Highway acted as decisively as the Association in 1914. Looking at the mounting cost of repairing motor-vehicle damages to the state's broken-stone roads, even to Page's favored bituminous-treated, macadam ones, New York's Governor Martin H. Glynn ordered an end to the construction of macadam roads in favor of brick- or concrete-paved ones. Earlier, New York had followed Massachusetts' lead and "adopted a system of macadam roads eminently fitted to light horse traffic but utterly unsuited to the wear and tear of the heavy automobile. The result has been that an ever increasing motor traffic is pounding the roads of New York apart almost as fast as they can be built. What the State has saved in the initial cost of its roads is more than lost in the great and increasing cost of maintenance."[8] The governor declared to the federal-alliance partner in his state, "I shall expect and shall exact results from the Highway Department." But he could only cajole the federal-road officers to whom the state highway department looked for leadership. AASHO, of which New York was a member, and the Lincoln Highway Association were naturally disappointed with the Federal-Aid Road Act of 1916. Still, the governor and the Association were not deterred from building more hard-surfaced roads.

VI.B Trucking Along to Victory

Automobility was on the national doorstep by 1916, even though the BPR did not acknowledge it. The more than three and a half million motor vehicles, including a quarter of a million trucks, that Americans registered in 1916 previewed the coming era. Soon after the First World War began in Europe in 1914, orders poured in from abroad for American trucks and cars. Ford's Model T became a vehicle of choice. Over one hundred and twenty-five thousand Ford T's served the Allied cause by the war's end. Indeed Lawrence of Arabia traded in his camels for Model T's. To meet European orders, American truck production rose from just under twenty-five thousand units in 1914 to one hundred and twenty-eight thousand in 1917.[9] The large and heavy chassis used in peacetime for luxury cars soon undergirded ambulance and truck bodies for export, and four-wheel drive vehicles served the military overseas.

Strong European demand for a broad range of American products, including motor vehicles, pushed the U. S. rail system close to its limits as goods began to pile up in east-coast warehouses and on boxcars awaiting shipment across the submarine-infested Atlantic Ocean. The United States' entry into the war in April of 1917 added American arms and soldiers to the movement east upon the already-taxed transportation system and enlarged the existing bottlenecks at east-coast ports. By late 1917, filled boxcars lined the rails for one hundred miles west of the port of New York.

The snarled rail system offered Midwestern truck manufacturers an unparalleled opportunity to demonstrate the viability of their product under the banner of national defense and with the aid of an organization of their own, the Lincoln Highway Association. As trucks rolled off assembly lines, their manufacturers began to load some of them with freight and ultimately had thirty thousand of them driven to east-coast ports largely over the only highway route laid out, constructed, and marked for motorized cross-country travel, the Lincolnway.[10] War allowed the automotive industry to demonstrate in a season, and beyond argument, that trucks could efficiently substitute for trains and boats for highly valuable cargo on long hauls. "Long-distance express [truck] service," the *ENR*'s editor noted in late 1918, was "beyond the expectations of even the most enthusiastic highway advocates five years ago; today even the man in the street regards them as commonplace."[11]

Trucks also began to supplement the increasingly snarled north-south seaboard rail system in 1917. They brought milk and produce from farms to city markets in a number of regions, carried widely variable intrastate freight, and moved increasingly into regional interstate hauling. Trucks fulfilled a prophesy the *ENR* editors had been making periodically since the turn of the century that "the commercial motor vehicle" would ultimately carry freight more cheaply and efficiently than animal-drawn vehicles and successfully compete with railroads for intermediate-length hauls (up to forty miles).[12]

VI.C Just in Time for the Automotive Express

The wartime explosion in truck traffic had predictable consequences upon the roads and upon the stature of the BPR. Trucks destroyed in a season many of the "good stone roads" built over the previous decades to BPR standards for wagon and carriage traffic.[13] Circumstances beyond the ken or control of the BPR – American participation in a world war, the snarling of the railroads, and the resulting vast surge in the use of their roads as truck thoroughfares – conspired to highlight spectacularly the federal ostrich. The world war made the automotive lobbyists look like planning experts and the federal highway engineers seem like bumpkins locked away too long in the Department of Agriculture.

Before the war, the *ENR* had been quite skeptical of federal aid for state and local highway construction and had opposed proposals for a national highway system built and maintained by the central government. The editor regarded state and local roads as most appropriately the business of state and local government, and he dismissed a national highway system as a subsidy for "a few wealthy pleasure seekers at a time when the masses of the people are crying out in protest against the increased cost of living."[14] But with the wartime explosion of motorized trucking, a new *ENR* editor reversed the journal's course and favored the creation of a national highway system under a national highway commission.[15]

(INDOT)

Stalled-out and ankle deep.

Editor E. J. Mehren did not merely imply criticism of the BPR within his advocacy of national highways under a national commission. First, he noted, "it would take upwards of one hundred and fifty years at the present rate of building to bring it [federal-aid highways] up to a unit" or system. Next, he volleyed across the BPR's bow:

> It is too much of course to expect from Washington, either as a war or a peace measure, that strong type of leadership which befits one of the greatest constructive programs the country has ever known. The federal road administration has too long been in a subordinate capacity, lost in a department – that of Agriculture – whose main interests are foreign to road work. The time has clearly come for setting up a Federal agency commensurate with the highway future of this country – a commission with the autonomous standing of the Interstate Commerce Commission, the Federal Trade Commission or the Tariff Board. No other plan is adequate for the leadership which the federal government should furnish in the tremendous highway development we face.

Later, Mehren claimed "the BPR ... surely cannot be charged with anything but the utmost conservatism." Finally, he reported, "lack of vision and of sympathy with new conditions have been the chief deficiencies in the bureau in the immediate past."[16]

Two leading umbrella organizations critical of current operations – the Highway Industries Association and AASHO – called for "a great highway convention" in December 1918 to address the direction of postwar highway policy and administration. Behind the scenes, Logan Page worked to divide his opponents by meeting AASHO's minimum demands and at least to keep the federal-state alliance together. "The Page bill," which AASHO ultimately endorsed, recommended amendment of the federal-aid act to appropriate significantly more money for construction, eliminate post-road limitations that militated against regional and interstate routing, and raise the per-mile ceiling on federal aid to allow equal matches with the states for building substantial hard-surfaced roadways.[17]

Page's efforts at division worked, but his proffered compromises came as too-little, too-late to mollify the automotive interests and engineers persuaded by the *ENR*. A majority of the convention delegates voted for the creation of a new National Highway Commission which would inevitably sideline the BPR and its federal-state alliance, and AASHO dissented. Although unplanned and beyond his control, Page made one last contribution to the BPR: he suffered a heart-attack and died just before the highway convention opened.[18]

The unexpected opening at the BPR provided the agency a reprieve and offered a new head, Thomas MacDonald, time to work with the automotive interests, the *ENR*, and other critics before they had indicted his leadership. When he took the BPR helm in 1919, he faced rough political seas. But, as Daniel Luten had learned, one should not have underestimated MacDonald's ability to take charge. He organized his allies and compromised with, outflanked, and outlasted his critics. In the end, the 1921 Federal Highway Act kept the BPR securely in control of the nation's highway development.

VI.2 TOO LATE TO BACK IN, INDIANA DRIVES STRAIGHT AHEAD

There are some advantages to being the last state by which the new is tried. Unlike the BPR, which had to back into automobility, Indiana entered the federal-state highway alliance so late and with such a considerable range of powerful and vocal automotive interests present that the state planned and operated with those interests right on hand from the commission's beginning.[28]

Indianapolis alone had seven automobile plants in 1905, which was more than Detroit could then boast. The chorus of Hoosier carmakers for whom a state-highway system might translate into greater sales rose to over one hundred rooted in several dozen towns and cities from 1905 to 1910, including known names like Apperson, Cole, Cord, Duesenberg, Haynes, Marmon, Perry, Studebaker, and Stutz. Although after 1910 these manufacturers lost their predominance to Michigan-based Ford and General Motors, which had adopted mass production and high-volume sales, Indiana car-makers exercised their still considerable, economic and political clout until the Thirties; and they wielded that clout on behalf of good, hard-surfaced roads.

Detroit could not quickly upstage Hoosier promotional activities for automobility. By driving a vehicle of their own make from their Kokomo factory to New York City for a Brooklyn physician in 1899, Elwood Haynes and Elmer Apperson had with proper fanfare taken the first thousand-mile automobile trip in the United States. Even before Carl G. Fisher's venture with the Lincoln Highway, he and James A. Allison headed a group of local businessmen who opened the Indianapolis Motor Speedway for a "racing carnival" in September 1909. The serious accidents which resulted on the crushed stone surface led the promoters to pave the track with brick before December's race. Large crowds encouraged the promoters to organize the first five-hundred-mile race in 1911. The "500" quickly became a nationally recognized event despite the opposition of some of the "better families" of Indianapolis who regarded the race "as a vulgar spectacle."[29]

An irrepressible promoter, Fisher looked for a north-south route to supplement the east-west Lincolnway. He turned his attention to southern Florida where, envisioning Miami Beach as a winter vacation spot for Midwesterners, he speculated on real-estate. To get Hoosiers and others to the Florida sun by motor vehicle, Fisher proposed the creation of an interstate highway from Chicago to Miami. Indiana's governor, Samuel Ralston, quickly embraced the Dixie Bee Highway, and he organized a meeting of the governors of those states through which the highway would pass. In September 1916, Fisher and Ralston joined others in Martinsville, Indiana, to celebrate the inauguration of the highway from Indianapolis to Miami.[30]

VI.2A The Campaign to Commission Hoosier Speedways

The Indianapolis Chamber of Commerce organized a special conference for "friends of good roads" in mid-January 1914, at which Clarence Kenyon of the Western Paving and Supply Company summoned support for the organization of an Indiana state highway commission. Influential persons must have pledged their help, for Governor Samuel Ralston soon appointed a special commission to consider road conditions in the state.

The governor's advisory commission of five included two urban Democrats with substantial records for public works construction. Democrat Thomas Taggart had vigorously promoted Indianapolis as a City Beautiful when fellow Democrat and also a mayor, William H. O'Brien, had successfully spearheaded the building of a massive levy around Lawrenceburg after three successive Ohio River floods.[31]

The governor's three other commissioners were associated with the Purdue University-Department of Agriculture-BPR network. Republican Addison C. Harris had served in the lower house of the state legislature before President William McKinley appointed him U. S. Minister to Austria-Hungary. Alonzo Harris, president of the Purdue University Board of Trustees, may have influenced brother Addison's views about the need for a proper state-highway system.[32] R. L. Sackett, the fourth commissioner, was a professor of Sanitary and Hydraulic Engineering at Purdue, a more advanced position in the same department in which Daniel Luten had once served. The last commissioner, farmer Leonard F. Clore, seems at first glance to come from outside the core of good roads interests. But as the state's first county Agricultural Agent, he also operated within the Purdue University-Department of Agriculture network.[33]

The advisory commission concluded quickly that "the time has come when a more equitable and uniform system of making and maintaining the public roads in the state should be provided."[34] The commissioners called upon the legislature to "establish a state department of roads" under the guidance of a board consisting of the Purdue University dean of civil engineering, the state geologist, and a third person selected by the governor. The board should, with the governor's approval, employ "a competent and qualified highway civil engineer" to administer the highway department and pay its employees from the State Automobile Fund. The proposed funding acknowledged a direct initial relationship between the new road system and automobility.

As matters stood in 1914, only townships had the power to build improved roads, although they shared maintenance responsibilities with the counties. The governor's advisers complained that "the state does nothing towards the construction or maintenance of our public roads," not even providing standard specifications or quality-testing materials. Apparently presuming that road-building required no professional knowledge or skill, the state legislature had laid down only one requirement for the superintendent of construction – that he live in the township where the road was being built.

Why change things? First, "the ordinary gravel road is found insufficient to resist the stress and force of automobiles, trucks and the like: and so a new problem is presented." Second, "eighty percent of the travel is upon … ten to twenty percent of the mileage. These much traveled roads are expensive to make and maintain, under the use of automobiles, trucks and other … vehicles." Third, "if there is a main traveled road between two or more counties, or cities, crossing several townships in its course, some parts of the road may be made of gravel, some of stone, some of cement or brick …, and other parts not even improved. It cannot be otherwise until there is some general state system provided requiring more unity and concord of action along the lines of that road."

The federal government provided a fourth important reason to change. The federal House of Representatives had already approved a federal-aid road bill, a measure which "in some form will soon become a law." To share in the federal subsidy, Indiana needed to create a state highway commission similar to what the governor's advisory group with the help of a federal – probably BPR – adviser now suggested. Consistent with BPR guidelines, the proposed commission would provide "suitable standard specifications" for roads and bridges and approve all specifications "tendered by builders or others." Surfacing materials also needed testing under the commission's direction "in the laboratory at Purdue University or some other laboratory within the state." Finally, the governor's advisers recommended that the counties and townships be allowed to employ as road superintendents only those whom the new state highway commission certified as professionally qualified.

The governor's advisory commission requested "suggestions and advice" from about twelve hundred local government officials as to "whether a state system of main roads should be built and kept up by general state tax and federal aid." The commission's request went largely ignored. Less than a third of the officials contacted appear to have answered the commission's letter. Under twenty percent of the whole expressed support for a state highway system, and under ten percent opposed it. One Howard County official articulated a middle ground that a number of the silent majority and some of the other respondents may have occupied:

> The three mile road law [under which township road control operates] … has done wonders for individual communities. Instead of connecting county seats and larger cities to be used

(INDOT)

A "community road."

for automobile speedways, the farmers have petitioned for the improvement of roads leading to their local grain market, to their churches and consolidated school. No other one thing has done so much toward improving social conditions in rural communities. ...

In constructing these three mile roads, or community roads as I prefer to call them, very little attention has been given to connecting the roads with those of other communities, consequently the automobile speed fiend taking long trips across the country finds 'connections' very poor, and consequently desires a system of main roads across the state.

I, as well as many of my neighbors, run automobiles, but we are opposed to hav[ing] the present law changed to build speedways across the state. I am heartily in favor of main roads being built over the state by state tax and federal aid, but allow the state commission the supervision of these main roads only and leave the other roads in the care of county road superintendents and other local officers.[35]

Automobility, these Hoosiers argued, should be partly accommodated but without embracing the federally promoted centralization of highway administration.

VI.2B Introducing a State Highway Commission

Only after the President signed the 1916 Federal-Aid Road Act authorizing direct and annual funding for those with a BPR-approved state highway commission did a majority of Indiana legislators accede to the rudiments of what Governor Ralston's advisory group had recommended earlier. The legislature finally approved the creation in March 1917 of a four-person, bipartisan, highway commission in which professional engineers selected and designed the main roads, while at the same time the counties remained significantly involved in their operation.[36] In the spirit of shared governance, the law also gave the state more say over local or "community roads" than the Howard County official of 1914 and some others wanted.

The law authorized the state highway commission [ISHC] to recommend to the governor up to two thousand miles of "continuous and connected highways" as "main market roads." In selecting these roads the commission needed to consider not only "lines of travel connecting main market centers" but also "the kind and volume of traffic." State and federal aid would be available for up to half the cost of constructing new surfaces and bridges that followed ISHC specifications and plans on these designated roads. Otherwise, roads and bridges remained under the control of the counties through which they passed. County authorities decided when to seek funding for building or rebuilding roads and bridges, administered the bidding and contracting processes subject to approval of the ISHC, and paid half the bill for improvements.

If the 1917 act allowed for more local meddling in the state highway system than the BPR wanted, it also permitted more state influence over county road and bridge matters than probably most local officials preferred. Should fifty or more freehold electors petition, or county officials on their own volunteer, then the specifications and plans for any county road or bridge project estimated to cost over two thousand dollars would have to be submitted to the ISHC for approval. Once ISHC became involved, the county could award a contract based only on specifications and plans that the state's engineers had already approved, thereby giving the ISHC a veto over materials and design.

VI.2C Boilermakers at the Helm

The state highway engineer – "a competent civil engineer, experienced and skilled in highway and bridge construction and maintenance" – stood at the center of the ISHC apparatus. The state commissioners nurtured the network established with Purdue University by naming two of its alumni, William S. Moore and William H. Rights, as its lead engineers.

When William S. Moore became chief engineer for the highway commission, he was already well known in important professional circles in the state.[37] The Indianapolis native had won acclaim as a halfback for the Purdue football team in the mid-1890s. For his academic efforts, he received a bachelor's degree in civil engineering in 1898 and a master's of civil engineering in 1910. After graduation, "Billy" Moore moved to South Bend, where he played recreational football with Notre Dame graduates and Studebaker car-makers, married, and served in the city engineering departments of South Bend and Mishawaka. He had designed several substantial City Beautiful bridges before opportunity took him out of state briefly.[38] Moore was working as city engineer for Grand Rapids, Michigan, when the ISHC called him back to Indiana in 1917.

William H. Rights became Moore's assistant engineer at the ISHC. After earning a bachelor's in civil engineering from Purdue in 1890, Rights worked as Columbus' city engineer from 1897 to 1913, sometimes served as Bartholomew County surveyor and consulted as a civil engineer before joining the newly organized state commission.[39]

(INDOT)

An early state road survey.

Moore and Rights explored four thousand miles of Indiana roads during the summer of 1917 in their official automobiles. By the end of September, they had identified eight hundred miles of "Main Market Highways," which could also be called interstate since they reached from one state border to the other and articulated with similar roads in adjacent states. Because "all of these will be hard-surfaced roads," they were also automobile-based. Moore reported to a joint meeting of engineers from four Indiana societies "that it is the interest of the state to build the roads for motor trucks, and not the trucks for the roads."[40]

The ambitious program that the ISHC engineers had launched came to a halt in May 1918, when a Hamilton County judge declared the highway act unconstitutional and the state supreme court adjourned for a three-month vacation before ruling on the commission's appeal. The entire ISHC engineering staff was laid off, and all construction stopped until after the supreme court overruled the county court in the fall of 1918. W. S. Moore headed north to South Bend, W. H. Rights took an assignment with the YMCA in Europe, and the *ENR* editor shook his finger at a state supreme court that assigned its summer break a higher priority than highways.[41]

VI.2D Roaring Through the Twenties on Two Thousand Paved Miles

The trauma over the ISHC's constitutionality pulled apart more than the commission's staff. Indeed, the legislature substituted a new highway act in the spring of 1919 different in some substantial ways from its 1917 effort.[42] First, a director who was not necessarily an engineer would administer the revised commission organization. The chief engineer now headed the division of construction. Second, the legislature expanded the number and scope of "Main Market Highways" by requiring the connection of all county seats and cities with over five thousand inhabitants to "trunk highways from other states." Third, shared governance became a one-way affair with control emanating from the state to the local level. Now the main roads would be "constructed, reconstructed, repaired and maintained by the state highway commission out of the state highway funds." County taxes no longer accounted for half the construction and maintenance costs of the Main Market Highways, and county officials lost all superintendence of them. Although county participation in the state road system disappeared, the 1917 mechanism for occasional state control over county design did remain.

The new commission at first selected a transition team headed by experienced state administrators to restart highway operations. It named the chairman of its predecessor, L. H. Wright of Indianapolis, as director, and he, in turn, selected John D. Williams as his chief clerk.[43] Williams, who arrived at the ISHC with considerable experience in the state auditor's office and insurance department, became director in 1922 and served in that post through 1928. Not surprisingly, the new ISHC leaders regularly confronted partisan demands for appointments. Williams successfully fended off the Ku Klux Klan, which promoted legislation in 1925 to bring the ISHC's patronage within its orbit. But neither he nor his successors sought total victory over mainline Republican and Democratic demands for jobs for the partisan faithful.[44]

William J. Titus

Only the ISHC's engineering staff appears to have remained largely untouched by partisan appointments. The politics there were of a different order. Anxious to see that the 1917 sputtering start did not cost Indiana its share of federal patronage, Wright added sixty percent to W. S. Moore's old salary to corral a recent BPR district engineer, H. K. Bishop, as the ISHC's Chief Engineer. Wright also appointed Chesleigh E. Gray and William J. Titus, respectively, Chief of the Bureau of Contracts and Chief of the Bureau of Bridges.

Bishop quickly brought state practice into nearly complete accord with his former employer's policies. The new chief engineer "standardized" all highway work "to comply with the requirements of the U. S. Bureau of Public Roads."[45] His term, though, was short-lived. Wright dismissed Bishop after sixteen months when plans for the next year's work had been completed and the state tax board had cut the ISHC's budget. Apparently convinced that relationships with the BPR were shored up enough, Wright saw no need to retain a high-priced engineer. Once more the *ENR* editor scolded Indiana, this time for being overly parsimonious.[46]

Hoosier-born and bred, Chesleigh Gray succeeded Bishop as chief engineer. A native of Shelburne, Gray had graduated from Rose Polytechnic Institute with a degree in engineering and served as a U. S. Army Corps of Engineers captain during the First World War before joining the ISHC. Gray acted as chief engineer until 1925, when the director dismissed him for disagreeing with the commission's rejection of Marion County's specifications concerning road rock.[47]

A second and even better-connected Hoosier succeeded Gray, and that successor remained on the job until 1933, when he became a Senior Highway Design Engineer for the BPR. William J. Titus requires little additional introduction.[48] Born in Indiana, student at two state universities, sometime employee of Daniel Luten, interpolator of Luten's ideas about artistic design in George Hool's *Reinforced Concrete Construction*, and designer of concrete bridges for a Midwestern railroad company, Titus returned to Luten's employ as an assistant engineer in Indianapolis before moving across town as Chief Bridge Engineer for the revived state commission.

The staff shakedowns within the leadership of the revived ISHC never touched the commission's ties to automobility. At its second beginning, the ISHC accepted the offer of the national government to give it surplus Great War trucks. "We had one source of equipment to draw on and that was the Government trucks. The availability of these trucks caused us to motorize the entire organization from the beginning."[49]

The relationship between the Indiana affiliate of the AAA – the Hoosier State Auto Association [HSAA] – and the ISHC was exceedingly close in the early Twenties. M. E. Noblet, HSAA's manager and secretary in Indianapolis, peppered the ISHC administrators

"BETTER ROADS AND BRIDGES" FOR "BY FAR THE GREATEST NATION ON EARTH"

Only a short time ago in America men drove their carts, wagons and carriages on muddy, uncertain roads, and looked to the fast moving railroads when they wished speed or to 'make time.' Today the automobile carries the larger portion of the traveling public at railroad speed. Every highway is a railroad, and the hard type highway is more important than the steel railway because there are vastly more moving vehicles upon them in proportion than there are trains on the road bed. …

How we congratulated ourselves upon the establishment of post offices, which make it possible for easy and expedited correspondence and all very economically. How much better it is now when long distances are eliminated and citizens of far distant communities crank up an old car and swiftly and with pleasure speed away to personally see and visit friends. And don't forget that much of the commerce of the nation is hauled overland on huge freight autos that annihilate time and distance in this newest transportation service.

Good roads are the same to our nation as clean, wholesome circulation to the human body. Would we consider good health expensive? Never, even though we pay a huge amount to medical skill to restore it, in case it becomes necessary. There is nothing in the world dearer to all than good health. Good highways permit free circulation to citizens, commercial, agricultural and industrial enterprises.

– *A. V. Burch, Vice-chairman,*
Indiana State Highway Commission[50]

and engineers with letters prodding the commission to seek wider rights-of-way, build smoother and straighter roads, install better signs, and keep the motoring public informed about detours.[51]

The ISHC staff responded promptly, courteously, and responsively to the HSAA on most occasions. When, for instance, Noblet reported that the state road between Angola and Waterloo had claimed the rear axle of a member's auto, A. H. Hinkle, the ISHC's maintenance chief, responded that he would turn the lethal stretch of road into "a real race track" that summer. He promised to make it "as suitable to drive over … as the `Indianapolis Speedway.'"[52]

The HSAA reciprocated. When, for example, a farmer in Rush County refused to cede to the ISHC the wider right-of-way that it and the automobile club wanted, the HSAA delivered the farmer's consent to the commission. On another occasion, ISHC director Wright wrote to Noblet about a bill that had been introduced in the legislature which would turn some of the state's automobile fees over to the counties. Noblet and HSAA moved quickly into the political trenches to fight the move. The HSAA did not equivocate on its commitment to the ISHC: "we believe that every cent of money available from the motoring public of the State of Indiana should go the State Highway Commission road funds and into no other fund," the HSAA and the Indiana Auto Trade Association declared in a leaflet.[53]

The relationship between the auto interests and the ISHC did have some rocky moments, as, for example, when the national manager of the AAA reported to the press in detail about two dreadful road stretches – one in Nevada and the other in Indiana – that he had experienced in a cross-country journey. The director of the ISHC was offended, and he let the AAA national manager know it. Manager Ernest N. Smith responded apologetically: "The work of the present commission has been so uniformly excellent that I should hate above all things to have any harmful remark made about it, particularly from a representative of an organization such as our own, which stands primarily for the building of good roads as well as for the type of personnel and activity covered by your excellent commission."[54]

Even as inter-agency and inter-organizational matters were being worked out, the ISHC had its sights set on the highest or most basic automotive priority – paving the highways. The commission did not need to hear from private automotive lobby groups to know the importance of hard road surfaces. ISHC's own employees vividly reported to their superiors the trials and traumas they underwent driving federal-surplus trucks from the State Fair Grounds in Indianapolis, where they had been shipped by rail, to their stations in different parts of the state. Trucks that had apparently managed battle-scarred European fields could not always navigate Indiana's dirt and gravel roads: "Many of our trucks `stuck in the mud' on their way to their destination."[55]

Not surprisingly, the ISHC phoenix arose with a five million dollar construction program – almost all of which was for paving – in 1919-1920. Indiana's hard roads, which measures thirty-seven miles in 1920, stretched to over twenty-one hundred miles within ten years. By 1927, the ISHC could report that "most. . .trunk lines have been paved."[56]

The rapid increase in Hoosier ownership and use of motor vehicles drove and justified the ISHC construction programs. Indiana residents had already registered more than two hundred and eighty-six thousand motor vehicles by 1920. By 1930, the number had jumped to about eight hundred and seventy-six thousand – more than a three-fold increase within a decade.[57]

Although automobiles made up the bulk of motor vehicle registrations, the growth of "heavy traffic" most affected highway design. Truck routes for "express and light freight" to Indiana cities had been established by 1922. They carried, for example, livestock and milk into Indianapolis and transported household goods from places as distant as St. Louis and Cleveland. Bus lines also mushroomed across the state. One hundred and twenty bus lines clocked

(INDOT)

(INDOT)

"Two of the many reasons why it costs more to build and maintain good roads today than in former years." – A. H. Hinkle, ISHC (1924)

more than sixteen thousand miles per day by 1922. Within two years, the number of bus lines had nearly doubled to two hundred and twenty and their total daily mileage had more than tripled to fifty-two thousand and eight hundred. "Horse-drawn traffic has almost disappeared from our main highways," the ISHC reported in 1925, and provided only "the local farm traffic."[58]

With concrete instead of gravel or mud under their wheels, motor vehicles sped along hard roads at rates well beyond what animal-drawn vehicles could achieve. The dramatically increased speeds that automobiles, trucks, and busses quickly achieved on concrete surfaces heaped a range of new demands upon highway designers. Sharp corners and twisting roads that caused no great difficulty for horses and buggies became safety hazards for fast-moving motorized vehicles. As the ISHC staff recognized, "the increased speed of operation of our motor vehicles" put "an increased burden" on state roads. "The rate of speed has been doubled and the distances traveled materially lengthened, all making a heavier load upon our present state highway system."[59] The ISHC responded by straightening and widening roads and broadening shoulders.

(INDOT)

"All that was left of a team and wagon after it was hit by a truck at night; the wagon carried no light." – A. H. Hinkle, ISHC (1924)

The internal-combustion revolution swept through Indiana in the Twenties as it did through most of the rest of the United States and established a "car culture." Robert and Helen Lynd, who studied Muncie in 1925, saw the revolution growing triumphant. When the Lynds asked a resident to pinpoint the major source of change in Muncie, the answer was as clear as it was decided: "'I can tell you what's happening in just four letters: A-U-T-O!'" The Lynds agreed, concluding that "ownership of an automobile has now reached the point of being an accepted essential of normal living."[60]

Hitched to automobility through its political network as well as its own motorized wheels, the ISHC willingly joined the celebration of the internal-combustion revolution. "The automobile has enlarged the living conditions of all the people. Better roads have given increased postal service, furnished transportation for the consolidation of our public schools, and have provided the means for a wider social view of life as well as a wider interchange of the products of the farm and work shop."[61] In short, the ISHC rhapsodized at the end of the Twenties, "the public highway of today is a track for commercial and pleasure transportation, upon which a million automobiles and trucks operate for gain and for pleasure. This means probably six hundred million vehicle-miles ... over our highways each year."[62]

(INDOT)

One ISHC sign for automobility.

VI.2E Fast Lanes for the Wheels of Democracy: From the Great Depression to the Second Great War

The Thirties introduced a decade of turbulence for the ISHC that made the false starts, Klan forays, and occasional tight budgets of the Twenties seem like superficial ripples on mostly-calm waters. The new decade opened with the lengthy and deep depression, which hobbled private enterprise and increased federal intervention in the lives of citizens and bureaucrats alike. Seizing the opportunity in adversity, the ISHC expanded its control over road development throughout the state without effective opposition.

The Thirties closed with preparations for the Second World War by a President determined not to repeat the mistakes of the First. Neither President Roosevelt at the national level nor the ISHC at the state would underestimate the progress of the internal combustion revolution. The President insisted on preparing the nation for a war in which motor vehicles would play a critical role. The ISHC, in turn, continued to serve the evolving demands of automobility within Indiana.

At first, the growing unemployment of the early Thirties touched the ISHC only indirectly. While Indiana counties and cities increasingly transferred tax and bond revenues from construction projects to relief for feeding, clothing, and housing the many new poor, the ISHC's priorities remained the same – to pave roads and build bridges. "There has been no business depression in the state highway departments," the *ENR* editor reported in late 1930. Indeed, the ISHC had more money to spend in 1930 than in 1929.[63]

As the depression deepened in 1931 and 1932, effective control over more of Indiana's roads and bridges shifted from the counties to the state. Indiana counties began to eliminate maintenance as well as construction in order to apply every available penny to the survival of the growing ranks of unemployed residents. St. Joseph County, for instance, announced the abandonment of its road program for 1932. The ISHC, in turn, annexed almost another thousand miles of county roads, and it doubled the maintenance division in the winter of 1931 as part of the governor's "program of relief through employment." The depression was not bad news for the ISHC in other ways, too. Depressed prices during those hard times enabled the state to pave an extra hundred miles of roadway beyond budgeted estimates in 1932 alone.[64]

The continuing downward spiral of economic activity finally galvanized the federal government into action. Highway construction seemed like such a good sponge for unemployment relief that it became a prime conduit for both President Herbert Hoover's and Franklin D. Roosevelt's federal-aid programs. With the federal-state alliance in place, Republican and Democratic administrations instinctively expanded highway activity without significantly modifying the priorities of agency policy. President Hoover, for example, supported the Federal Emergency Relief and Construction Act of 1932, which pumped three million additional dollars of federal-aid funds into the pipeline for Indiana state highways.[65]

Roosevelt's New Deal generated a much larger number of programs with bigger budgets and more wrinkles, some of which ultimately expanded the ISHC's domain substantially. Under a Public Works Administration grant, the ISHC widened the shoulders of a thousand miles of state highway in 1933 and 1934.[66] The National Industrial Recovery Act produced even more largess and a wider reach for the ISHC. By dedicating forty-four percent of a ten-million dollar grant in 1934 to building roads and bridges in Indiana's urban areas, the federal government projected the ISHC into areas that state law had previously precluded. But, even though it entailed more external regulation of design and construction, cities by and large welcomed the new federal-state intervention, since urban officials were otherwise largely without funds either to hire the unemployed or to build needed public works. In time, federal aid came for county road and bridge building, too, again accompanied by external regulation of design and construction.[67]

The lure of extraordinary federal grants only strengthened ISHC ties with Purdue University and further intertwined their federal-agency networks. Prescient "rumors" in 1935 that Congress might pass "a large Federal appropriation bill for highway work" led the ISHC to expand its road and bridge planning. Purdue "cooperated by making available for drafting purposes members of its senior class prior to graduation. Supervisors were sent to Lafayette from the highway department's central office in Indianapolis and many plans were developed by the students who utilized the facilities available at the University."[68]

(INDOT)

Road-widening with Works Progress Administration labor north of Batesville in 1936.

Purdue's support of the ISHC extended beyond emergencies to include considerable collaboration on research projects and sponsorship of extension services patterned after farm-related Department of Agriculture activities. Purdue had begun in 1914, for example, to host an annual "road school" where county, city, and consulting engineers might hear directly from their academic and governmental counterparts as well as meet with industry spokespersons, including the automotive interests. Soon after Indiana joined the federal-state highway alliance, the ISHC's staff began to play a large and regular role in the annual event along with representatives periodically invited from the BPR and other Midwestern state highway commissions. For most of these Hoosier engineers, road school provided an informal reunion of Purdue graduates.[69]

The ISHC used the lower bids and larger governmental grants of the Thirties to press its agenda. Hard times slowed the rate of growth but did not stop increases in the number of motor vehicles used in Indiana. Auto registrations reached more than a million by the time of American entry into the Second World War.[70] Straighter, wider, and longer paved roads only encouraged more journeys of greater length and at higher speeds. The ISHC oversaw at least sixty-seven hundred miles of paved roadway within its system by the end of 1942 – almost three times as many miles as a decade earlier and over one hundred and eighty times more than in 1920.[71]

With its trunk highways already paved before 1930, the ISHC turned to widening lanes from nine to ten feet each and adding more of them for increased and faster-moving traffic. The state ordered its first three- and four-lane pavements for roadways near major cities about a decade after the Boston Post-Road had been paved "to carry four lines of vehicles" and after Wayne County, Michigan, had built two parallel four-lane pavements separated by a strip for street cars from the county line to Pontiac.[72] The Lincoln Highway was no sooner widened to four lanes than the state regretted its decision because of the dramatic increase in the number and seriousness of accidents. The multiple-lane roadway reportedly tempted "drivers to speed their cars, provoking impromptu races that result in cars traveling at excessive speeds cutting into lines of opposing traffic."[73]

A means to contain wrong-way speedsters was only about half a decade old. Milwaukee, Wisconsin, first experimented with "boulevard construction" – travel ways separated by a median. As the *ENR* editor noted in 1935, "safety lies in separating the two pairs of travel ways by a parkway that will absolutely prevent cross-over travel. This of course is not new thought but practice had been slow to commit itself for the greater right-of-way and grading costs; accident experience has been necessary to bring conviction that the extra expense is justified if safety of travel is to be assured in four-lane road design."[74] The ISHC expedited its decision in 1935 "to build no more four-lane roads without a dividing parkway," and it quickly developed a strategy for converting old two-lane roads into divided or dual lane highways. The commission planned to build a pair of ten-foot lanes on each side of an existing trunk highway and then to replace the original lanes with sod. The state's first dual-lane pavements were constructed in 1937.[75]

In late 1940, M. R. Keefe, the ISHC's chief engineer, announced preliminary plans for the state's first "superhighway," a cross-country, limited-access road connecting Chicago on the west to the Pennsylvania Turnpike on the east. Keefe's plan for an express highway across the northern part of the state offered a toll-less version of what later became the Indiana Tollway.[76] Again, the nature of the "super-highway" had been a matter of national discussion for some time. In 1926, the *ENR* editors defined super-highways as including at least four-lane pavements with separation of intersecting road and railway grades. To this they had added in 1935 the dividing parkway or median.[77]

Even as the ISHC under BPR encouragement outlined a twenty-year program, President Roosevelt began to insist on a "national defense" agenda which by 1940 cut across BPR-ISHC priorities. Indeed, Roosevelt at length vetoed business-as-usual highway appropriations. Concerned about American involvement in another world war, the President did not want to find a transportation system snarled as it had been in the Great War.[78] Vetoes caught the BPR's attention and encouraged the agency to acknowledge, if not always to accede to, the Commander-in-Chief's priorities.

At both the national and state levels, creative administrators repackaged a lot of old projects while they did some new strategic thinking. The ISHC's widening of U. S. #40 – the Old National Road – with dual lanes became part of "Highway `Rearmament'," the bypass planned for Indianapolis turned into a "defense road," and the twenty-year blueprint became a "post-war plan." The master plan addressed the needs of a motorized Indiana by emphasizing the widening and improvement of road surfaces, the extension of dual-lane mileage, the relocation of roads around congested areas, the improvement of state routes in cities, and – not least – the construction and rehabilitation of bridges and overpasses.[79]

(Putnam #1835, *ENR*, 1941)

U.S. #40 over Deer Creek which by-passed an open-spandrel arch bridge with a two-lane roadway that the state had designed and built less than ten years before (Putnam #237).

NOTES

[1]For a fuller discussion of the BPR and federal action from 1916 through 1921 – to which much of the analysis here is owed – see Bruce E. Seely, *Building the American Highway System: Engineers as Policy Makers* (Philadelphia, 1987), 36-65. See also Stephen B. Goddard, *Getting There: The Epic Struggle between Road and Rail in the American Century* (New York, 1994), 43-64, 102-19.

[2]"Annual Meeting of the Indiana Engineering Society," *ENR* (21 January 1909) 61: 77.

[3]"An 'Accelerated Test' of Road Wear by Automobile Traffic," "Concerning the Wear of Road by Automobiles," *ENR* (8 July 1909) 62: 42-43, 46.

[4]"The Construction and Preservation of Macadam and Bituminous-Macadam Roads," "The Possibilities of Portland Cement as a Road Material," *ENR* (9, 23 December 1909) 62: 639-40, 710.

[5]W. Stull Holt, *Bureau of Public Roads: Its History, Activities and Organization* (Baltimore, 1923), 10.

[6]For materials on Fisher and the Lincoln Highway, see: Lincoln Highway Association, *Lincoln Highway: Story of a Crusade That Made Transportation History* (Sacramento, California, 1995 reprint of 1935 ed.), 1-28, 125-137, 253; Austin F. Bement "The Lincoln Highway Tour in 1916," *Complete Official Road Guide of The Lincoln Highway* (Detroit, Mich., 1916), 31-35.

[7]Gray, *Alloys and Automobiles*, 173-76; Peter T Harstad, "Harry Ostermann and the Lincoln Highway: He's Gone on Ahead," *Traces of Indiana and Midwestern History* (Spring 1994) 6: 28-29.

[8]"Roads of Brick or Concrete Instead of Macadam for New York State Highways," *ENR* (2 April 1914) 71: 752-53.

[9]"Motor-Vehicle Factory Sales and Registrations, and Motor-Fuel Usage: 1900-1957," *Historical Statistics of the United States* (Washington, D. C., 1961), 462; James J. Flink, *Automobile Age* (Cambridge, Mass., 1990), 73-78.

[10]The Lincoln Highway Association's field secretary spent the 1917 winter exploring possible routes from the Midwest to the east with army and BPR engineers. Except for Ohio, where some alternatives were available, the Lincoln Highway provided the route of choice. *Lincoln Highway*, 95-96, 109, 255-56.

[11]"A New Era in Highway Transportation," *ENR* (28 November 1918) 81: 967.

[12]"Some Engineering and Economic Results of the Commercial Motor Vehicle," *ENR* (22 June 1911) 65: 762-64; "Hard Roads, Marketable Produce," *ENR* (9 August 1917) 79: 243.

[13]"War Orders Divert Freight from Railroads to Highways," *ENR* (23 August 1917) 79: 366.

[14]"A Project for 51,000 Miles of National Highways," *ENR* (16 May 1912) 67: 937-38.

[15]"*ENR* in the past has been opposed to a national system, but the new conditions introduced by motor trucking have so changed the situation as to leave no doubt of the necessity of completing the highway net of the country by adding to our county and state systems a national system tying the various states just as efficiently as rail lines now do for the bulk of our transportation." "Cabinet Attitude on a National Highway System," *ENR* (5 December 1918) 81: 1010.

[16]"Wanted – A Connected National Highway System," "Post-War Highway Drive Begins," "A Suggested National Highway Policy and Plan," "Vacant Directorship in BPR," *ENR* (3 October, 14 November, 19, 26 December 1918) 81: 608, 878-79, 1112-17, 1155.

[17]"A Highway Issue," "A New Era in Highway Transportation," "National Highway System Gets Strong Backing," "Representative Highway Congress in Chicago Unit[e] for National System," *ENR* (28 November, 19 December 1918) 81: 965, 967, 1108, 1145-46, 1149.

[18]John M. Goodell, "Logan Waller Page," *ENR* (12 December 1918) 81: 1095, 1098.

[19]Thomas H. MacDonald, "The Federal Aid Road Law and Changes Suggested by its Practical Operation," *Public Roads: A Journal of Highway Research* [*Public Roads*] (January 1919) 1: 3-6; "New Roads Official Names," *Public Roads* (March 1919) 1: 3.

[20]"Federal Highway Act (1921)," as reprinted in Holt, *BPR*, 86-93.

[21]Bruce E. Seely, "The Scientific Mystique in Engineering: Highway Research at the Bureau of Public Roads, 1918-1940," *Engineer in America*, ed. Terry S. Reynolds (Chicago, 1991), 309-42; Seely, *Building the American Highway System*, 100-35.

[22]T. R. Agg, "Comprehensive Investigations in Highway Engineering Needed," *Public Roads* (August-September 1919) 2: 35-36.

[23]A. R. Hirt, "Memorandum on Mr. Agg's Article," *Public Roads* (August-September 1919) 2: 36.

[24]Anston Marston, "A National Program for Highway Research," *Public Roads* (January-February 1920) 2: 34-37.

[25]"The Organization of the Advisory Board on Highway Research," *Public Roads* (September 1924) 5: 13.

[26]Agg, "Comprehensive Investigations," *Public Roads* (August-September 1919) 2: 35; A. C. Rose, "Highway Research Board Holds its Seventh Annual Meeting," *Public Roads* (January 1928) 8: 243.

[27]"Road Research – The Future," *ENR* (19 December 1929) 103: 953. The *ENR* editors correctly find the results of the Research Board and of the BPR's scientific research as meager. Both, though, overlook the political dimensions inherent within the research programs. Seely, "Scientific Mystique in Engineering," *Engineer in America*, ed. Reynolds, 309-42.

[28]Wallace S. Huffman, "Indiana's Place in Automotive History," *Indiana History Bulletin* (February 1967) 44: 15-19; "100 Years of Automotive History," *Traces* (Spring 1994) 6; Ralph D. Gray, *Alloys and Automobiles: The Life of Elwood Haynes* (Indianapolis, 1979), 67-116; Clifton J. Phillips, *Indiana in Transition: The Emergence of an Industrial Commonwealth, 1880-1920* (Indianapolis, 1968), 263-70; James H. Madison, *Indiana Through Tradition and Change: A History of the Hoosier State and Its People, 1920-1945* (Indianapolis, 1982), 208-13;

James H. Madison, *Indiana Way* (Indianapolis, 1986), 274-5; John B. Rae, *American Automobile Industry* (Boston, 1984), 21-30.

[29]Church and veterans groups almost succeeded in stopping the use of Memorial Day for racing. Only the governor's veto prevented a statutory prohibition in 1923. Madison, *Indiana thru Tradition*, 344-45.

[30]Roy Boomhower, "Carl G. Fisher: The Hoosier Barnum," *Traces* (Spring 1994) 6: 24-27.

[31]George I. Reed, "William H. O'Brien," *Encyclopedia of Biography of Indiana* (2 vols., Chicago, 1895), I: 331-33.

[32]Reed, "Addison C. Harris," *Ency of Biog of Ind*, I: 27-29; obituaries, *Indianapolis Star* (3 September 1916), and *Indianapolis News* (4 September 1916).

[33]Obituary, *Indianapolis Star* (7 October 1946).

[34]For the findings of the commission, a number of which are quoted in the following paragraphs, see *Report of the Indiana Highway Commission to Governor Samuel M. Ralston* (1914) and *Bulletin of the Indiana Highway Commission* (1914), copies of which are located in the Indiana State Library, Indianapolis [ISL].

[35]P. C. Stineman to IHC (14 August 1914), as printed in *Bulletin of the IHC* (1914), 13.

[36]"Dobyns-Duffey Highway Commission Law," *Laws of the State of Indiana: 1917* (Ft. Wayne, 1917), 253-73.

[37]"Indiana Highway Commission Appoints Engineer," *ENR* (26 July 1917) 79: 138; "W. S. Moore Dies at Age 94," *South Bend Tribune* (21 October 1969), 2nd sec: 1, 19; Purdue University, *Alumni Directory, 1875-1934* (West Lafayette, 1934), 345.

[38]See I.3 above.

[39]Purdue, *Alumni Directory*, 158; "W. H. Rights to Locate in Denver," Columbus *Evening Republican* (18 September 1919).

[40]"Report of State Highway Commission," *Year Book of the State of Indiana for the Year 1917* (Indianapolis, 1918), 480-87; "State-Aid Hard Roads for Indiana," *ENR* (15 November 1917) 79: 918; "Four Indiana Engineering Societies Meet Together," *ENR* (7 February 1918) 80: 283.

[41]"Report of the State Highway Commission," *Year Book of the State of Indiana for the Year 1918* (Indianapolis, 1919), 521-24; "Indiana Lower Court Holds State Highway Law Invalid," *ENR* (16, 23 May 1918) 80: 976, 1016-17; "Indiana Highway Commission Law Declared Constitutional," *ENR* (23 January 1919) 82: 205-6.

[42]"An Act Creating a State Highway Commission," *Laws of the State of Indiana: 1919* (Indianapolis, 1919), 119-40; Madison, *Indiana thru Tradition*, 182-92.

[43]Lawrence E. Lyons served as director in 1921. On John D. Williams, see *Book of Indiana*, ed. Frank M. "Kin" Hubbard (Indianapolis, 1929), 187, 355; obituary, *Indianapolis News* (27 May 1948), 2nd sec: 167.

[44]Madison reports successful Republican and Democrat patronage assaults on the ISHC during the Twenties and Thirties in *Indiana thru Tradition*, 93-103.

[45]"Report of the State Highway Commission," *Year Book of the State of Indiana for the Year 1919* (Indianapolis, 1920), 559-73; "Indiana State Highway Commission Selects Chief Engineer," *ENR* (15 May 1919) 82: 988.

[46]*ENR* (7 October 1920) 85: 677.

[47]"Indiana Highway Commission Engineer Dismissed," *ENR* (6 August 1925) 95: 234; obituary, *Indianapolis News* (7 November 1949): 2. Gray's case against the ISHC is aired in the Indiana Association of County Surveyors and County Engineers, *Some Correspondence of the Indiana State Highway Commission* (1925), and J. M. Henry, *A Short History of the Indiana State Highway Commission* (Indianapolis, 1926).

[48]For a description and analysis of the Titus text in Hool's third volume, see III.2 above. William J. Titus, deposition (8 August 1916), District Court of New Jersey, *Daniel B. Luten vs. The Dover Construction Co.: Complainant's Record (Supplemental)* (Indianapolis, 1916), 77; Purdue, *Alumni Directory*, 187; obituary, *Indianapolis Star* (16 July 1950), 2nd sec: 5.

[49]"Report of the State Highway Commission," *Year Book of the State of Indiana for the Year 1920* (Indianapolis, 1921), 1145.

[50]News Release, ISHC Correspondence (1921), Indiana State Archives, Indianapolis [ISA].

[51]"Hoosier State Auto Association," ISHC correspondence, ISA.

[52]A. H. Hinkle, to M. E. Noblet, 2 May 1921, ISHC correspondence, ISA.

[53]L. H. Wright to M. E. Noblet, 20 January 1921, ISHC correspondence, ISA; *The Automobile Owners and Dealers of Indiana* (1922).

[54]Ernest N. Smith to J. D. Williams, 20 October 1925, ISHC correspondence, ISA.

[55]"Report of the SHC," *Year Book for Indiana for 1920*, 1145.

[56]"Indiana Spending $5,000,000 on Roads," *ENR* (4 September 1919) 83: 488; "Report of the State Highway Commission," *Year Book of the State of Indiana for 1927* (Indianapolis, 1928), 1063; L. S. Bowman, "Indiana State Highway Commission: Accomplishments and Costs" (1945), ISL.

[57]"Report of the ISHC," *Year Book of the State of Indiana for the Year 1929* (Indianapolis, 1930), 1117.

[58]"Report of the SHC," *Year Book of the State of Indiana for the Year 1922* (Indianapolis, 1923), 1013-14; "Sixth Annual Report of the SHC," *Year Book of the State of Indiana for the Year 1924* (Indianapolis, 1925), 1237-38; "Seventy Annual Report of the SHC, *Year Book of the State of Indiana for the Year 1925* (Indianapolis, 1925), 1151.

[59]"Report of the ISHC," *Year Book of the State of Indiana for the Year 1928* (Indianapolis, 1929), 1172.

[60]James J. Flink, *Car Culture* (Cambridge, Mass., 1975); Robert & Helen Lynd, *Middletown* (New York, 1929), 251-63.

[61]"Report of the ISHC," *Year Book of Indiana for 1928*, 1172.

[62]"Report of the ISHC,"*Year Book of Indiana for 1929*, 1114-15.

[63]"Some Current Highway and Highway Bridge Activities," *ENR* (25 September 1930) 105: 502.

[64]"Indiana Plans Winter Road Work on Short-Day Basis," and C. H. Parrish, "Winter Road Work for Unemployed," *ENR* (26 November, 10 December 1931) 107: 861, 930-31; "Entire Road Program of Indiana County Abandoned," "Low Costs Mean More Paved Roads in Indiana," *ENR* (21 January, 5 May 1932) 108: 108, 668.

[65]William J. Titus, "Report of the Division of Construction," *Year Book of the State of Indiana for the Year 1932* (Indianapolis, 1933), 1141.

[66]"Indiana Widens Shoulders of 1,000 Miles of Road," *ENR* (21 June 1934) 112: 808-9.

[67]"Report of the Bridge Department," *Year Book of the State of Indiana for the Year 1933* (Indianapolis, 1934), 495-96; "Report of the ISHC," *Year Book of the State of Indiana for the Year 1934* (Indianapolis, 1935), 622-23; "Report of the Engineer of Design," *Year Book of the State of Indiana for the Year 1939* (Indianapolis, 1940), 747.

[68]"Report of the SHC of Indiana," *Year Book of the State of Indiana for the Year 1935* (Indianapolis, 1936), 479.

[69]"Men at Work," *ENR* (27 July 1939) 123: 118; "Purdue to Have Road School," *ENR* (15 January 1925) 94: 130; "Engineers Plan Purdue's Annual Road School," *ENR* (11 January 1940) 124: 66.

[70]Hallie Myers, *Engineering Safety and Convenience into Indiana Highways* (Indianapolis, 1939), 39.

[71]Bowman, "ISHC: Accomplishments and Costs," ISL.

[72]"Broad Highways," *ENR* (6 August 1925) 95: 205-6; "The Trends in Highway Development," *ENR* (6 January 1926) 96: 4.

[73]"Four-Lane Road Hazards," *ENR* (26 September 1935) 115: 445; William J. Titus, "Report of the Division of Construction," *Year Book of the State of Indiana for the Year 1930* (Indianapolis, 1931), 1243.

[74]"Shall Superhighways Be of Single or Dual Pavement Strips?" *ENR* (14 November 1929) 103: 756-57; "Four-Lane Road Hazards," *ENR* (26 September 1935) 115: 445.

[75]"Four-Lane Road Hazards," *ENR* (26 September 1935) 115: 445; John W. Wheeler, "Progressive Road Widening Planned for Indiana," *ENR* (5 December 1935) 115: 773; "Report of the Chief Engineer, ISHC," *Year Book of the State of Indiana for the Year 1937* (Indianapolis, 1938), 914; M. R. Keefe, "Indiana Divided Lanes," *ENR* (13 January 1938) 120: 74-75.

[76]"Indiana Road Head Tells of Superhighway Plan," *ENR* (28 November 1940) 125: 2; "Indiana Plans 20-Year Highway Program," *ENR* (13 February 1941) 126: 212.

[77]"The Trends in Highway Development," *ENR* (6 January 1926) 96: 4; "Four-Lane Road Hazards," *ENR* (26 September 1935) 115: 445.

[78]Seely, *Building the American Highway System*, 165-77.

[79]"Speeding Highway 'Rearmament'," "Indiana to Start Defense Access Roads Construction," *ENR* (2 January, 3 April 1941) 126: 63, 505; "Indiana Road Commission Develops Post-War Plans," "Belt-Line Defense Road Planned for Indianapolis," *ENR* (31 July, 25 December 1941) 127: 144, 913.

(Ripley #1234)

SR #101 over Hogan Creek.

(Knox #690)

VII

Spanning the Motor Speedways

(INDOT)

The old along with the somewhat newer. A covered bridge still carried the highway now paved for the automobile. A companion metal-truss structure served light-rail traffic, and a three-span plate girder took a regular railway across the parallel road and light-rail routes.

On facing page – George Rogers Clark Memorial as framed by a center span of the George Rogers Clark or Lincoln Memorial Bridge. (Photo by William E. Kelly, Graphics Engineering, INDOT)

Besides requiring straight, wide, and hard-surfaced roadways, motor vehicles generated more replacement of highway bridges within a few decades than had any previous change in American transportation. No machine on the roadways dealt more death and destruction than the automobile, and none so linked bridges with safety in dramatically new ways. Before the automobile came onto the American scene, bridges were considered safe when they bore the traffic loads of the day over a solid deck and between hub guards and railings. With the motor vehicle, however, load-bearing demands escalated dramatically, the speed of vehicular missiles required more room in case of driving error, and persons and structures on the road needed special protection. Safety applied to more dimensions and assumed a larger role in bridge design. When linked with the expanding federal-state highway system, its bureaucracy, and its commitment to scientifically-based or rational design, the automobile dramatically altered the nature of spans as well as their number.

Change swept across the Hoosier landscape unevenly. Street railways had forced urban officials to replace a number of bridges designed for wagons well before 1920.[1] In cities, motorized transport added many more streets to the roster of those needing greater load-bearing bridges. Unsullied by trolleys, the counties faced the consequences of motorized transport somewhat later and more sporadically. Still, in the countryside too, heavily loaded, fast moving trucks ultimately tested and increasingly destroyed bridges designed for wagons. Unlike their city counterparts, though, most county authorities could sidestep at least some of the consequences of motor vehicles, since the Indiana State Highway Commission [ISHC] gradually incorporated the most heavily traveled roads into the new state system and took responsibility for the bridges on them.

VII.A Hoosier Bridges Falling Down ...

Between 1920, when the state assumed control over its first thirty-two hundred miles of county roads, and 1941, when it maintained more than ten thousand miles of highway, the ISHC inherited four thousand bridges mostly built for horse-drawn traffic.[2] The mere act of designating a road as a state highway inevitably attracted those with heavier and faster-moving vehicles seeking the shortest routes between major and often long-haul designations. The new routing inevitably guaranteed the greatest destruction of bridges that the state had ever witnessed.

The Allies had tried out armored trucks as tanks towards the close of the First World War. The automobiles and trucks that Hoosiers drove might not have been plated like those ersatz tanks, but the bridges they hit were not shielded by trenches and embankments either. "Reports that one or more of our old bridges have been destroyed by heavy truck loads, by traffic accidents or by floods are received in our office almost every week of the year," noted the ISHC's chief bridge engineer around 1923. The number of bridges destroyed mounted over the years: nineteen bridges "were wrecked by speeding motor cars" in 1929 alone.[3] Truss bridges suffered most.[4] Since trusses typically bore their loads alongside and above the roadway, they were especially vulnerable to automotive ramming. All too often, breaking an endpost, taking out supporting members, or pushing a truss off the substructure damaged the whole beyond feasible repair.

If, as the ISHC's head design engineer asserted, "no highway is stronger than its weakest bridge," then the additional and heavier automotive traffic on state roads required that at least some of the shakiest horse-and-buggy spans be replaced.[5] But, even if the ISHC had presumed that wagons would continue to dominate highway transport, it would have replaced a large number of inherited bridges as quickly as possible. They represented a smorgasbord of design and construction to whose considerable diversity almost a hundred governmental units, dozens of designers, and scores of fabricators and contractors had contributed. Maintaining such a giant melange of bridges would have bedeviled any group and inevitably inspired some reduction in their variety. The diversity must have been especially galling, moreover, for a government department prone to standardization.

Motor-vehicular safety became the byline of much bridge replacement. In the "ideal state," the *ENR* editors announced, highway bridges would "afford the same degree of safety to traffic as other portions of the road." The automotive public of the early Twenties, though, still lived with "thousands upon thousands of horse-and-buggy bridges:"

AUTOMOBILE SCALES TRUSS OF BRIDGE EAST OF CITY

Ellis Culbertson's Machine Strikes Abutment and Runs up Angle of 45 Degrees to Top, Hanging in Perilous Position.

Ellis Culbertson, driving a practically new five-passenger automobile, was not hurt this morning when he lost control of his machine at the entrance of the bridge over the pond between the mill race and the river bridge east of Rushville.

The car dashed to the left of the road and struck the cement abutment which projects about a foot above the surface of the highway. This deflected the car which was headed for the water twenty feel below when the left front wheel hit the endpost of the north truss of the bridge. Even this was not enough to stop the machine and it continued in its course, the two left wheels running up on the truss, which slants at an angle of 45 degrees, until the top of the truss was reached.

When the right rear wheel left the ground, the car toppled over on its side and Mr. Culbertson fell out on the floor of the bridge. The top was the only thing which prevented the car from toppling over on Mr. Culbertson and seriously injuring him.

Mr. Culbertson was driving east, returning to his home after having purchased a lawnmower, which was the indirect cause of the accident. The mover was in the back seat. The handle flopped over and struck Mr. Culbertson in the back. He let loose of the steering wheel with his right hand to push the lawnmower handle back and evidently pulled the car to the left with his left hand. 'I had no time to jump,' said Mr. Culbertson just after the accident. 'It all happened in a flash and before I knew it the car was hanging on the side of the truss and I was lying on the road.' ...

Mr. Culbertson apparently was going rather fast because the rubber tire nicked the cement abutment and cracked the iron at the end of the truss when the machine started its perilous climb.

– *Rushville* Daily Republican[6]

(INDOT)

Trucks and trusses.

(INDOT)

> Accordingly we are troubled today by great numbers of narrow bridges, slow-speed bridges, bridges located on curves or at impossibly awkward jogs of the road, trusses or girders so placed as to invite endwise collision, low roadway curbs, unprotected truss members, and weak railings – not to mention such minor botherations as caving approach banks, abrupt jumps at entrance and exit, and bad spots in the bridge floor that force traffic out of line.[7]

The editors treated "horse-and-buggy" bridges as a class and found them generally unsafe for motor vehicles primarily because of the ever-growing differential in speed between animal-powered and horseless carriages. By 1928, Indiana's 1918 speed limit of twenty-five miles per hour had been raised to forty. But, as the Bureau of Public Roads [BPR] acknowledged, "a great many users of the highway travel at speeds much higher than the legal limit." "The models of motor cars produced in recent years are all capable of relatively high speeds, and the general trend toward increased braking capacity and greater riding comfort has given drivers a sense of greater security at high speeds."[8] As a consequence of the higher speeds which motor vehicles attained, traffic safety ruled bridges designed for animal-drawn vehicles off the federal-state highways.

Bridges now needed to be road-aligned. While sharp turns on roadways did not disturb the pace of horse-drawn vehicles, they added new risks to travel for car and truck drivers. Failure to slow down increased the danger of hitting a vehicle coming from the other direction, of running off the road, or of ramming a bridge. Safety for the motorist and passer-by as well as protection of surrounding property offered new incentives, therefore, for road-aligned bridges on straight highways. Straight-arrow roads both encouraged the building of some spans in new locations and put a premium on the construction of more bridges which angled across a river. Skewed bridges "required more difficult features of design and construction" and used more materials than spans which crossed a river on the perpendicular and left all compensating turns in the roadway.[9]

Automobiles also needed wider bridges for broader roads than did wagons to provide the same margin of traveling safety. The periodic and discrete programs to widen roads and bridges typical of the Twenties mushroomed in the middle years of the depression when the federal and state governments sought labor-intensive public works projects for the unemployed. In 1933, for example, the ISHC inaugurated a program to widen eighteen hundred miles of state highways and three thousand associated bridges and culverts.[10]

Periodic ratcheting up of load-bearing capacity in wider bridge structures did not affect all existing structural types equally. Just as motor vehicles destroyed more truss bridges than other kinds, so they also pushed existing timber and metal truss structures towards the top of the ISHC's list for replacement by increasingly testing their designed limits. Removal ranked well ahead of rehabilitation for the older timber and metal truss spans, since most of them could neither be significantly strengthened nor widened without taking the whole structure apart.

As the demands of automobility evolved, the federal-state highway alliance kept modifying what spans it would clear off or abandon on the landscape across the state and nation. Taken together, the introduction of multiple lanes, the addition of dual-lane and some divided highways, plus the general increments to proper roadway and shoulder width, all accumulated to such a point that by the late Thirties the ISHC found itself even bypassing or selling-off some bridges of its own design constructed only a decade earlier.

(INDOT)

A now-narrow span.

VII.B Grade Separations: Saving Time and Lives

The automobile neither first caused nor did the federal-state highway alliance first raise the question of separating the grade of roadways and railways. The issue initially surfaced in Indiana late in the Nineteenth Century largely in urban areas where animal-drawn vehicles, trolleys, and pedestrians competed at crossings with steam locomotives. Indiana's major cities lobbied the railroads for rail elevation, and the rail companies responded with programs to build subways under their tracks or overpasses above them. But, with its presence assured by 1920 and its advocates well organized and highly placed, the automobile added significantly new dimensions to the public discussion, and the highway alliance extended the call for grade separations across the countryside.[11]

(INDOT)

Deferring to the Iron Horse crossing SR #6.

The public and the highway alliance leaders focused intently on the outcome of the crossing contests between motor vehicles and locomotives. By 1922, motor vehicles provided eighty-nine percent of the challenges to locomotives, as reported in a Pennsylvania Railroad study of crossing accidents. The contests favored the train over the car or truck, the latter machines and their occupants almost always suffering greater damage and bodily harm. Motorists bore all seventy-one deaths and one hundred and thirty-one injuries reported in the Pennsylvania Railroad study. The federal Bureau of the Census reported a total of nearly eighteen hundred crossing deaths in 1934 which, while it accounted for less than five percent of all automotive fatalities in that year, occupied more than that proportion of the public's and the alliance's attention.[12]

Highway alliance leaders discussed separating the grades of highways from railroads largely under the rubric of safety. But the data from the Pennsylvania Railroad study suggest that in actuality impatience with waiting sparked the motorist's interest in grade separations. Seventy-seven percent of the automobile accidents with the Pennsylvania Railroad occurred, for example, after the motorists had run around crossing gates or disregarded guards. The city engineer of Chillicothe, Ohio drew a similar conclusion in 1925. A majority of the residents voted for a bond issue to fund the city's share of the cost of a grade separation where no serious accident had ever occurred and about which "traffic convenience" was the only major topic of conversation. The city engineer calculated that closing Main Street at the railroad during the average work day cost motorists and pedestrians "no less than twenty-five hundred man-hours" of waiting.[13]

(INDOT)

"They failed to 'Stop, Look, & Listen,' until it was too late." –A. H. Hinkle, ISHC (1924)

The move to separate highways from railroad tracks increased demand for new bridges in Indiana, too. The ISHC, as usual, started slowly, and Indiana before 1928 was not among the top ten states in the construction of grade separations. In 1928, however, Indiana rose to fourth among the states with the building of twenty-four separations. Viewing "with horror the increasing list of fatalities and complete destruction of property due to accidents at such crossings," ISHC planners had by 1940 gradually accumulated well over a hundred overpasses and underpasses separating rail lines from highways.[14]

VII.C Build Them Up With Steel and 'Crete ...

Replacement, relocation, and response to new demands encouraged the ISHC in its first three and a half years to design nearly one hundred and fifty bridges and to contract for their construction at a cost of over one and a half million dollars. Soon the state's annual bridge construction ranged between one and a half and two million dollars, a rate which placed Indiana "in the front rank in highway bridge expenditure in 1924 and again in 1925."[15]

The ISHC's new bridges differed from those they replaced, sometimes in location, often in material, and usually in form. The ISHC, for example, never replaced a covered bridge nor a stone arch with one of similar material and form. And its considerable number of steel-truss structures represented an ever-smaller portion of the system's spans. For many reasons, reinforced concrete emerged as the material of preference for automotive spans, although it had effective challengers.

During the Twenties, the state spent between fifty and sixty percent of its bridge construction dollars on metal trusses. The proportion dropped considerably in the Thirties, as metal trusses accounted for around thirty percent of the bridge construction dollars at the decade's start and slid to between ten and seventeen percent at the end. Steel trusses always retained certain advantages for particular sites. Navigable waterways virtually required the wide and high clearances that the longer truss spans most readily provided. For other rivers and streams, a single metal-truss span could sometimes bridge a crossing that might take multiple spans

ISHC BRIDGE CONTRACTS FOR SELECTED YEARS: Steel and Reinforced Concrete Structures[16]

	Number of Spans	Total Span Length in Feet	Average Span Length in Feet	Total Contract Dollars	Average Cost per Span Foot
1923					
STEEL (ALL)	41	3,512	82	659,549	188
TRUSSES	41	3,512	82	659,549	188
CONCRETE (ALL)	108	3,392	31	519,720	160
ARCHES	16	987	62	203,946	207
BEAMS	53	1,830	35	231,272	126
SLABS	39	575	15	84,502	147
1929					
STEEL (ALL)	44	4,889	111	874,965	179
TRUSSES	44	4,889	111	874,965	179
CONCRETE (ALL)	101	3,731	37	699,081	181
ARCHES	39	1,903	49	354,678	186
BEAMS	52	1,680	32	319,515	190
SLABS	10	148	15	24,888	168
1933					
STEEL (ALL)	58	4,965	86	780,284	157
TRUSSES	27	3,503	130	503,199	144
BEAMS/GIRDERS	31	1,462	47	277,085	190
CONCRETE (ALL)	153	6,054	39	1,139,239	195
ARCHES	62	2,858	46	623,186	218
BEAMS	82	2,748	34	485,822	177
SLABS	9	160	18	30,231	189
1939					
STEEL (ALL)	84	5,586	67	1,383,320	248
TRUSSES	8	1,318	165	227,578	173
BEAMS/GIRDERS	17	970	57	399,895	412
CONTINUOUS BEAMS/GIRDERS	59	3,298	56	755,847	229
CONCRETE (ALL)	62	2,219	36	794,564	314
ARCHES					
FILLED	19	791	42	430,019	544
OPEN SPANDREL	3	230	77	46,040	200
BEAMS	33	1,068	32	273,016	256
SLABS	5	94	19	40,448	430
CONTINUOUS SLABS	2	36	18	5,041	140

of concrete with some in-stream substructure. To eliminate vehicular damage to trusses and to permit widening, designers could in a few cases run the roadway over the tops of the trusses. But placing the load-bearing structure below the deck decreased waterway clearance and courted a washout from floods. Thus below-deck trusses were limited to sites with especially high banks.

The dramatic decrease in the ISHC's use of metal-truss bridges during the Thirties was both absolute as well as relative when measured in total span-feet and construction dollars. The switch away from steel trusses, however, did not generate a significant and lasting increase in concrete construction. Indeed, the ISHC largely switched from one metal bridge form to another, especially from short trusses to steel beams. Unlike the trusses, the steel beam substitutes offered additional design simplicity at competitive construction prices, in part because improvements in the state highway system itself made feasible the transportation of larger, heavier beams to sites not within practical reach a decade earlier.[17]

Although only sometimes capturing the largest share of ISHC bridge contracts, reinforced concrete played a significant role in state design and construction. All its forms – arches, beams, and slabs – were automobile-friendly. They did their carrying below the roadway, out of reach of motor-vehicular ramming and amenable to later widening. During the Twenties, the ISHC dedicated between forty and fifty percent of its total span-feet and contract dollars to concrete structures. Those percentages increased dramatically to between fifty and sixty during the first half of the Thirties, when job creation ranked ahead of economic efficiency in public policy. For awhile, officials more often approved concrete construction because it involved a lot of low-skilled labor. With the easing of the economic crisis by the end of the Thirties, though, the ISHC's percentage of total span-feet and contract dollars dedicated to concrete dropped back significantly to between thirty and forty percent.[18]

VII.1 THE ISHC'S JOB SHOP FOR DESIGN

> Highway system development, state and federal financing, centralized administration, traffic management, structural permanence – all that is distinctive of modern public roads – are the creation of a far shorter period than the life of *Engineering News-Record*. …
>
> Highway engineering has become … a specialty. It is not the task of the general practitioner. Indeed as the business of highway building has developed under state and federal aid there is little occasion for the independent practitioner.
>
> – *C. S. Hill, Associate Editor,* ENR *(1924)*[19]

The ISHC followed the common practice within the federal-state highway alliance of doing its own bridge design. The ISHC's own staff prepared "practically all" of the plans for the bridges constructed on Hoosier state highways. Only in very rare cases – where, for example, a major Ohio River bridge was being designed – might the state use non-government engineers as consultants. Never is the ISHC known to have accepted an "independent practitioner's" design introduced through the bidding process. With a small handful of exceptions, some engineer on the state payroll designed virtually every one of the twenty-one hundred and fifty-five bridges – nearly thirty-seven miles of spans – which the ISHC built before the Second World War.[20] An engineer who resigned from the ISHC in a major tiff with the commissioners in 1925 characterized the system as "the complete socialization of the industry of bridge engineering in the State of Indiana."[21]

"Centralized administration" meant more than design by state-employed engineers. It also meant design consistent with national "standards" as well as BPR approval of the plans proposed for each structure slated for federal aid – the vast majority of ISHC bridges.[22] Such processing review and approval of plans consummated federal authority through funding. Centralized administration, in short, involved government control at both the state and the federal levels.

The BPR's claim to the possession of special and scientific expertise, which applied to bridge as well as to roadway design, buttressed the bureau's authority. Even before Thomas MacDonald took its helm, the BPR strongly favored rational over empirical bridge design. MacDonald's emphasis on bureau research and the BPR's informal superintendence of the Highway Research Board, including its standing committee on bridges, elevated the bureau's scientific mantel.[23] Well publicized claims to preeminence in highway matters, when added to its funding power, must have wrapped the federal bureau in a magisterial aura which promoted deference from state highway bridge designers who ranked lower in the governmental chain of command.

VII.1A The Complications of Reinforced Concrete

Bridge designers in the BPR system could achieve scientific correctness in some materials and forms more easily than in others. Since the 1850s, engineers had computed the stresses of bridge members in wooden and metal bridges.[24] Reinforced concrete, however, presented a range of challenges to designers armed only with nineteenth-century theory and its accompanying mathematical formulas. Concrete's properties had not been fully determined, the combination of two materials – concrete with steel – added a thorny twist to calculations, and the fluidity of concrete before setting encouraged the use of structural forms where members joined, thereby further complicating analysis. At the edge, designers of some kinds of reinforced concrete structures relied on the sorts of empirical formulas and educated guesses out of favor with the BPR.

The bureau demonstrated its preferred methodology for designing bridges by funding research projects to address particular aspects of reinforced concrete slab, beam, and arch design.[25] Each research project report specified the theory being applied, attempted to separate and control the variables being studied, and sought quantitative data often generated with specially built devices. The researchers aimed, where possible, "to formulate a mathematical basis for design," often recommending some adjustments in the design formulas used in accordance with established theory.[26]

The static analysis which engineers conventionally employed with timber and metal designs worked for computing most of the stresses associated with concrete slabs and beams. The stresses in reinforced concrete arches, however, could not be computed from these principles alone, and arches were therefore considered "statically indeterminate." With the assistance of elastic theory, though, arch stress could be calculated and thus scientifically-correct design achieved. But the calculations were so complicated and time-consuming that they had led Daniel Luten and others to "empirical formulas" or shortcuts. In the early Twenties, Professor George Beggs of Princeton University advanced empirical knowledge of the elastic arch through modeling. He developed a much-touted system for computing the reactions of elastic structures by measuring the deflections in a paper model.

Despite its strong preference for rational design, the BPR bowed to the practical with arches. It evaluated the systems of empirical analysis available and finally prescribed a set of formulas which it expected the states to follow.[27] Where possible, however, the BPR dressed its judgments in scientific robes and publicly engaged in scientific ceremony. In a highly-publicized, symbolically-significant spectacle, the bureau carefully planned and elaborately measured the destruction of a North Carolina open-spandrel arch bridge. Among its tests of some of the more prominent practices and presumptions of how arches acted, the BPR took a close look, for example, at the much talked about Beggs system of empirical analysis. The bureau ultimately gave the professor and his method its qualified blessing. In addition to written reports, the BPR produced a twenty-minute motion picture which featured "the method of loading, test apparatus in place, measurements being taken, and close[d] with scenes taken during the destruction of the bridge with explosive[s] in experiments by the War Department."[28]

The BPR's commitment to the scientifically correct or to a quite narrow range of heavily fenced empiricism limited acceptable design and restrained invention far more than would empirically-based planning. Occasionally, however, a state highway designer sought BPR support for an innovative design, and sometimes the bureau responded affirmatively. For instance, Conde B. McCullough, by now a leading engineer for the Oregon Highway Commission, successfully proposed BPR participation in the erection of a concrete arch by the Freyssinet method which had been used in France since 1908.[29] But change would most often come within the system, particularly within Indiana, in small increments along largely predictable paths.

Instead of seeing a new bridge project as an opportunity to experiment with innovative design, alliance engineers more often began with the shoe-horn, seeing how much a conventional, already approved design had to be altered to fit a new location and its roadway. Indeed, planning that passed through two sieves of government bureaucracy and academic certification encouraged a resort to standardization. At the least, imitation was a hedge against insanity for the ISHC's bridge engineers, who on average had to turn out plans for more than a hundred separate bridges yearly from 1920 through 1940, and in the early years each set of plans had by law to include one design for steel and one in concrete.

VII.1B Just-In-Case Design

Other factors prompted the ISHC's addition of a just-in-case wrinkle – a belt-and-suspenders approach – to its mostly conventional design. Automobility, for example, changed so many bridge design requirements that over-design for the present might anticipate near-future needs.[30] "The development in the number, the weight, and the speed of motor vehicles during the past decade has been beyond the belief of those whose duty it was to anticipate the demands of traffic, when designing and building the present bridges," rued the ISHC's Bridge Engineer in 1923. His designers now "arbitrarily" extended current requirements to "the limit for which we can reasonably and economically provide in present design and construction." Even then, successors in the bridge department might "look back on our standards as the products of short-sighted men."[31]

Safety, which ranked high in the priorities of BPR system engineers, added to the pressures for just-in-case bridge design. A design failure on a roadway can produce an outpouring of complaints from those who were jostled by potholes, whose tires get punctured by sharp concrete edges or pieces of reinforcing, or who suffer delay by detours. But such complaints are minor when compared with the public outcry which the loss of life and vehicles from a bridge failure can generate. Driven by the whip of competition in the marketplace, the private bridge designer may well rank cost-saving efficiencies somewhat higher in relation to risk than would a government designer on salary and not subject to design competition. A government designer would more assuredly lose his job over a design that failed than over one that prescribed a surplus of materials. In many cases, the publicly-employed engineer succumbed to the strong temptation nurtured within his agency's environment to build redundance into design.

Detailed, just-in-case design also provided a hedge against construction variations. While the BPR-ISHC system eschewed the marketplace for design, it favored a highly competitive but also circumscribed arena for construction. The government's engineers wanted a large number of construction companies to operate within a quite prescribed field. This drove the design engineers to produce highly detailed plans "to clearly show the exact construction contemplated, because we believe that carefully prepared plans are first necessary in order to secure carefully constructed bridges."[32]

To enforce its detailed standards, the state relied heavily upon construction monitors. The ISHC proudly reported that "careful inspection has attended the construction work – not an inspection which contents itself with the setting of a few line and grade stakes and a final inspection when the work is complete, but an inspection which tests and measures all materials which enter into the work and continually checks the foundation depths and all other dimensions of the structure while it is being built." Securing "qualified" inspectors, a Hoosier highway commissioner reported, is as important as getting "the best engineers possible," if the work is to be "properly done."[33]

The BPR-ISHC system's penchant for detailed contracts reduced the public's risk and at the same time limited its benefits. It discouraged contractors from experimenting with alternative materials or techniques, some of which might have proven inferior while others could have increased efficiency. Presuming highly-regulated building by the lowest common competitive denominator of contractor, the ISHC specified micro construction standards rather than harder-to-measure performance criteria. Usually, the quality of the contractor exceeded the state's most basic assumption. Occasionally, it did not.

THE BRIDGE FAMILY

The owners and operators of small American enterprises have typically regarded their own lives as too ordinary to record. Indeed, the minimal records of small building contractors who competed successfully for ISHC bridge contracts during the Twenties and Thirties have mostly discarded and the social life of the workers has largely passed without written reflection. Mary Ann Clark Moore, however, has provided some snapshots into that nearly forgotten world. She saved some photographs and wrote about growing up in an extended Hoosier "bridge family" between the world wars.[34]

During these decades, Robert H. King operated a moderate-sized bridge construction business out of Danville. He concentrated on smaller projects that a "gang" of six to eight experienced men, doubled or tripled in size as needed with labor hired on site, could manage. In a year with a bumper crop of contracts, King fielded two gangs with trusted foremen, and he visited each construction site weekly. King won one of the first ISHC contracts in 1920 and secured at least three dozen others during the decade before the Great Depression set in.[35]

Besides a willingness to perform hard manual labor during long work days, it took skill and networking to become a core members of a King bridge gang. At the outset a candidate had to be familiar with carpentry and known as hard-working and reliable. The practice of communicating and evaluating qualifications through word of mouth generally limited employment to King's and his valued workers' kin

The bridge gang at Charleston with the Clark brothers on the truck's hood.

or neighbors. Paul Clark, who with his brothers Earl and Beuford became core members of King's gang in the early Twenties, joined after returning to civilian life from the Great War.

Although King concentrated his bidding efforts on structures in and near Hendricks County, both ambition and circumstance led him to contract for construction projects in almost all parts of the state over

(Continued on next page)

Left – Paul Clark placing reinforcing rods.

time. His successes required the gangs to live away from home for months at a time. Married to Vonele in 1922 and father to Mary Ann and Pauline, Paul Clark lived with his immediate family in more than two dozen places beyond their North Salem base between the world wars.

For Mary Ann Clark, "family" included more than her father, mother, and younger sister. It encompassed at a social distance the King family in whose Danville barn, where bridge machinery was sheltered and repaired, and over whose lot, where lumber was stored, she frequently played while her dad worked. Occasionally she "visited with Mrs. King and enjoyed looking at her bird books" in the large house nearby. More immediately and intimately, family included her father's brothers and their wives along with the other core members of the gang with whom she traveled and lived in different parts of Indiana and of Illinois.

Mary Ann Clark grew up in the age of automobility. Paul owned a car – indeed, always a Ford – to which he periodically attached a two-wheeled trailer with wooden side racks during the Twenties and early Thirties. On moving days, the Clarks packed the trailer with a coal stove for heating and a kerosene one for cooking; metal bed springs with a set of attachable homemade wooden legs for the parents and canvas cots for the kids; folding chairs and a card table; wooden packing boxes loaded with supplies which, when emptied, stacked, and covered with print curtains, became kitchen cabinets; a sewing machine for Vonele to make and mend the family's clothing; and a tent to supplement the new temporary residence.

The bridge families that descended on a small Indiana community for a few months "couldn't be choosey about living conditions." An empty house available for rent was quite rare. More often, the family occupied a spare room or two. Sometimes a "summer kitchen," shed, or garage became home. Almost always, the Clarks used the outdoor privy and well, leaving plumbed indoor facilities for the landlord's family.

A garage that became a bridge-family home in Eaton.

Mary Ann attended two and sometimes three different schools in a year and worshiped in churches of quite different denominations. Outside of school and church, she played near her latest new home with her sister and other bridge-family kids, sometimes inventing stories around a buggy found in an old barn, an automobile abandoned without tires on a back lot, or with her companion doll. On weekend evenings there was often much kibitzing among the adult bridge-family members who gathered together from their separate perches for chili suppers and to play checkers, dominoes, and euchre.

As hard times sharpened ISHC bidding and eroded King's competitive edge, he turned in 1930 to construction in Illinois to sustain business. That market, too, soon turned sour, and King hobbled along with only an occasional ISHC contract in each of the next several years. With King's business slowed to a trickle, Paul Clark was fortunate to get a job with the Hendricks County highway department in 1931, and the family settled back into its North Salem base for several years. When New Deal pump-priming rejuvenated his business by 1935, King reemployed Paul Clark – this time as foreman of one of his gang's.

Once again, the family took to the road, although in improved circumstances on a foreman's pay. In 1938, for example, the Clarks purchased one of the first house trailers put on the market. Their twenty-seven-foot "Covered Wagon" included a convertible couch-bed, an ice box, a gas stove, a closet, and a few cabinets. Water still

(Continued on next page)

Mary Ann and her younger sister, Pauline, in front of their "Covered Wagon." The girl's tent was not included in the photo.

had to come from a nearby well and the out-house remained a necessity. "Roosevelt toilets," a New Deal innovation of concrete stools with wooden seats in the out-house, however, sometimes provided another happy improvement in the daily lives of the bridge families. To offer some privacy for their parents, the girls continued to sleep in a tent when weather permitted.

Now somewhat older, Mary Ann could explore the new communities by bike, attend town baseball games, go to the movies, and play dominoes and cards with the adults. She could also participate at the edges of the bridge work world. She accompanied her father, for example, on many of his nightly trips to the temporary bridges or "run-arounds" where she helped the foreman light the kerosene lamps which warned motorists of a bridge detour. On cold spring or late-fall evenings, she helped to tend the "salamanders" – wood- or coal-filled barrels fired near setting and curing concrete to keep it from freezing. As her education gradually surpassed her father's, she began to make other small contributions to his work. She started to draft – with a script more handsome than her father's – the weekly time sheets that Paul submitted over the weekend to Mr. King who prepared paychecks which he usually personally delivered to the workers on the job some days later.

Vonele Clark posed atop a large cement mixer at Seymour as she had perched some evening before, although then with a shotgun. Determined to stop nightly siphoning of gasoline out of the construction crane, Paul and Vonele had laid in wait and had caught the culprits in the act. Vonele held the unlucky ones at gunpoint while Paul telephoned the local authorities.

The Second World War brought an end to this bridge family saga, but not to Robert H. King's and Paul Clark's bridge building. Both joined the Seabees, naturally at different ranks, and constructed spans for the military. Mary Ann Clark, in turn, closed her "gypsy" years by marriage in 1942 without forgetting the excitement of having been "a bridge builder's kid." She would "always remember all the places I called home, all those wonderful experiences and the many friends – all over Indiana."

VII.2 ARCHES FOR THE STATE MOTORWAYS

VII.A Filled Spandrels

Filled-spandrel arches accounted for nearly one-fifth of the spans which the ISHC designed and had built before the Second World War.[36] William J. Titus, who designed the first ones, applied his considerable training and experience to design and placed his stamp firmly on the state's arches before he took a BPR post in 1933. His proteges, Fred Kellam and Frank Y. Hardy, carried on the Titus traditions in the ISHC's bridge department into the Second World War.[37] Quite predictably given his background and his institutional setting, Titus promoted conventional and low-risk design laced with touches of his mentor Luten's style.

Titus, Kellam, and Hardy favored the three-centered arch shape which approximated the semi-ellipse akin to Luten's preferred "linear" form.[38] By 1920 the three-or-more-centered ring – rather than the circular or segmental shapes – had become the conventional form with which designers began their planning for a particular site. Only with an arch placed under fill so deep that its weight put unusually heavy pressure on the ring would the ISHC engineers sometimes return to circular forms. The state engineers did often design each end span of a series to be unsymmetrical – that is, the ring extended further around in an additional arc towards the abutment than at the pier – to deliver some of the lateral thrust of the series vertically to the substructure, allowing for a more trim abutment.

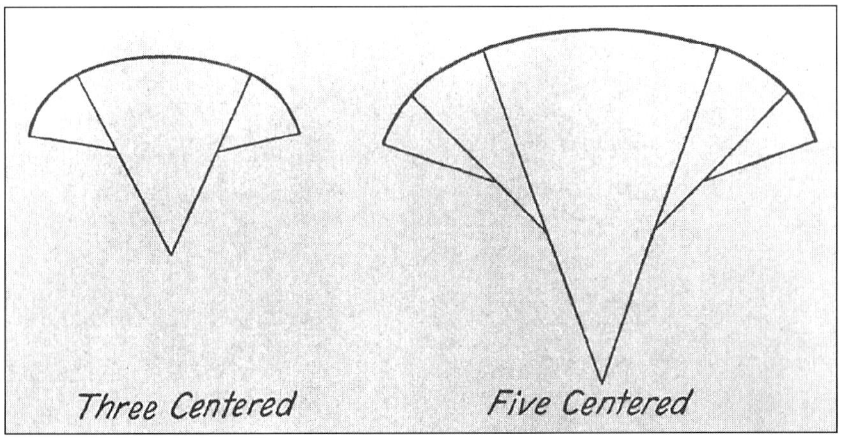

Titus made use of some of Luten's reinforcing practices and not of others. Like Luten, Titus treated piers as integral extensions of the rings. This design decision permitted the piers as well as the rings to be dramatically slimmed, thereby both reducing construction costs and increasing waterway clearance. Titus abandoned Luten's ring system, however, for the federal-state alliance system convention of two-tier reinforcement. Here is a good example of just-in-case design: "the general practice is to carry both rows of steel throughout the entire span" George Hool had noted approvingly, "thereby eliminating any possibility of failure due to an inadequate provision for tensile stresses."[39] Besides virtually doubling the amount of steel over what Luten would have used with his single and continuous set of rods, the state's use of two tiers of ring reinforcing also carried some rods through regions of high compression, where they might in consequence pop out of the concrete. To "prevent buckling" and to keep the inner and outer sets of bars properly spaced during the pouring of concrete, Titus followed Hool's advice to add a third set of reinforcing members or stirrups which tied the upper and lower reinforcing rods together.

Titus designed his filled-spandrel arches cautiously in other ways, too. The ISHC prescribed shorter and therefore more spans rather than longer and fewer ones for a given structure length. The flatness of considerable portions of Indiana's terrain gave an edge to short-span filled arches which minimize the distance between the roadway and the watercourse. As a span lengthens, the distance from the waterway to the crown or center of an arch ring necessarily grows and accordingly provides a higher roadway than with a shorter span. Terrain, however, does not appear to have dictated

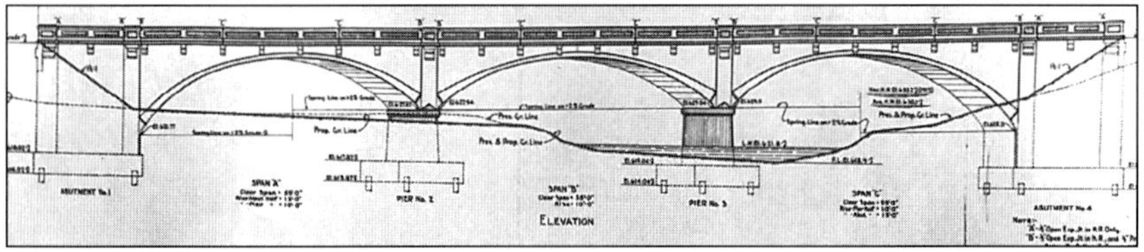

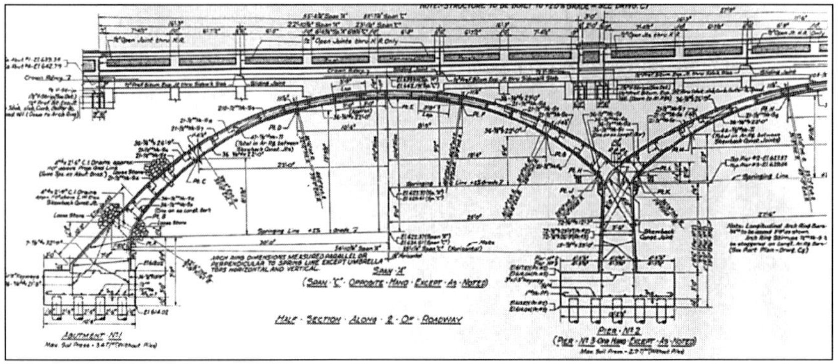

(Whitley #1584; INDOT)

S.R. #5 over Eel River, South Whitley. Note (above) the two-tier reinforcing with stirrups in the ring extended well into the pier and abutment. The ring on the outer span is unsymmetrical: it "springs" or enters the substructure much lower at the abutment than at the pier. The additional arc at the abutment shifts more of the structure's lateral thrust to vertical pressure, thereby reducing the size of the abutment needed to address it. The drawing also shows the rail panels roughened to create a contrast in texture and color with the posts and copings.

(Whitley #1584)

(Marion #1087)

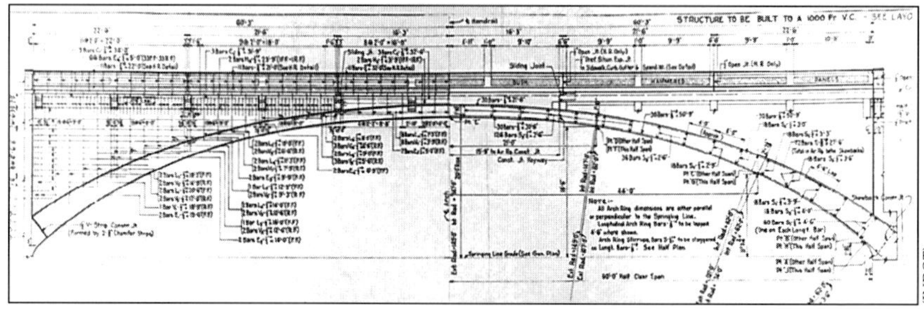

(INDOT)

(Marion #1087)

Meridian Street over the White River, Indianapolis. The Meridian Street Bridge contains one of the very few one-hundred-and-twenty-foot ISHC spans at its center to survive (above, bottom photo). Since INDOT has reportedly not retained its original plans for this bridge, the design for the Pittsburg Bridge over the Wabash River of spans of similar length has been provided as a substitute here (above, middle). The ring is nearly segmental and carries the usual two-tier longitudinal reinforcing with stirrups. In the Meridian Street structure the considerable lateral thrust which its central segmental ring develops is addressed by the three-centered spans which flank it and by the unsymmetrical ones at the structure's end.

every case where the state called for short spans. That short-span arches also required less construction skill, permitted more routinized building, and allowed a greater margin for unknown future requirements may have sometimes individually or collectively proved decisive in design decisions. In any case, the ISHC only occasionally called for arch spans of over eighty feet, and on average planned spans in the mid-forty-foot range.

Longer spans did have advantages. They could sometimes be built more cheaply and, with fewer piers in the river, were less exposed to failure from structural undercutting or debris-generated water pressure. The state did design and construct at least eight spans at one hundred and twenty feet each, its very longest filled arches. In the mid-Thirties, Frank Hardy incorporated six of these eight long spans into the ISHC's lengthiest filled-spandrel structure over the Wabash River at Pittsburg in Carroll County.

VII.2B Artistry and Applied Psychology

Titus applied to ISHC bridge decor some of the implicitly structuralist principles he had drawn from his apprenticeship with Luten.[40] Both Luten and Titus expressed their admiration for the lines of three-centered or linear arch rings. While Luten had regularly emphasized those lines by having the spandrel walls finished with a different texture than the ring, Titus made less distinction between the two. He simply cut the ring's upper line into the concrete to distinguish the ring somewhat from the spandrel wall.

Titus did take pains to provide a craftsman-like finish to the bridge surface. A few hours after pouring the concrete, the builder would remove the forms from the sides and smooth the surfaces with wooden floats. Once the concrete had hardened, workers polished the sides with carborundum brick and water but no wash. "The result is a smooth, white surface which presents a very fine appearance."[41] A slightly-raised vertical pilaster or post over the pier or abutment bounded the lines of the ring at span end.

Titus' rails contained the main decoration found on the typical ISHC bridge. He contrasted the white surfaces of his filled-spandrel rings and walls with the design and finish of his parapet rails. The rails were bordered with base and cap copings and with posts extending above the pier and abutment pilasters. Depressed panels "roughened with a bush hammer" emphasized "the rougher and darker aggregates beneath the surface." His rails became a state trademark. The same pattern appeared on the parapets of hundreds of beams and slabs as well as of filled and open-spandrel arches built all over Indiana.

Little matter to the motorist that the rails of a bridge might be the only decorated part of a bridge or that all forms of concrete spans carried exactly the same style of rails. Speeding ahead ideally in a straight line, the motorist rarely saw the side of a bridge, and so would not observe its overall form, let alone variations in the finish of its parts. If the roadway passed over a bridge without interruption or change – a standard the ISHC adopted early in the Twenties – then the motorist might speed across a span without even taking notice of its rail.

By 1930, "psychology" played "an important part in the design used on Indiana bridge approaches." The state's engineers extended approach rails beyond a bridge's entrance and exit. From the outer edges of the roadway's shoulder outside the load-bearing structure, the rails curved gently inward to the span's regular trough. Flared rails, the state engineers believed, funneled the motorist unconsciously into the span and then sent him/her back to land-bound pavement. Psychology seemed to work: "Such structures

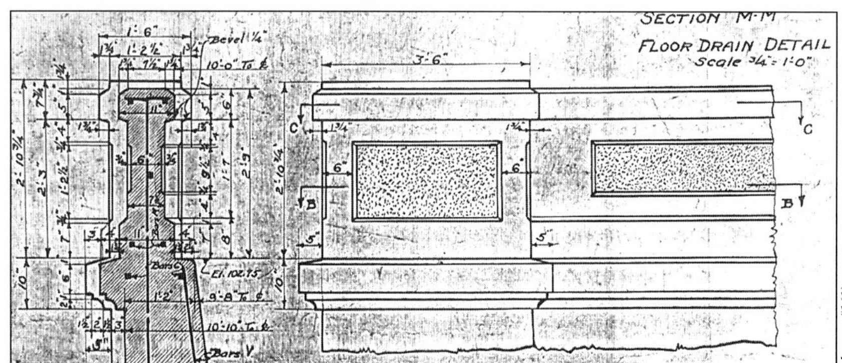

(Morgan #161)

Rail design with bush-hammered panels on old S.R. #37 over Little Indian Creek.

("Blue Book of Personal Greetings," INDOT, Vincennes)

"Another wreck—nobody hurt."

have figured in virtually no collisions, whereas the old straight handrail bridges have a serious accident record."[42]

In time the national standard as articulated by the *ENR* called for bridges wide enough to incorporate the roadway's shoulder, rather than those that disguised its partial absence. "Every bridge narrower than the approach highways," the editors stated in 1935, reduced capacity "very greatly" and thereby created "an economic problem of major proportions." The bridge designer needed, as "an essential principle," to provide "such a profile, width, surface, and approach transition that the bridge will not reduce either capacity or safety during its life." The designer must consider "the chance of wheels running on the shoulder, the tendency of some drivers to veer away from truss, girder or curb, the obstruction of visibility, the deflecting effect of entrance roughness or expansion-joint arrangement, and finally the adaptation of the structure to widening or other road revision."[43] The ISHC gave up rail psychology for the national standard well before 1939: "Modern highway bridges are of sufficient width to accommodate all traffic on the road and do not require reduction of speed of the vehicles traveling over them."[44]

VII.2C Open Spandrels

The siren of monumentality called to the ISHC's bridge designers when a particular location made an open-spandrel arch sensible. The open-spandrel arch carries the forces from its roadway to the ring through columns or secondary arches rather than through fill and spandrel walls. The absence of a curtain wall highlights the shape of the ring and columns against the sky. Since the ring does not need to contain a load of fill, it can be reduced from a barrel to two or more ribs. The open-spandrel only multiplies the delight of engineers and others who also find filled-spandrel arches appealing.

Except in special settings, the more appealing open-spandrel, however, costs more than the filled one to build, for more scaffolding and form work are required to erect columns or open secondary arches than to build spandrel walls upon a ring. Not surprisingly, the ISHC built fewer than a dozen open-spandrel arches as compared with approximately one hundred filled-spandrel ones in the Twenties and Thirties.

The open-spandrel arch, however, was sometimes the sensible design choice, especially where a considerable vertical distance separated the waterway from the road. Here the weight and width of fill would so increase the cost of the required ring and spandrel walls that a fill-less or open-spandrel structure became an economic as well as an aesthetic choice. With substantially less structural weight to carry, open-spandrel arch spans could also be very long. Their central spans could cross large rivers and minimize the number of high and therefore sizeable piers. In the case of the ISHC, open central spans ranged up to one hundred and eighty feet, sixty feet beyond the state's longest filled spandrels.

The ISHC's and other BPR-system engineers brought certain conventions to the shape and the reinforcing of their open-spandrel arches. They traditionally started with a parabola rather than an ellipse as the model for the curve of their rib-rings in order to more nearly equalize the forces placed upon each section through the columns, forces which, under fill, would generally have been diffused.[45] For his ring reinforcing, the designer of the filled-spandrel

THE DEER CREEK BRIDGE ON OLD U.S. 40

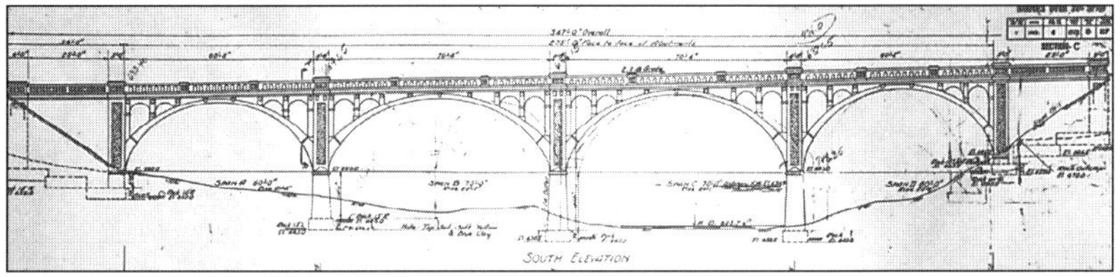

(INDOT)

(Putnam #237)

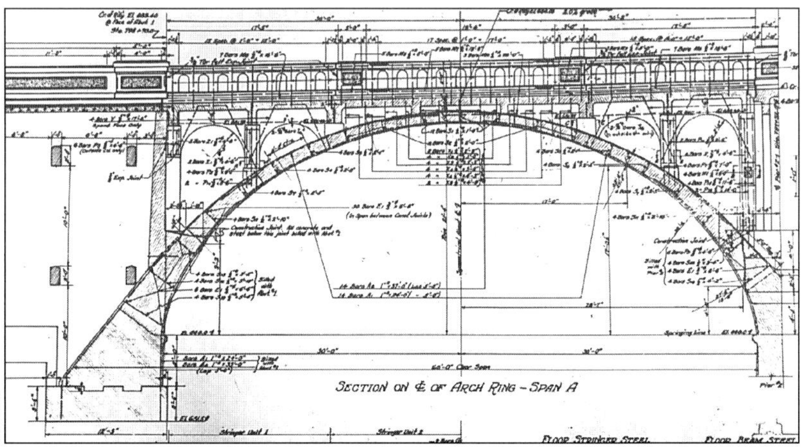

(INDOT)

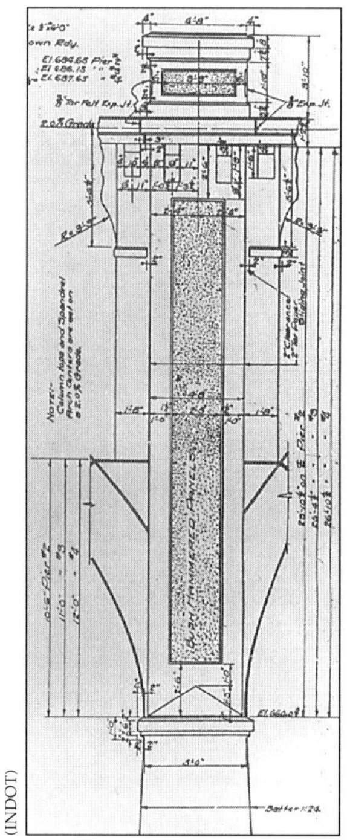

(INDOT)

(Putnam #237)

Old and classic, this ISHC open-spandrel arch has been saved in good condition as a consequence of bypass at the outset of World War II.

arch paid most attention to the weight of the fill and of the bridge structure. Without the shield of fill, however, the open-spandrel designer needed to take quite seriously the impact of moving loads upon his bridge. Here the ISHC's use of a row of reinforcing at the top and bottom of each rib with stirrups tying the rows together became quite appropriate. Luten also increasingly followed the same pattern of reinforcing with open-spandrel ribs.

Although firmly committed to a utilitarianism that allowed little more than a bow to structuralist decor, the ISHC's bridge designers could not resist all temptation to accentuate the soaring lines of open spandrels. The ISHC extended to the open-spandrels' pier pilasters the bush-hammered finish reserved on most of its bridges solely for parapet rail panels, thereby using color and texture to accentuate the structural framing which the tall, vertical piers provided for the graceful parabolic rings they bordered and rooted. Instead of depressed, solid panels in the parapet rails, state designers also used open-arched balusters with a hint of City-Beautiful style to them. By reporting the construction of almost every one of its open-spandrel structures to a national audience through the *ENR*, the state-employed engineers bragged a little about their modestly celebratory spans.[46]

On at least one occasion, the ISHC's engineers designed an avowedly memorial structure. It was, of course, an open-spandrel bridge. About 1923 the Illinois state highway authorities opened discussions with the ISHC about replacing the old metal spans across the Wabash River at Vincennes, and the two state commissions settled on a new high-level steel-truss structure located about a block away from the old one. Just as financing was being concluded in 1927 on the new bridge, the Indiana legislature appropriated funds for the erection of a memorial to the explorer, George Rogers Clark, on the riverside just south of the old bridge in Vincennes. Later, the federal Congress appropriated funds for its share of the memorial. Since the Clark Memorial included the Indiana end of the proposed bridge, sponsors urged that the new spans harmonize with the memorial.

(INDOT)

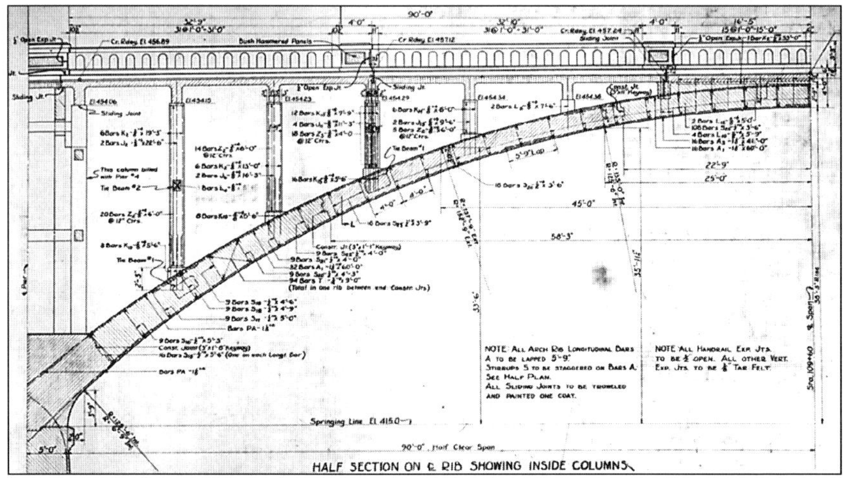

(Knox #690, INDOT)

(Knox #690)

George Rogers Clark or Lincoln Memorial Bridge under construction (above, top) with the old metal bridge in place to the north. The "Half Section" of the end-span (above) shows the parabolic ring and its two-tier longitudinal reinforcing with stirrups extended well into the abutment. Also note the open-arch rail panels.

A differently-designed "memorial structure" would, the ISHC understood, cost more than the proposed steel trusses – in fact, fifteen percent more. But, having decided to build a quite special bridge, the Indiana authorities set William J. Titus and Fred Kellam to work on an unusual open-spandrel arch (Knox #690).[47] As designed, the entire grade line of the George Rogers Clark Memorial Bridge followed "a parabolic vertical curve" of twelve hundred feet. In addition to its aesthetic statement, the dominant curve allowed the span-rise ratio to remain the same for spans of different length, thereby avoiding unequal thrust on different sides of piers and providing adequate navigational clearance under the one-hundred-and-eighty-foot central span. The line of each span's rib-rings visually resonated with the overall parabolic grade-line curve.

Titus and Kellam decorated the Memorial Bridge as they had other ISHC open-spandrel structures. The memorial, though, would not be confined to the spans which Titus and Kellam designed, for the bridge approached the Wabash through an eighteen-acre park on the waterfront in Vincennes. Architects from Bennett, Parsons & Frost of Chicago appended an entrance to the engineers' Memorial Bridge which, in sum, provided more decoration than would be associated with any other ISHC bridge built before or after. The Indiana approach contained a landscaped plaza, statuary, and stone-facing reminiscent of City-Beautiful decor.

While expressly proud of their spans, Titus and Kellam remained remarkably quiet about what they thought of the architects' contributions to their work. The state engineers' silence about the beautification may have foreshadowed the apparent ease with which the ISHC changed the name of their bridge from the George Rogers Clark to the Lincoln Memorial by 1938.[48] Renaming the bridge distinguished the engineers' structure from what the architects had erected on the Vincennes waterfront. But, by not walling off the architects' large, carved, entrance pylons which evoked the George Rogers Clark era from the engineers' spans, the bridge's memorial message became more mixed than transformed.

Pedestrian approach to the George Rogers Clark or Lincoln Memorial Bridge from the north park (above) and the entrance pylon (right).

(Knox #690)

VII.3 SLABS AND T-BEAMS: THE COMMON CARRIERS

The government-design system worked especially well for the smaller and simpler concrete beam and slab spans. Indeed, the ISHC very quickly adopted "standard" plans for beams and slabs. They did not have to design these standards from scratch, and they revised them only incrementally and occasionally. Being the last in the nation to establish a BPR-approved state highway department provided the Indiana engineers with a larger pool of federally-certified plans from which to borrow as well as with more seasoned BPR network advice from which to profit. The ISHC belonged to AASHO, its chief engineer sat on the Highway Research Board, and BPR publications, including *Public Roads*, circulated widely through the commission's offices.

For the latest in conventional wisdom, the ISHC's engineers could also depend upon reports issued by the ever-more-important and influential national professional associations of engineers and trade groups like the American Society of Civil Engineers, American Concrete Institute, and Portland Cement Association. George Hool and Nathan Johnson reprinted many of these reports, for example, in the *Concrete Engineers' Handbook* (1918), a periodically revised text that BPR officials cited and to which the ISHC's design engineers apparently turned often for authoritative opinion.[49]

Unless some very special feature dominated a site for one of these smaller-span structures, the ISHC designer sorted among his sets of standard plans for the best match and then copied the essential if not the entire design for the bridge. Here off-the-shelf design reached its epitome with bridges of a given form being cloned and then dressed in the ISHC's ubiquitous rail design, submerging even family resemblance in the state uniform.

T-beams and slabs provided the only concrete forms which the ISHC used for short bridge spans, and both did their load carrying through or partly under the deck. T-beams and slabs were first cousins in another way, too. Engineers recognized their structural kinship by routinely analyzing slabs as rectangular beams.[50] The ISHC, however, did not regard the cousins equally. T-beams accounted for more than a third of the spans the state constructed in the Twenties and Thirties. The ISHC ordered about half as many slabs in the Twenties and less than a tenth as many in the Thirties.

VII.3A T-Beams

The ISHC's sets of standard T-beam spans changed in several ways over time. First, the spans became longer. Hool and Johnson recommended T-beam spans up "to about" fifty feet "from the standpoint of economy" in 1918.[51] The ISHC's bridge department started out cautiously in 1922 with four sets of standard designs for between twenty and thirty feet. Within a year they had added a pair of standard plans up to thirty-eight feet, and by 1927 they had a forty-foot model. The next addition to span length did not come until the end of the prewar period. In 1938 the department added a forty-five foot standard and capped the process with one for fifty feet only in 1941.

Three other general, partly interrelated, changes occurred in the standards. The ISHC began to experiment with larger and more reinforcing rods in its beams in 1936. By 1940 the typical number of

(INDOT, Vincennes)

Three new forty-foot spans of ISHC T-beams completed in Washington County in 1934.

T-BEAMS

The lower reaches of the beam make up its stem, and the upper slab provides the flanges or crossing of the T. A row or more of steel rods is set in the concrete towards the bottom of the stem to address tension, which is greatest near the center of the span. Usually, several of the longitudinal bars are bent upwards from thirty to forty-five degrees as they approach each end of the beam where tensile stress decreases and shearing stress increases. In general, the longer the span, the greater the stresses, with increased tension calling for the use of heavier and larger numbers of reinforcing bars and added compression requiring deeper stems. Yet the use of stirrups, or U-shaped hangers, to hold all the reinforcing in place, ties stem and flange together, effectively making the deck an upper part of each beam. The deck's double duty allows for the stems to be smaller, which, in turn, provides less structural weight in need of support across the span.[52]

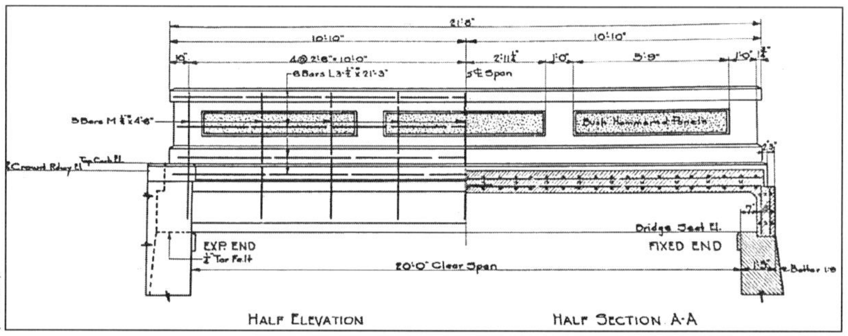

(INDOT)

The "Half Elevation and Half Section" (above) for an ISHC standard T-beam (1922) shows the longitudinal dimensions, the end at which the beams are "fixed" or attached to the substructure, and the appearance of the side. The "Section" (below) describes the placement of the longitudinal reinforcing and stirrups largely in the stem. The dots along the upper reaches denote transverse reinforcing in the flanges.

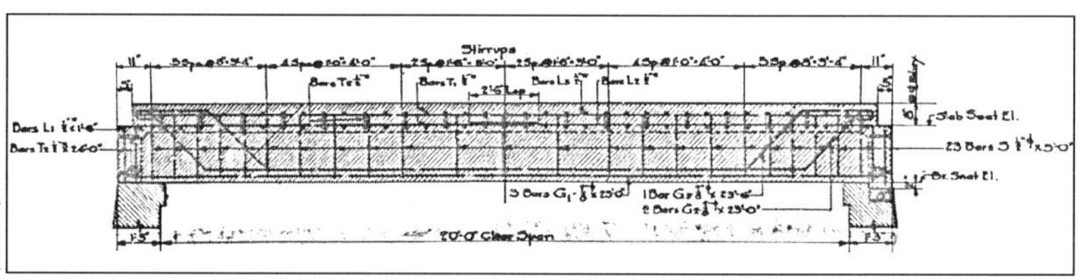

(INDOT)

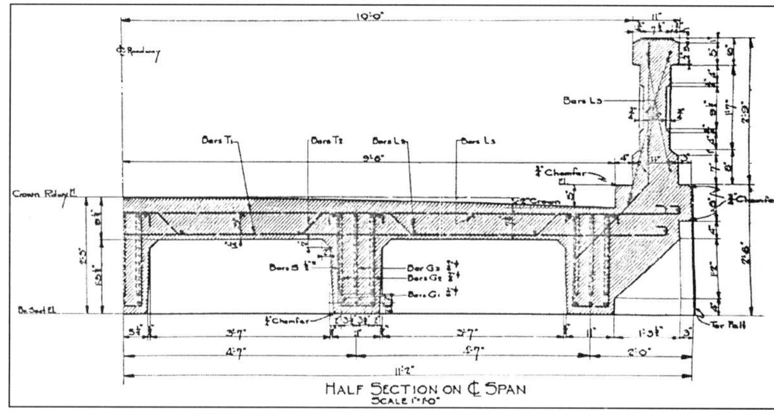

(INDOT)

The "Half Cross-Section" of a standard ISHC T-beam (above) highlights the transverse reinforcing in the flanges and rails. Here the dots low in the stems denote longitudinal rods. The 1929 photograph (below) of a state span under construction shows the hooked upper ends of the stirrups draped over longitudinal rods at the prospective tops of beam stems as well as some of the transverse bars running across the stem tops and then bent down towards the bottom of the flanges.

(INDOT, Crawfordsville)

longitudinal reinforcing rods had risen from ten to twelve and the diameter of each had moved from less than to more than an inch. Next, stem widths gradually increased over time. Foot-wide stems for a standard twenty-four-foot span in the early Twenties spread out half again as much by the Second World War. Finally, the ISHC also extended its flange widths. During the Twenties and early Thirties the tops of the T's reached between twenty and twenty-four inches from stem to stem. Designers began to experiment in 1936 with thirty-inch flanges which by 1940 had grown to thirty-six inches in some cases.

Here as elsewhere, the ISHC's changes were incremental and within the BPR system's paradigms. The ISHC's standard T-beams of the Twenties and Thirties followed most of the design rules that Hool and Johnson laid down in 1918.[53] Three of their stipulations illustrate ways of staying securely on the safe side of design by keeping the beams compact: First, T-beam stems should not be more than twice or three times as deep as they are wide in order to keep the stem-flange connection strong. Only one of the ISHC's standard plans had a stem even twice as deep as it was wide. Second, flanges should be no more than three times as wide as the stem. The ISHC never went beyond half the recommended maximum ratio. Third, flange thickness should be at least one-third of the stem's depth. The ISHC followed this rule for its standard plans for spans of twenty to thirty-four feet, but transgressed the line in the Thirties for spans of forty feet and longer.

T-beam spans appear on average to have cost the ISHC less per span-foot than competing forms. They were cheaper than concrete arches or slabs as well as steel beams, although they also proved somewhat less durable. Because the T-beam's range of span length was more limited than arches, trusses, and, by the late Thirties, steel beams, long structures required more spans of T-beams and consequently more supporting piers than would arches or trusses, eroding and often eliminating the Ts economic edge. Beyond use on relatively short crossings of minor streams, the ISHC's engineers prescribed T-beam spans of between thirty and forty feet for use only where heavy currents did not threaten simple piers and where they could be easily constructed because the site was often dry or in shallow water.

(Lake #149)

(INDOT)

ISHC T-beams over the Pennsylvania Railroad. With the bents constructed, Rieth-Riley Construction Company workers moved on to prepare the framing for the beams and deck (top) and then to complete the whole for use in 1934 (center). Even today, the tell-tale signs of smoke from steam locomotives remain on the outer beam's fascia (bottom).

BOTTOM-LAND TRESTLES

A number of the Indiana's rivers traverse wide, shallow valleys subject to rises which periodically blanket the bottom lands with overflow waters for several days to weeks. Before the ISHC and motor vehicles, the counties and their residents had gotten along by accepting road closure which bottom-land flooding occasionally and temporarily induced. The county bridged the river's channel with trusses erected on substructures built high enough to keep their investment safely above any sea of flood water which spilled over the valley. With animal-drawn wagons, local residents could ford water-covered bottom lands much sooner and more extensively than could those driving motorized vehicles, a difference which essentially shortened road closures for most local residents.

Because the ISHC serviced motor vehicles with minimal capacity for fording, bottom-land inundations presented the state with "an emergency condition."[54] Flooding confronted drivers, accustomed to traversing wide dry stretches quickly, with unanticipated stoppages in unfamiliar country. To ameliorate the automotive emergency, the ISHC in its early years designed and constructed on occasion a two-track, single lane timber trestle across the bottom lands to the old county bridge to serve motorists until the state could build a proper structure of its own.

In some cases, the ISHC might in time erect a filled causeway over shallow valley areas. But across major river bottom lands, the state needed to allow for waterway clearance in order to keep its roadway from damming flood water. For such sites, T-beam viaducts of "three-span units" became the ultimate design of choice.

The spans rested on piers of capped two-post bents seated on a single concrete footing above timber foundation piles. The state designed a depression in the central piers of the unit to hold the superstructure's "keyways" or feet in place and allowed the unit to expand at its ends. The contractor built a reusable set of forms for three spans with a removable side to allow early finishing of the exterior concrete. The builder erected the unit forms, placed the beam reinforcing as though each span were simple or disconnected from the others, and then poured the concrete into the forms as though all three spans were a single unit. Once the concrete in the unit had cured adequately, the contractor moved the forms to the next location for reuse.[55]

(Continued on next page)

(INDOT, Vincennes)

An ISHC trestle approaching the White River near Washington.

The Big Four highway-overpasses incorporated some constants.[8] They were all designed to carry unusually heavy traffic, specifically an interurban engine. Most are five spans with a center span of thirty-one feet over the two railroad tracks, flanked by two shorter spans on each side. Every span uses two outer-edge and one deeper and wider central T-beam with three-and-a-half-foot flanges.

Sometime before the First World War the Pennsylvania Railroad also built a series of what appear to be continuous highway overpasses, a few of which remain in Wayne County.[9] The Railroad's engineers relied on what seem to be continuous girders, which do their load-bearing above the roadway, rather than on T-beams, which carry below it. As with the Big Four, most of the Pennsylvania's grade separations were five-span structures, with a center span of thirty-seven feet over two tracks flanked by a brace of shorter spans on each end. The girders are supported on two-column bents apparently integrated with them except at the ends of the structure.

CONTINUOUS BEAMS OVER THE BIG FOUR

The T-beam's longitudinal reinforcing would end over the pier for the simple or jointed span. Some of the stems' reinforcing follows the lower edge of the beam across center span and is displaced to near the upper edge of the deck towards the substructure. To make the structure continuous, the railroad's designers extended the upper edge rods of one span across the top of the substructure and connected them with those of the adjacent span, making a single system of longitudinal reinforcing. Furthermore, the designers added a set of rods which run diagonally from the top tier of longitudinal reinforcing to about three feet into the two-column bents, tying the beams to all the bents except for those at the ends of the structure. Only the end bents or abutments allowed for expansion.

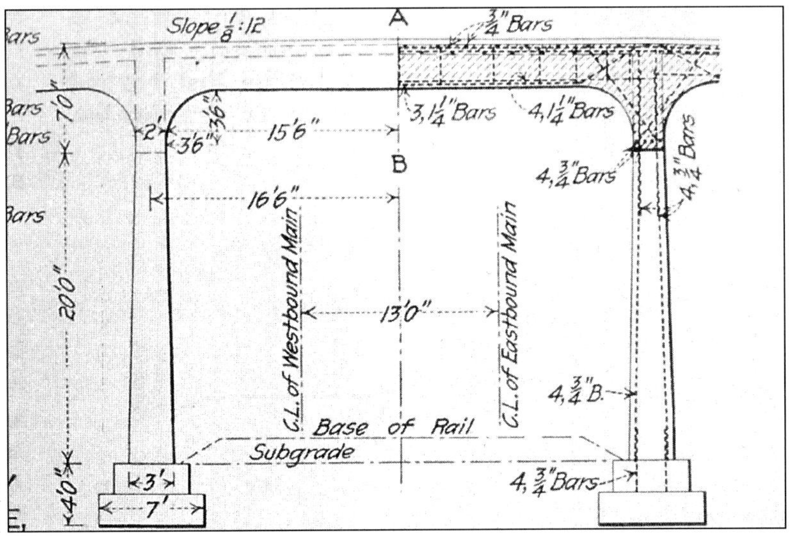

(ENR, 1907)

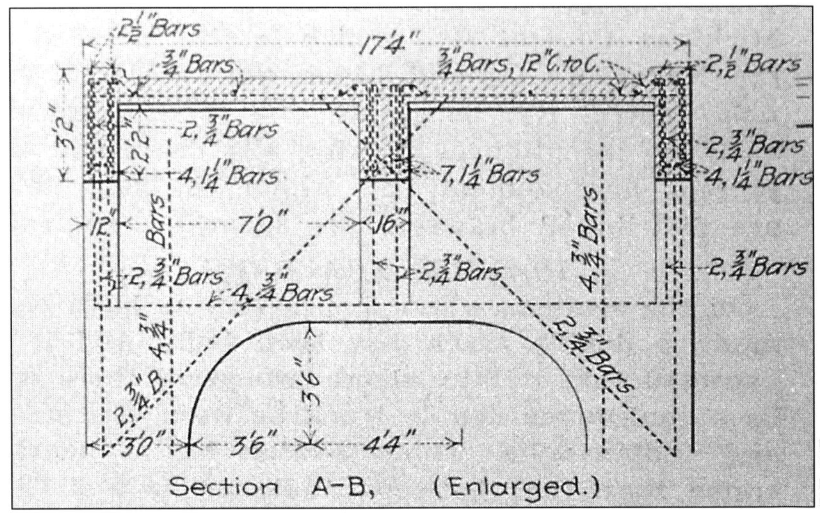

(ENR, 1907)

If Big Four plans for a grade separation at Shelbyville serve as a model of its continuous-beam structures, then the railroad used two-tier longitudinal reinforcing with stirrups in its beams (left) with each tier displaced over the bent. "Section A-B" (right) shows how some of the transverse reinforcing was also embedded in the bents.

The Pennsylvania Railroad's continuous-girder separations in Wayne County. All, like the Johnson Street overpass in Dublin (upper), carry nicely-indented panels. All transfer deck load to the girders and bents through floor beams as illustrated in Dublin's Davis Street separation (lower).

(Marion #2511)

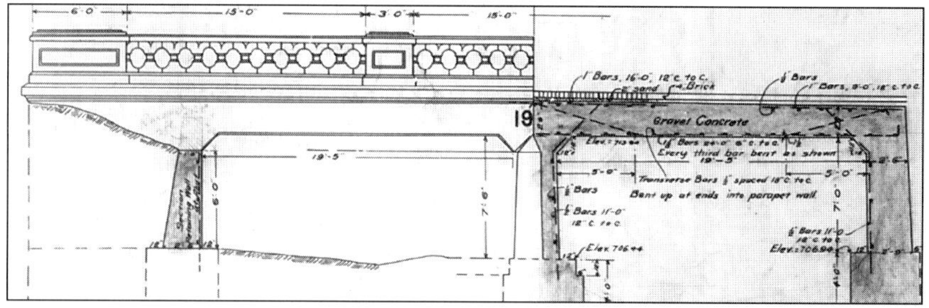

The removal of the open-circle panels of the rails and the burial of the rail posts in concrete block, along with some spalling of the slab which incidentally uncovered stirrups (upper), have deprived Klausmann's Dorman Street Bridge in Indianapolis of some of its charm over the years. According to the "Half, Longitudinal Section" (below), most of the longitudinal reinforcing is concentrated towards the bottom of the slab until transposed upwards over the pier to join rods from the adjacent span. Some rods do run longitudinally near the top of the slab, starting a short distance from the substructure and then angling down deeply into it.

The advantages of continuous structures for urban circumstances also enticed some city engineers into this form of largely empirical design. Streets lined with pre-existing buildings ruled out the option of raising the level of the roadway to achieve added waterway clearance, putting a premium in a number of such cases on spans with superstructures as slim as possible. Since continuity provided more carrying for less girth, Indianapolis city engineers occasionally turned to small continuous structures before the First World War. Indeed, some were framed, or at least partly framed, as well as continuous.

Indianapolis city engineers favored slabs and experimented with wide, boxed structures consisting of rectangular tubes through which water flowed. In 1910, for example, H. W. Klausmann designed a continuous, reinforced-concrete slab structure with two nineteen-and-a-half foot spans in which the slab reinforcing ex-

(Marion #2510)

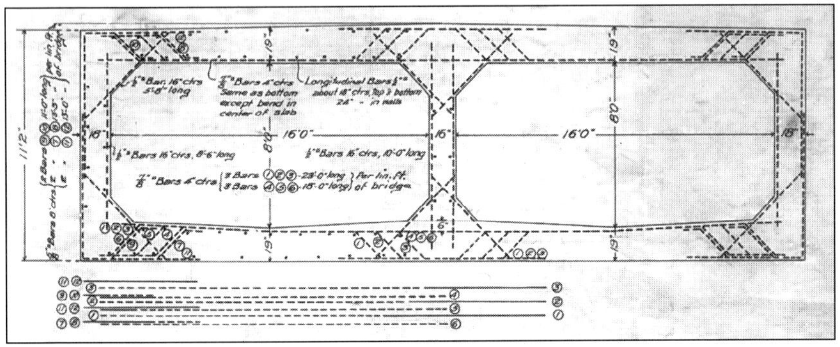

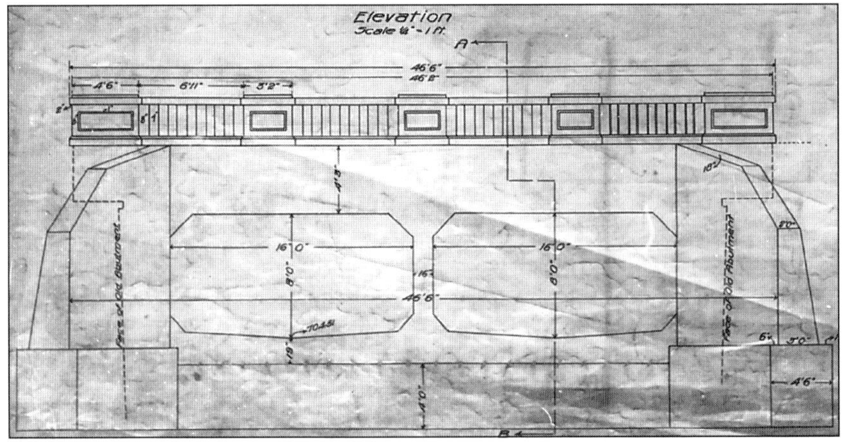

Jeup's Michigan Street Bridge has suffered in much the same manner as the Dorman Street spans. The open panels of its rails have been filled in, and the concrete has spalled (above, left). The designer treated his box tops like slabs (below, left), concentrating his longitudinal reinforcing towards the lower edge and angling it towards the upper edge only towards the sides of the box. The longitudinal rods of the box tops join over the shared inner side. Another tier of rods runs diagonally from near the upper edge of the box top towards the most distant edge of the closest side, where the rods are bent to parallel that side until turned another forty-five degrees to run diagonally close to the lower edge of the bottom.

tended into the pier and abutments (Marion #2511). Klausmann's successor, B. J. T. Jeup, drew the plans for a two-span boxed bridge in 1916, which contained a pair of concrete rectangles, the reinforcing of which continued around all four sides. Built with sixteen-and-a-half foot spans each, the boxes are quite wide (Marion #2510).

VIII.1B The Concrete Frame

When Arthur G. Hayden, design engineer for New York's Bronx Parkway Commission, presented the rigid frame to the national engineering profession in 1923, he made history. At least Carl Condit, whose preeminence among historians of American structure has been widely recognized, has thought so. Condit ranked Hayden's rigid frame as the second "most important innovation in concrete bridge design" in the twentieth century and the most cost-saving method offered before the Second World War.[10]

The Bronx Parkway Commissioners set out in 1913 to create a system of limited-access parkways which became a prototype for later expressways.[11] Parkway spans needed to be both efficient and beautiful. To address structural and economic efficiency, Hayden designed seventy-four, rigid-frame, grade separations between 1922 and 1930. To achieve beauty apparently required the services of architects.

Hayden did not think either "the architect or engineer alone" could best meet all the "structural requirements" and "rules of beauty" which the Park Commission sought. "Scientific minds and artistic minds," he observed, "are rarely combined:"

> The development of engineering mechanics and mathematics has been so rapid in the past twenty years that even the practicing engineer is hard pressed to keep pace with the engineer mathematician. The architect on the other hand has centuries

> of artistic development to study and digest beside being faced with the problem of adjustment to new structural developments. Either task is formidable in itself. Specialization is a compelling force, and co-ordination is increasingly urgent.

The rigid-frame form gave Hayden a considerable range of structural flexibility that the architect might exploit by changing lines, erecting entrance buttresses, or sheathing with stone or brick.

Hayden made the structure's slab or beams so thoroughly continuous with the abutments that the terminology of superstructure and substructure seemed anachronistic. The thoroughness of Hayden's elastic treatment of slabs and beams came much closer to Luten's merging of an arch's ring with its piers or abutments than to the Big Four railroad engineers' extension of T-beam reinforcement a few feet into piers. A rigid frame's abutments were cast right along with its slab or beams. Hayden arched the beams or slab slightly and increased their depth some to strengthen the working materials close to the abutments, tops, or knees, where stress would be most pronounced and complex. He designed the abutments as vertical extensions "which share in absorbing the work" of the system. The abutment walls looked like wedges tapered downward toward the footing where stress was less pronounced.

For the slabbed, rigid frame, reinforcing rods extended longitudinally close to the top and to the bottom of the slab, and then both tiers angled sharply downward to run along the outer and inner sides, respectively, of the abutments. Since the bottom edge of the slab was arced to increase the deck's depth towards the knees, Hayden tied the two tiers of reinforcing together with stirrups near midspan to keep the lower rods under tension from straightening and whipping out of the concrete. Where T-beams carried the deck, most of the longitudinal rods ran near the lower edge of the stem over midspan and were displaced at different points to the top towards the knees, as would typically be the case with a simple or a continuous beam design. The rods then curved downward to extend along the outer edge of the abutment.

Hayden was able to put his basic designs through some "tedious" and "complicated mathematical analysis" and yet keep the cost per unit relatively low because he designed a great many spans that were structurally alike. This practice provided him with a number

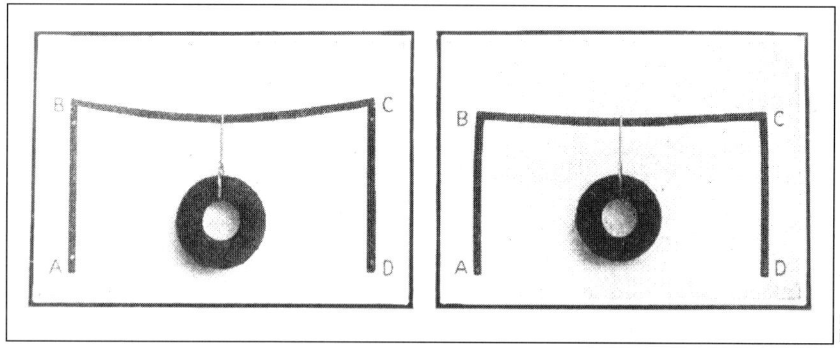

(ENR, 1926)

Simple beam (left) and rigid frame (right) of equal dimensions and under the same load. Note how much more the simple beam, which is separate from its columns, deflects than does the frame, in which the beam and the columns are integral.

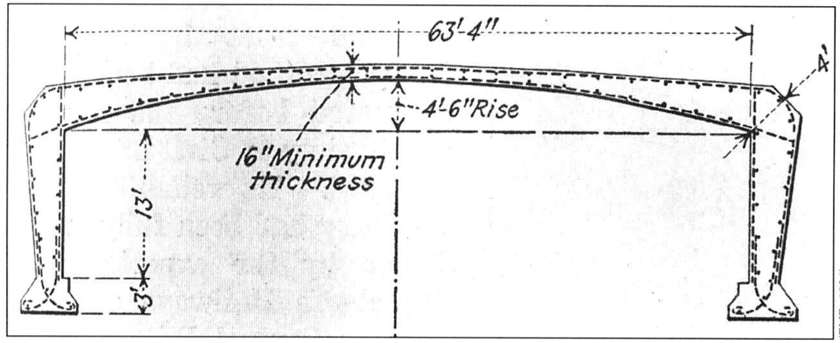

(ENR, 1926)

Hayden's solid-slab design of rigid frame.

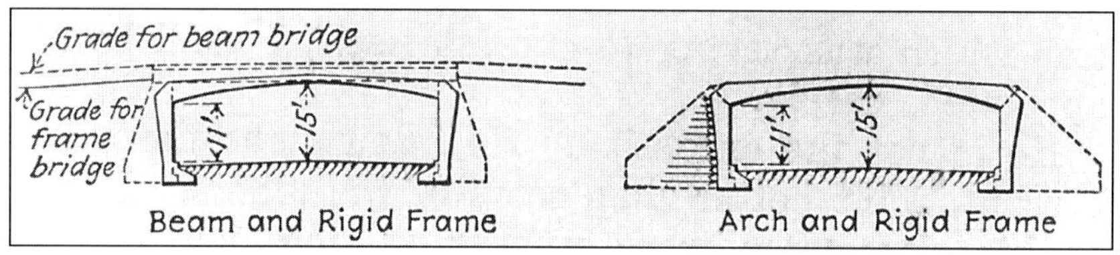

(ENR, 1926)

Beam, arch, and frame. Beams need more floor depth than frames, thereby requiring a higher roadway grade and often greater approach embankments (left). The arch requires either greater rise or heavier abutments than the frame (right).

of rationally-analyzed standard designs. Where appropriate and trustworthy equations did not exist, Hayden turned to other methods. Unsure, for example, of the impact of stresses within the knees of their frames – where the reinforcing and the concrete curved sharply from the slab or beams into the abutments – the Park Commission hired Columbia professors to test several sample models in the university's engineering laboratory. In 1926, furthermore, Hayden tried another way to cut free of the albatross of mathematical analysis. He endorsed and the commission purchased a set of the "deformeter gages" that Professor George E. Beggs of Princeton University had developed to measure the deflections created by load changes in models made of elastic material.

National recognition of Hayden's achievements came quickly. In 1923, Professor Beggs had himself publicly praised Hayden for introducing substantial efficiencies with his "elastic continuous structure." "The work that a structure must do in supporting its load is not only divided among all members of a rigidly connected structure, but the total work to be done is less by reason of the rigidity." The editors of the *ENR* were equally impressed. Hayden's design, they argued, had indeed brought "continuous structures to the fore."[12]

VIII.2 A LITTLE CONTINUITY WITH THE MOSTLY DISCONTINUOUS IN INDIANA STATE DESIGN

VIII.2A Watching and Waiting in the Twenties

When the Indiana State Highway Commission [ISHC] started building bridges in 1920, the Bureau of Public Roads [BPR] expected most statically indeterminate design plans to meet the same analytical standards as proposals for statically determinate ones. The bureau's strong emphasis on rational analysis required the completion of those same complicated, tedious, and time-consuming calculations that had driven most private designers of statically indeterminate structures to rely on practical experience and empirical rules-of-thumb. As Albin L. Gemeny, the BPR's long-time, senior structural engineer, would say somewhat later, designers needed to stay alert to new developments in their field, "while retaining in general the conservatism necessary for safety of such important structures as bridges."[13]

The BPR's insistence on scientifically-correct analysis appears to have slowed down the construction within the highway alliance in the Twenties of most statically indeterminate forms, except for the arch, whose pedigree reached back into antiquity. The experience with continuous design could be measured at least in decades and that with rigid frames in a handful of years. A continuous structure predictably passed through the highway alliance system first. The *ENR*, for example, published an article that the Engineer of Bridges for the Wyoming Highway Department had prepared concerning a continuous, inverted T-beam structure erected in his state in 1923. No rigid frame was built on an alliance highway, however, until 1927.[14]

BPR policy reinforced what the ISHC's chief bridge engineer already thought. George Hool had advised William J. Titus to be wary of continuous design when Titus was still a student at the University of Wisconsin:

> The designs of most of the continuous girder bridges built monolithic with supports are conservative and safe, but still there are some bridges of this type, especially those with exceedingly high piers or bents, which are classed by many engineers as dangerously extreme considering our present knowledge of reinforced concrete. The fact that light sections in concrete construction have quite often developed cracks ... has made these engineers hold to the conservative idea that

> the difficulty in providing intelligently for movement due to temperature, change in moisture content of the concrete, etc., is so great as to make light construction something to be avoided wherever possible.[15]

Titus mixed prudence with deference to several authorities.

Some elements in the ISHC's T-beam viaducts might be mistakenly identified as continuous. The state's engineers did design "three-span units" as a strategy to reduce construction costs, not to enhance structural efficiency. The contractor built a frame for three spans and poured the concrete as a unit. The beam reinforcing, though, was not linked between spans nor mingled with that of the substructure. Neither the amount of concrete nor the amount or placement of steel reinforcing used in three-span units differed from what the ISHC would have designed into three simple spans of the same length.[16] There was far less continuity here, in short, than the railroad engineers had used much earlier in their Hoosier highway overpasses.

Continuity's allure, though, did not diminish with the years. Its benefits were so hard to deny especially with concrete slabs, and the dangers from depending upon empirical knowledge with such a relatively simple structural form so minimal, that the ISHC's engineers could not finally resist at least thinking about a slight risk. In 1929, they produced a two-span continuous slab standard plan which moved beyond their earlier three-span T-beam unit by integrating the longitudinal reinforcing across the spans and by reducing slab depth about twenty percent over simple spans of the same length. Continuity was not extended to framing, however, for the slab was set into the center bent through a keyway, not through the integration of reinforcing. Design was one step; construction another. The ISHC never built a span with its 1929 continuous slab design.

Discouraging continuity remained professionally respectable in the Twenties, especially within the highway alliance. The net balance for continuous as against discontinuous beams remained a matter of discussion in the national engineering press. Continuity generally got saluted for stress management and sometimes for economic advantage.[17] Debate aside, though, "indeterminate structures [apart from arches] were rarely used in this country" before their stresses could be easily and quickly calculated. That requirement would soon be satisfied.[18]

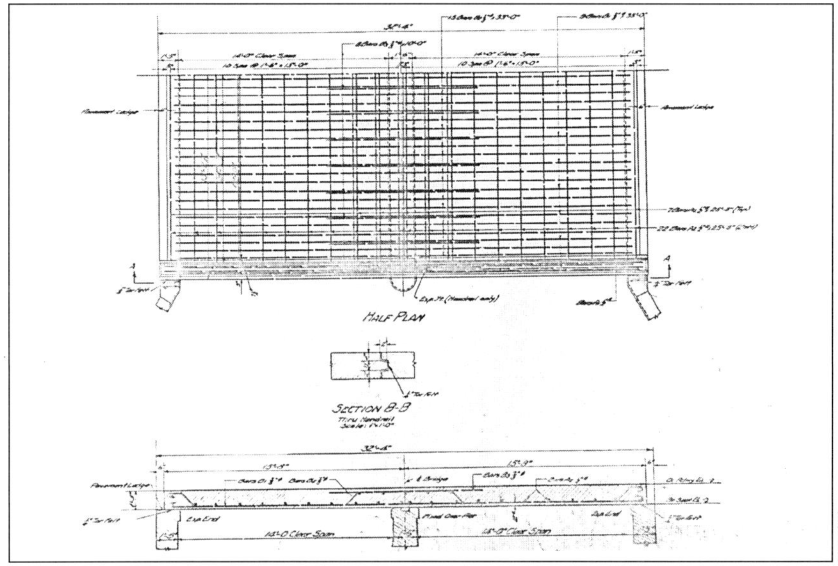

(INDOT)

ISHC continuous-slab standard (1929). "Section B-B" shows the slab as fixed through a keyway over the piers. If this were two simple slab spans, the longitudinal reinforcing over the pier would look like it does at the abutments. But instead of having disconnected, hooked ends abutting above the pier, the rods are transposed from near the lower to near the upper edge and extended over the support to join the rods from the slab of the adjoining span, making the two continuous. The darker lining of the reinforcing over the pier than at the abutments in the "Half Plan" denotes some of the reinforcing that the ISHC designers added for stress in continuous spans.

VIII.2B A Nod in the Thirties to Yesteryear's Design Innovations

Distinguishing the "analytical procedures" needed as tools of research from those for design, Hardy Cross, a professor of structural engineering at the University of Illinois, developed in 1930 a relatively simple method to certify arithmetically the rationality or scientific correctness of continuous design. Cross made "the continuous beam an article of commerce" largely by hurdling the last obstacle which the BPR placed in the way of federal aid for the construction of continuous design within the highway alliance. Even then, the ISHC was not quick to jump on the long-delayed design bandwagon.

The first Hoosier trial of any of the new forms came in 1935, when the state designed and contracted for its lone rigid frame (Carroll #1788). Unusual circumstances probably precipitated this exceptional ISHC structure. The state had just begun to erect bridges inside city limits, and this was a rare case where the ISHC designed and built a structure to carry a local street, not a state highway, as part of a grade separation. Substantial homes already lined North Street in Delphi, and the dip through which a state highway passed had been bridged for over two decades so the state had little option but to maintain the street as a thoroughfare at its existing grade-level.

Every foot of additional clearance the state wished for its highway had to come from below North Street. That meant blasting rock out of an extended cut or reducing the depth of the bridge's superstructure. Since it had to cross a twenty-eight-foot highway flanked by city sidewalks, the bridge needed to span just a little under forty feet, almost double what the ISHC allowed for its concrete slab spans. While state arch and T-beam designs could handle that span-length, the ring of the filled-spandrel arch substantially limited clearance towards the abutments, and, even at thirty-nine or forty inches deep, T-beams offered too little clearance as well. A solid-slab rigid frame, though, could span this distance with a deck thickness at center of less than half of the T-beam, which provided the best but still unsatisfactory conventional alternative. Furthermore,

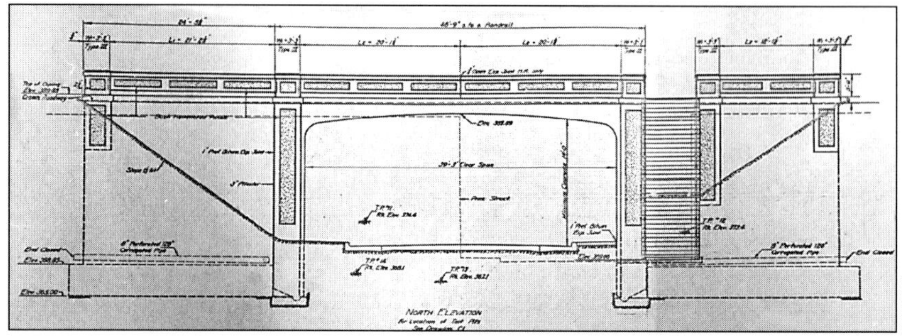

(INDOT)

ISHC's only rigid frame: North Street over U.S #421, Delphi. The horizontal slab's two-tier reinforcing with stirrups bends at the knee and extends vertically through the frame or abutment into a pivot or hinge at the structure's base.

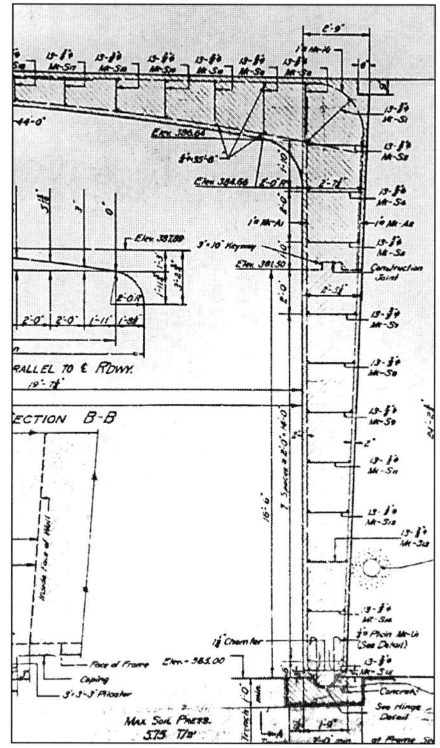

(Carroll #1788, INDOT)

both the arch and the T-beam would have required considerable excavation for their abutments that the rigid frame avoided.[19]

The ISHC's engineers designed their rigid frame with style but, unlike the Westchester County Park Commission, without the assistance of architects. The crossing was slightly skewed, which made the already more-complicated-than-usual analysis and reinforcement a bit more difficult. To allow pedestrians direct passage between the streets, the ISHC added a stairway between the two, and the bush-hammering that the state usually reserved for rail panels the ISHC used here on the pier pilasters as well.

`DELPHI, SHE'S BEAUTIFUL. LOOK HER OVER.'

"These ... beautiful structures, [which] we dedicate to public use today, will stand for years [as] monuments to the engineering skill and workmanship of the State Highway Commission. They beautify our city and make more real our slogan ... `Delphi, She's Beautiful. Look her over.'"
– William C. Smith, master of ceremonies, 1936 Delphi dedication of the North Street Viaduct and the nearby Deer Creek arch.[20]

(Carroll #1788)

North Street over U.S. #421, Delphi. The original viaduct (left) was built in 1908.

Still, neither decades of practice with the elastic treatment of beams and slabs across the nation nor simplified scientific certification made any serious inroads into ISHC construction before 1939. Then continuity gradually gained a respectable place within the state's bridge planning. The state constructed its first continuous-slab bridge a decade after the ISHC's earliest and unused standard plan for one. In 1939, the ISHC built a pair of twenty-foot spans (Dubois #3262) with a fifteen-inch deck (as compared with twenty-one inches for simple spans). Other contracts for continuous slabs followed in 1940, along with the state's first continuous structure of variable-depth beam. Six years after its first trial with rigid framing, the ISHC also made a modest return to it. In late 1941, the state partly framed a slab (Vanderburgh #3445).

VIII.2C Among the Last by Which the New is Tried

Whether the ISHC was less adventurous with its bridge design than the average state highway department remains unclear. Engineers often recognized caution as a professional characteristic, one that the BPR encouraged. Images of the consequences of a structural failure especially haunted the government-employed bridge designer.

The reluctance of the ISHC and of other agencies in the federal-state highway alliance to innovate arose in part from a tendency rooted in the system – the French engineering tradition imported into West Point early in the nineteenth century, transferred into the civilian engineering programs of colleges and universities over the course of the century, and carried by their graduates into govern-

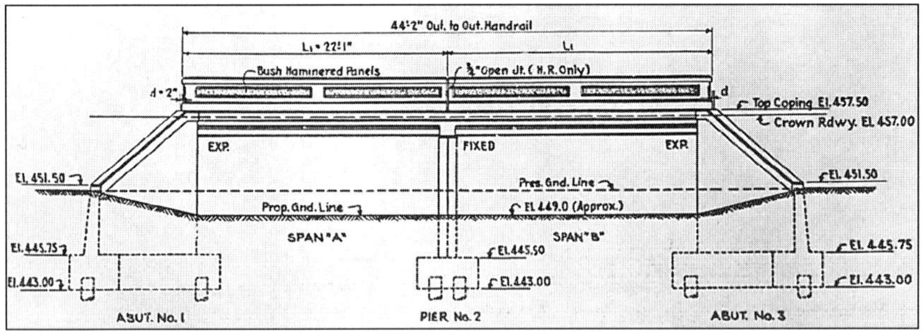

(INDOT)

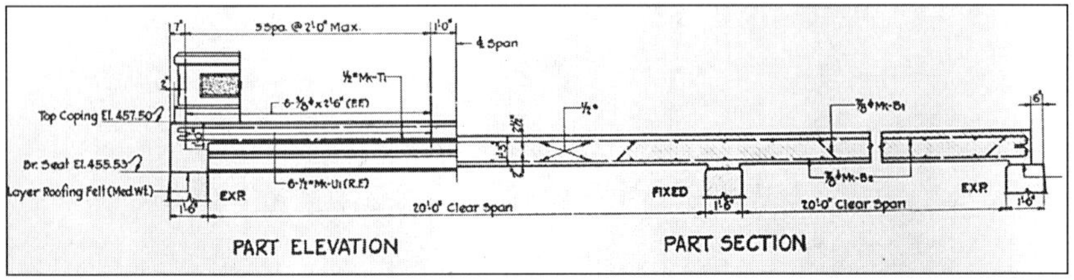

(INDOT)

(Dubois #3262)

S.R. #64 over Hunley Creek overflow. Although the pattern remains essentially the same, the ISHC did amend its 1929 continuous-slab standard for this set of plans (above), at least to incorporate two-tier longitudinal reinforcement.

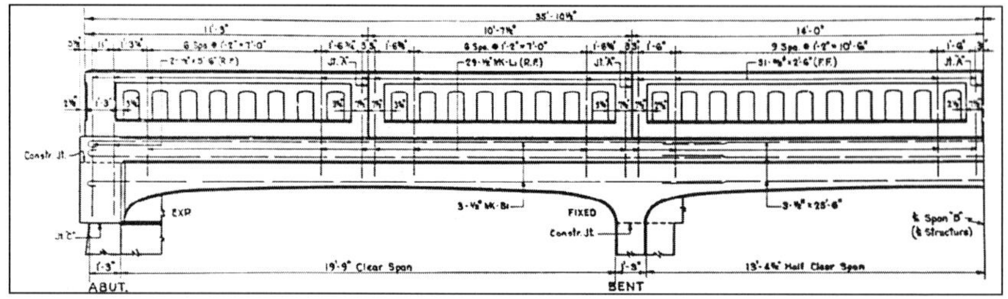

(INDOT)

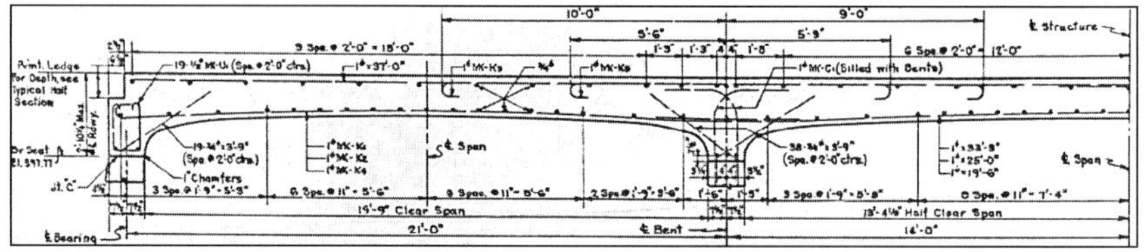

(INDOT)

(Vanderburgh #3445)

S.R. #57 over Big Blue Grass Creek. Fixed to the bents and allowed to expand over the abutments, the slab's two-tier longitudinal reinforcing received some framing assistance from vertical reinforcing in the bent arced into the upper slab as well as from other rods lodged diagonally downward from near the slab top into the bent. There was, in this case, no transposition of longitudinal rods from the lower to the upper areas of the slab over the bents. The bents are quite light, and the open-column rail panels promote an unusual substitute for the solid, bush-hammered standard, together helping to give the structure an especially graceful appearance.

ment bureaus in the twentieth century. To counter an earlier *ENR* article in lavish praise of British contributions to engineering, James K. Finch, a Professor of Civil Engineering at Columbia University, drafted "French Pioneers in Engineering" (1930).[21] The fifteenth to the eighteenth centuries, Finch argued, encompassed three hundred years of "undisputed French engineering leadership." In bridge work, the French "brought the stone arch bridge practically to perfection by the close of the eighteenth century." Indeed, the British called in a Frenchman, Labelye, to build Westminster Bridge across the Thames at London in 1739.

At the end of his article, however, Professor Finch addressed "why engineering in France suffered an eclipse with the advent of the nineteenth century" – the time in which, incidentally, Americans were patterning their academic engineering on the French. Finch attributed part of the decline to "a change in emphasis in French engineering work," a shift which occurred as the Academy of Sciences at Paris secured "a stronger and stronger grip on French engineering matters." By "1800 or 1810 science had the upper hand in the Ecole de Ponts et Chaussees [School of Bridges and Causeways] as well as in the Corps," leading a number of French engineers to make "notable and valuable contributions to the mathematical theories of structures" during the nineteenth century:

> These studies were essential to the changes which were gradually taking place in engineering technique – it was changing from an art to a science – but these men were primarily theorists who favored mathematics rather than construction and who found themselves in an environment that afforded little encouragement to construction. Theory grew apace, but actual construction was at a standstill. ... Thus engineering leadership passed from France to Great Britain and later took a long journey across the Atlantic to America. We should remember, however, that British engineering was founded on the truly remarkable achievements of the French workers of the seventeenth century and that the modern mathematical and rational methods of design were born in France, even if they were developed and extended by workers in other lands.

We should also remember that a preoccupation in the nineteenth century with theory mathematically expressed did not generate the innovation that had secured and maintained French leadership of engineering for the previous three hundred years.

Some twentieth-century American engineers within the academy have questioned whether an unqualified faith in scientific correctness best promotes the general welfare. Hardy Cross, the engineering professor who simplified stress analysis for continuous designs, expressed his concern about a professional preoccupation with calculable correctness. After acknowledging his "heartiest sympathy" in 1935 "with the traditional conservatism of engineers," Hardy concentrated on the single-minded advocates of correctness. The "designer of engineering structures upon which the integrity of property values and the safety of human life depend has no right to accept theoretical results unless he finds for them some support in experience or in common sense."[22]

The theorist, the laboratory, and the calculator, Hardy believed, needed to be used as supplements to, and not substitutes for, experience. "Experience is a guide which may be miscellaneous, fragmentary, unsatisfactory, often second-hand, frequently inaccurate," Hardy acknowledged, "but no engineer will discount its tremendous importance as evidence. All nature is trying to tell something of how its forces act. . . . Men may try to duplicate her phenomenon in a laboratory, but we never exactly reproduce the true natural problem, never fully ferret out her secrets." In short, theoretical-mathematical scientists and empirically-based inventors need to respect and check each other.[23]

The *ENR*'s editors also both supported scientific training and research and warned about an obsession with it. In an early address at the dedication of a college of engineering building, Charles W. Baker had advised his listeners to balance "theory" with "practical application." As he saw it, "the mathematical side of the profession and even the research laboratory have been if anything overdone already."[24]

Baker believed the nation needed engineers experienced and bold enough to lead as public servants. But the environment of the large-scale organization, with its government agencies in which many worked, fostered "the frailties of humanity" and accommodated the limits of professional training more than it encouraged engineers to rise above them:

> Some of us are lazy, and will take the easy course and let things run on in the rut of routine rather than make the effort to prepare new designs to meet changed conditions. Some of us are arrant cowards, and rather than run any risks we will spend money like water – so long as it is not our own money. I have seen engineers of this type take credit to themselves for their conservatism, and prate in official reports about the high-class construction they had secured, when the fact was that they had spent thousands where hundreds would have satisfied every requirement.
>
> I am perfectly well aware that there are men who defend that type of engineering. They claim than an engineer's business is only to look after accuracy, strength, permanence and safety in construction, to use only the best materials and accept nothing but the highest class of work. It is engineers of this type who are responsible for the idea, still too commonly held, that to put an engineer in charge of work means a great increase in its cost. If we had less such engineering, there would be a greater demand for engineers.

Like most of their colleagues in the BPR system, ISHC design engineers put security – for the public and for themselves – well ahead of cost-reducing innovation. In the case of continuous design, innovation delayed for decades – largely because it lacked simple mathematical demonstration – amounted to innovation denied for a generation. The ISHC honored conventional design, generally avoiding the surprises inherent in trying recent fashions. From 1920 to 1942, Indiana built a substantial highway system with safe and structurally unexciting bridges well integrated into it.

WITNESSING TO EMPIRICAL PRACTICE

(Lawrence #172)

A continuous girder built in Lawrence County in 1909.

(Clay #139)

A hybrid continuous slab-girder erected about 1910 in Clay County.

(Monroe #127)

A continuous slab with decorated rails built in Monroe County about 1915.

NOTES

[1]"Continuous Structures to the Fore," *ENR* (29 April 1916) 96: 677.

[2]As the equivalent of a rectangular beam, the reinforced concrete slab provides a parallel example of this development.

[3]For a more extended discussion of T-beams, see VII.3A.

[4]"Continuous Structures to the Fore," *ENR* (29 April 1926) 96: 677.

[5]See "On Stress in Simple, Tied, and Arched Concrete Beams," II.2A.

[6]Henry S. Jacoby, "Statically Indeterminate Reactions and Stresses," *ENR* (12 January 1912) 67: 128-29. "Design calculations for such structures are difficult and tedious, and up to relatively recent years this fact was a very real hindrance." "Continuous-Span Bridges and Structural Efficiency," *ENR* (22 August 1918) 81: 346-47.

[7]Hardy Cross, Professor of Structural Engineering at the University of Illinois, offered this definition of common sense. "Limitations and Application of Structural Analysis," *ENR* (24 October 1935) 115: 571.

[8]"Improvements on the Chicago Division of the Cleveland, Cincinnati, Chicago & St. Louis Ry," *ENR* (16 May 1907) 57: 535-39.

[9]Parts of this evaluation of the Pennsylvania Railroad overpasses should be understood as tentative since no plans of the structures have been located.

[10]For innovation, Condit ranked the column-and-slab system as the most important. For economic efficiency or cost, he ranked the rigid frame second to prestressing which was not introduced into the United States until *after* the Second World War. Carl W. Condit, *American Building Art: The Twentieth Century* (New York, 1961), 213-14, 280-81.

[11]Arthur G. Hayden, "Continuous Frame Design Used for Concrete Highway Bridges;" "Arthur G. Hayden, "Tests of Knees for Continuous Frame Concrete Bridges," *ENR* (11, 18 January 1923) 90: 73-75, 108-10; Arthur G. Hayden, "Two Parkway Bridges Planned by Engineer and Architect," *ENR* (2 July 1925) 95: 16-19; Arthur G. Hayden, "Rigid Frames in Concrete Bridge Construction," *ENR* (29 April 1926) 96: 686-89; Jay Downer and James Owen, "Translating Theory into Practice in Regional Engineering," *ENR* (19 September 1929) 103: 442-48; Hardy Cross and Newlin D. Morgan, *Continuous Frames of Reinforced Concrete* (New York, 1932), 244-46.

[12]Professor Beggs, American Concrete Institute *Proceedings* (1923), as quoted in Arthur G. Hayden, "Rigid Frames in Concrete Bridge Construction," *ENR* (29 April 1926) 96: 686; "Continuous Structures to the Fore," *ENR* (29 April 1926) 96: 677. The editors reaffirmed their judgment a few years later: "With the progressive weakening of what remains of the old influence [i.e., simple or statically determined structural thinking] and with the growing number of continuous bridges as a constructive precedent, we may look forward to more rapid extension of continuous structures." "Continuous Structures," *ENR* (17 January 1929) 102: 89.

[13]The evidence presented here about the impact of BPR policy upon statically-indeterminate design in the alliance during the Twenties is circumstantial. Hardy Cross, "Limitations and Application of Structural Analysis," *ENR* (17 October 1935) 115: 535; Albin L. Gemeny, "Bridge Engineering in 1937," *ENR* (10 February 1938) 120: 233-34.

[14]J. F. Seiler, "Continuous-Girder Bridges of Concrete," *ENR* (29 October 1925) 95: 722-25; "Ten Years of Achievement with Rigid-Frame Bridges," *ENR* (27 April 1933) 110: 531-33.

[15]George Hool, *Reinforced Concrete Construction* (2nd ed., 3 vols., New York, 1916), III: 423. Within a couple of years, Hool was less cautionary. The ISHC's engineers were probably aware of this growing acceptance of continuity within their profession. George Hool and Nathan Johnson, *Concrete Engineers' Handbook* (New York, 1918), 603-12, 622-38.

[16]The ISHC's engineers never represented these "three-span units" as continuous. In time, tension cracks would run across the deck over the piers as each span deflected independently of its neighbors. See also "Bottom-Land Trestles" in VII.3.

[17]Marshall Findley, "Continuous Beam Design Practice Brought Up to Date," *ENR* (21 March 1929) 102: 467-69.

[18]Hardy Cross and Newlin D. Morgan, *Continuous Frames of Reinforced Concrete* (New York, 1932); Hardy Cross, "Limitations and Application of Structural Analysis," *ENR* (17, 24 October 1935) 115: 535-37, 571-73; "For Continuous Structures;" L. E. Grinter, "Direct Design for Continuous Beams," *ENR* (16 March 1939) 122: 374-75, 388-90.

[19]For the same factors at work in similar circumstances, see Arthur G. Hayden, "Continuous Frame Design," *ENR* (11 January 1923) 90: 73.

[20]"Program for the Dedication of the New Deer Creek Bridge and Viaduct," Delphi, Carroll County, (2 July 1936), Carroll County Historical Museum, as quoted in Camille B. Fife and Thomas W. Salmon II of the Westerly Group, "Delphi Bridge on U.S. 421," HAER IN-91: 6.

[21]James K. Finch, "French Pioneers in Engineering," *ENR* (24 April 1930) 104: 676-83.

[22]Hardy Cross, "Limitations and Application of Structural Analysis," *ENR* (17 October 1935) 115: 535.

[23]Hardy Cross, *Engineers and Ivory Towers* (New York, 1952), 105, as quoted in Bruce E. Seely, "The Scientific Mystique in Engineering: Highway Research at the BPR, 1918-1940," *The Engineer in America*, ed. Terry S. Reynolds (Chicago, 1991), 336.

[24]Charles Whiting Baker, "A Broader Training for the Engineer," *ENR* (20 May 1909) 61: 536-38.

IX

BRIDGING HOOSIER BYWAYS WITH PRIVATE DESIGN IN HARD TIMES

The New Era of Prosperity which followed the close of the Great War entangled Indiana's private bridge designers in more rules and provided fewer opportunities for profit than before the war. The federal-state highway system became an intrusive reality, despite the Indiana State Highway Commission's [ISHC] sputtering start. The federally directed system increasingly took control of the best traveled roads in the state and made design of their bridges a government function. Hoosier cities and counties, however, still provided a significant arena within which private designers might compete, especially during the Twenties, when local government had monies to commit to construction projects.

The scientific correctness that dominated the federal-state highway alliance had not yet become a requirement for designs competing in the marketplace. Competition among private designers continued to encourage them to adopt empirically-based innovations, even as the market shrank. Among Indiana's private design entrepreneurs, Daniel Luten remained the most prominent in the Twenties, both within the state and across the nation. In turn, William S. Moore, who contributed something to every major design phase in concrete within Indiana before the Second World War, closed the period with the most significant, private-design achievements in the state during the Thirties.

IX.1 FROM CHECK TO CHECKMATE: LUTEN'S INNOVATIVE REPRISE

ARCH STUDY AND PRACTICE

The engineer's art in the field of arches is in the most flourishing condition of development that it has seen for decades. ... Of these developments, all of which are significant, the most far-reaching is the study of deck participation. ...

According to the results of all these studies, the arch rib does not act alone, but forms a unit with the spandrel columns and deck in resisting the deforming action of the load. This being so, our long-established view of division of function between arch rib and superstructure falls to the ground, and methods of analysis as well as methods of design require prompt revision. ... The composite arch is in effect a spandrel-braced arch, in which the columns and decks serve as spandrel bracing ... .

– *editors,* Engineering News-Record *(1929)*[1]

IX.1A Containing the Design Entrepreneur

A recorded social exchange underscores the importance of the patent system to Daniel Luten's success and decline. When in front of others Luten once criticized Elmer Goldsmith for being "a narrow field patent lawyer," the lawyer parried with a reported "Grand Slam:" "'All right about patent lawyers,' bristled Elmer. 'Hadn't been for them friend Dan over there would still be a poor school teacher.' And the bridge expert laid down his cards."[2] The removal of patents from Luten's entrepreneurial arsenal undoubtedly reduced the private designer's clout in the marketplace.

The biggest blows to the sale and use of Luten-design bridges, however, came from the almost total boycott of them, along with other private design, by the same and ever-growing federal-state highway alliance that helped to shear Luten of his patent advantage. In Indiana, the ISHC probably controlled between a third and a half of the highway bridge construction dollars expended during the Twenties. Luten neither expected nor received exemption from the federal-state alliance's nearly exclusive reliance on government design for all but the very largest or most unusual crossings.

That a protégée of Luten, William J. Titus, dominated ISHC bridge design in the Twenties did not give Luten a home-state niche. The ISHC hired the National Concrete Company, the Indiana arm of Luten-design construction, when the company submitted the lowest bid to build ISHC-designed arches or concrete substructures of metal spans. But the ISHC apparently never contracted for a Luten design nor hired him as a consultant.

Exclusion of private design from the federal-state highway system, however, was not enough. The alliance leaders wished to influence, if not to dominate, the road and bridge building of local government, too. Both the 1917 and the 1919 acts establishing the ISHC created a mechanism which, when triggered, required state approval of the specifications and plans for a county road or bridge before construction.[3]

When state control of county bridge planning had been invoked, the ISHC sometimes acted against Luten designs. In an early case that became engulfed in published charges of ISHC corruption, William J. Titus found his official role pushed to the edge, if not beyond the letter, of the law. Local residents, with an ISHC Commissioner in the lead, successfully petitioned for ISHC intervention concerning a new bridge to span the Wabash River at Attica between Fountain and Warren counties. Instead of just monitoring the design of others as the law allowed, Titus designed a concrete-arch structure for the site. When the Fountain and Warren County Commissioners opened bids in 1920, they found that the National Concrete Company offered two Luten designs, one which they proposed to construct at four percent and the other at six percent less than the lowest bid on Titus' ISHC design. Since the two Luten designs were not state-approved, the county authorities awarded a contract for $275,540 on the more expensive state design.[4]

Another rejection of Luten design was both more direct and revealing. The Franklin County Commissioners tried unsuccessfully to limit ISHC intervention in the planning of a bridge at Laurel over the Whitewater River (Franklin #17). In 1922, Titus reviewed plans which the county's engineer had drawn for the bridge. After comparing these plans "with our bridge at Brookville," Titus rejected them as well as some partial revisions offered in response.[5] The commissioners then tried to ignore the state by contracting for the construction of the Laurel bridge on the basis of the unapproved plans, but Titus threatened legal action.

Cut off at one pass, the Franklin County Commissioners next turned to the Luten design submitted with a bid from the National Concrete Company. Titus also rejected these plans, again largely on the basis of where Luten design differed from state practice. When he complained that Luten's plans were not as detailed as those of the ISHC, Titus ignored the differences between how the federal-state highway establishment related to builders and how Luten as designer related to the National Concrete Company. With decades of experience building almost only Luten designs, the National Concrete Company needed minimal written instructions for construction. As Titus also knew from his years in Luten's drawing room, his former boss used narrower and thinner arch rings and less reinforcing than the state's just-in-case designs. As ISHC representative, however, Titus now objected to Luten's lean rings. He likewise found the spandrel and other walls "very sparingly designed." It was exactly such efficiencies in design that gave Luten his competitive edge, and Titus must have understood this as he rejected it.

Even though excluded from about half of the bridge-building market in Indiana and across the nation, Luten and his construction affiliates continued to win a significant number of contracts where competition in design remained open. These entrepreneurs stayed prosperous in the Twenties as long as they exploited every opportunity. Luten designed at least two dozen structures which still survive in Indiana, including nine filled-spandrel and four open-spandrel arches.[6] The Luten interests succeeded even more markedly across states in the South.

IX.1B The Spandrel-Braced Arch

To compete effectively, the entrepreneurial designer needed to stay at the forefront of change. Combining intuition with training and experience, Luten tried to anticipate the changing needs of transportation and then to prepare bridge designs appropriate for the future. He foresaw, for example, some of the motor vehicular requirements for bridge design well ahead of the federal-state government engineers and many other private competitors. He grasped parts of the new reality of automobility before trucks imposed significantly on the nation's wartime transportation system, tore apart the good federal-state highways, and began to demolish their bridges in large numbers.

Luten addressed the Iowa Engineering Society in February of 1914 on "Hard Roads and Permanent Bridges." He sketched the changing role being assigned bridges in the emerging transportation system. Bridges, he noted, used to interrupt the road surface with a plank floor to carry traffic across a river, and signs warned travelers to cross the bridge no faster than they could walk. "The modern bridge," in contrast, "must carry the brick or macadam roadway over the river, with no interruption of its continuity and with no limitation of speed." "Hard roads, heavy loads, and high speeds," Luten concluded, had in ten years turned bridges from "luxuries" to "necessities."[7]

The design entrepreneur had to be more than a seer. He also had to introduce his insights to potential purchasers. The ideal was to be seen as both best and cheapest, which was a major challenge. In this situation, Luten's arches met the requirements of a modern bridge: they allowed the roadway to cross a bridge without interrupting its surface, reducing the speed of traffic, or having the bridge suffer from considerably heavier or faster moving-loads.

To the two phases of bridge design that he had articulated in early 1914, Luten added a third by the early Twenties. From this, he generated his last significant innovation. "Initially, we built bridges to carry traffic from one road end across an obstacle to another road end. Later, we improved bridges so that they could carry the road itself (whether paved or gravel) across the obstacle. Now, with most roads hard surfaced, we are beginning to design bridges in which the pavement itself becomes part of the bridge structure."[8] If with automobility bridges needed to carry an uninterrupted and paved roadway, why not in the name of structural and economic efficiency incorporate the roadway into the load-carrying work of the arch's structure? To transform necessity into virtue, Luten proposed a spandrel-braced arch.

Luten did not initiate the notion of a spandrel-braced arch any more than he developed elastic theory. His forte was in applying other people's ideas, sometimes in very transformed ways, to the needs of his time and circumstances. Proposals for an iron structure with an arched bottom chord and a horizontal top one connected with web bracing went back to at least the 1880s when Max Am Ende and Handyside and Company proposed to build such a bridge over the River Douro in Portugal. Malverd Howe, a Professor of Civil Engineering at Rose Polytechnic Institute, included a chapter on "the spandrel-braced arch" in his 1897 *Treatise on Arches*, a work which Luten had read.[9] In this case Luten kept the geometry of the old spandrel-braced arch's chords but put aside the metal structure from which a deck is hung and whose arches flank and therefore contain the roadway.

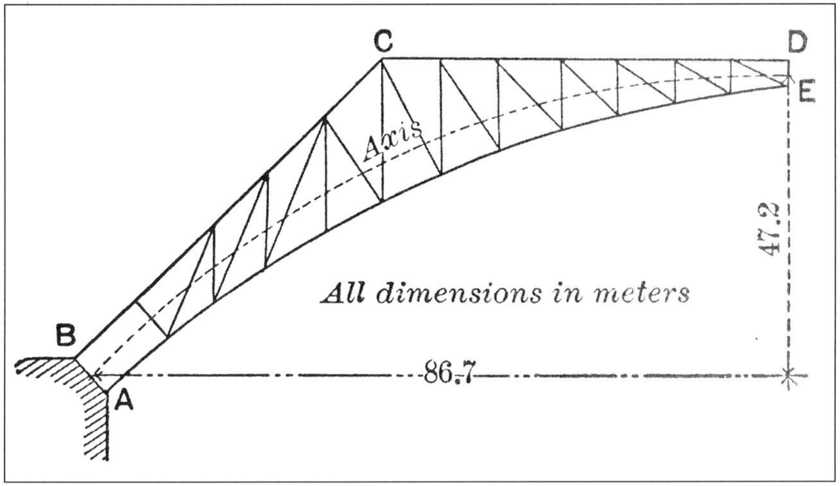

Howe borrowed from Messrs. Ende and Handyside for this figure of the spandrel-braced arch with the "Axis" showing the structure's center of gravity or line of pressure. The axis approximates that of a linear arch, Luten's favorite ring shape.

At the core of Luten's special arch sits a pair of reinforced-concrete spandrel braces, each arced at its lower edge, usually into a semicircle, with a horizontal upper edge to carry the slab deck. As is typically the case with reinforced concrete structures, the placement of the steel tells much about how a designer thinks his bridge works. The braces carry two tiers of longitudinal reinforcing rods, with one row close to the lower edge and a second generally paralleling some distance above, except over midspan where the upper row runs horizontally near the top edge. Without fill to dissipate the impact of moving loads, the nature, intensity, and regions of stress in the braces change significantly with traffic and so the two rows of longitudinal rods are tied together with stirrups. Essentially semi-circular and delivering most of their thrust vertically, the braces rest on modest concrete pads in which the longitudinal reinforcing is anchored. If the pads do not rest upon bedrock, then they are supported on timber piles. Luten cantilevered a slab deck over his spandrel braces and stiffened its ends with a thin vertical wall, which also contained the approach fill.

Following the logic of his early spandrel-stiffening to its conclusion, Luten abandoned the barrel ring for a pair of ribs which he merged with spandrel walls. Without the weight of a barrel ring and fill to address, and with the semicircle's advantages for handling stress in mind, Luten next essentially abandoned the abutment for a modest foundation pad. The slab deck, in turn, carried some load on its own, most of which it transferred vertically into the braces and to the span-end walls. In all, the spandrel-braced arch was remarkably light, required a minimum of steel, and embodied construction efficiencies. It could be built with reusable forms, its steel placed simply, and its concrete poured in large sections.

Developing a new design was one thing; marketing it, another, especially in the Twenties. Now effectively lacking patent protection against the purloining of his ideas in the marketplace, he needed to innovate and then to market defensively – that is, quietly. Afraid to advertise his spandrel-braced arch either through catalogs or in the pages of the *ENR*, Luten and his associates introduced the design without fanfare on a case-by-case basis at specific biddings. Even under these constraints, spandrel-braced arches began to dot the landscape. The Luten Bridge Company of York, Pennsylvania, successfully promoted the design in Ohio and West Virginia, erecting spandrel-braced spans from fifty to one hundred feet long beginning at least as early as 1924. Luten won contracts to build about a dozen spandrel-braced arches in Indiana from about 1927 to 1931.

Probably the best documented test of Luten's spandrel-braced arch occurred in New Hampshire, where the Luten Bridge Company successfully introduced the design in direct competition with a standard, state plan for a filled-spandrel arch on a trunk highway.[10] Of the three bids on the state design, the highest came in at just under fifty-one thousand dollars and Luten Bridge bid the lowest at just under forty-seven thousand and five hundred dollars, almost five percent less.

Apparently trying to crack the monopoly of government design for state highway bridges in New Hampshire, Luten Bridge sent in a second bid, this one based on a spandrel-braced arch. The

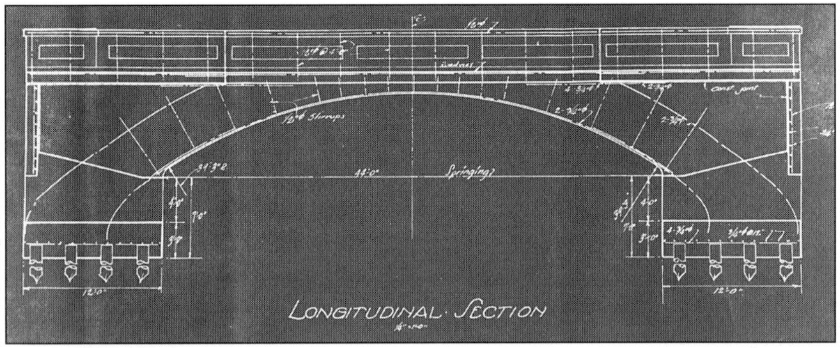

(Fountain County Commissioners Office)

(Fountain #124; Luten family)

Two Fountain County spandrel-braced arches illustrate the essence of Luten's design. In the Loeb Bridge (above left) and the MacDonald Bridge (above right) the spandrel braces act as arch ribs reinforced with two tiers of longitudinal rods and occasional stirrups, except over the crown where the horizontal deck cuts deeply across the brace and submerges the top arc of its longitudinal reinforcing into the horizontal rods of the slab deck. More than metaphorically akin to a top chord, the horizontal deck transfers load through the spandrel braces and, at span-end, through the light wall substitutes for abutments. Reinforcing rods angle diagonally downward from the near-bottom of the wall-abutments through the spandrel brace, and to the brace's lower longitudinal reinforcing at the springing line, where tied to complete a reinforcing system.

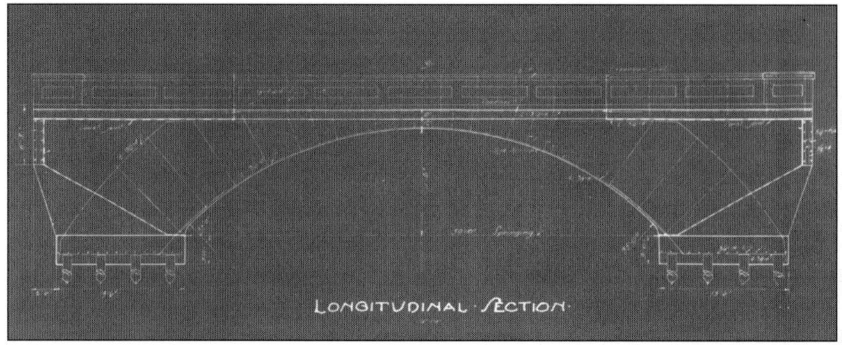

(Fountain County Commissioners Office)

(Fountain #130; Luten family)

company's bid of just over twenty-four thousand dollars on the Luten design amounted to about half of the average of the bids on the state plan.[11] Luten's spandrel-braced structure used at least one-third less steel than did the state's arch and only sixty percent as much concrete. Luten's design also allowed for the repeated use of some of the form-work, thereby reducing the cost of framing and labor.

In this case the "careful" New Hampshire commissioners broke ranks with the federal-state alliance's general practice and accepted the bid to build with private design. Although their decision saved taxpayers almost fifty percent, the commissioners kept this case an

Different Luten-design, spandrel-braced arches. Wide-spanned Roby Bridge, Harrison County,. West Virginia (above). Spandrel braces, themselves braced, raised on bents, and decorated with neoclassically balustraded rails in Waco, Texas (below).

(Luten Bridge Company)

exception to the general rule. Some New Hampshire townships and cities contracted for Luten's spandrel-braced arches, but the state never returned to the design.

As statically indeterminate and therefore without a scientifically certified pedigree, Luten's spandrel-braced arch would not be used

Design success and entrepreneurial fizzle. Main Street Bridge, Lebanon, New Hampshire. Note the use of four spandrel braces per span.

in other parts of the federal-state highway system either. How much formal analysis Luten performed on his new design remains shrouded as he stayed out of the national limelight. But he probably approached the spandrel-braced as he had earlier addressed the filled-spandrel arch, boiling down the formally correct equations into a series of quick and approximate – and therefore partly empirical – calculations.

The arch form could be shunned but not banished just because statistically indeterminate. Even the Bureau of Public Roads [BPR], which stood for scientifically correct research and design, had accommodated, albeit uneasily, to quasi-scientific analysis of arches. In its elaborate, instrument-monitored destruction of an open-spandrel arch structure over the Yadkin River in North Carolina, the BPR tested the arch analysis it had already endorsed and then re-endorsed the analysis with qualifications.[12]

The Yadkin River Bridge tests also provided support for some of the principles Luten had been following quietly in his spandrel-braced arch design but not for the ways he practiced them. The measurements revealed that the "arch with the floor system continuous over the whole span and rigidly attached to the tops of the columns" (or spandrel walls) became "in fact, a fixed, spandrel-braced arch." Although the *ENR*'s editors took special note of this finding, neither the finding nor the editorial acknowledgment of it translated into an implicit endorsement of Luten's particular design.[13] If the bureau's engineers and the magazine's editors had known of the Luten spandrel-braced arches which had been constructed here and there across the nation for nearly a decade, they might have complimented him for his innovation and then likely criticized him for inadequate analytical rigor. The "new results lay emphasis on the value of conservatism in arch design, and upon the need for skilled and thorough analysis before fixing the proportions of a proposed structure," the editors added. "Rule-of-thumb practice in arch design is definitely obsolete."[14]

By 1929, other factors also substantially limited Luten's days as an entrepreneurial design engineer. With the onslaught of the Great Depression, counties and cities on which Luten had increasingly depended for business transferred dollars from construction to relief for growing numbers of unemployed residents. Almost all the bridge-building done on highways during the Thirties occurred within the BPR system. For the next decade and a half, the boycotters of private and empirical bridge design had just about the only business to offer. One of Luten's associates compared the depression to a man who, when reaching the top of a ladder, "must back down or jump off. We all just recently jumped off, that's all." Luten demurred: "Somebody jerked the ladder out from under me."[15]

At age sixty-two in 1932, Luten closed his engineering offices and purchased a small broom factory, which he operated in Indianapolis until his death in 1946. More retirement activity than a second career, the Daniel B. Luten Manufacturing Company became as well known locally for its lunch hours as for its brooms. At midday individuals from across the city regularly assembled at the broom factory's offices to join Luten in a passion second only to bridge design, the playing of chess at which he was also sometimes a champion. On one occasion Luten challenged thirty SCIENTECH club members to simultaneous chess games. Although the outcome of the challenge went unreported, his colleagues continued to note the retired design engineer's zest for questions and his evident enjoyment at "watching the minds of others work."[16]

IX.2 FROM MOORE TO McANLIS: AT THE GOAL LINE

Daniel Luten, who called himself "Designing Engineer," may have been the only bridge entrepreneur in Indiana with a national reputation. Some other engineers in private practice, though, had a specialty in bridge design which they employed along with other lines of enterprise around a home base within the state. Not exclusively dependent on bridge design contracts for their living, these "Consulting Engineers" rolled along with the changing times. They managed through the Twenties when the rise of the ISHC halved the market for private design, and they scrambled during the Great Depression for the occasional private design opportunity that the federal-state highway alliance system had not engrossed. Of these consulting engineers, none seems to have worked as widely and as creatively as William S. Moore of South Bend.

"Billy" Moore, the Purdue University football halfback and civil-engineering graduate, designer of some City Beautiful spans for South Bend and Mishawaka, and first chief engineer of the ISHC, opened an office as a consulting engineer in South Bend in the mid-Twenties.[17] He advertised specialties in surveying, subdivision development, sewers, and bridges. Although Moore apparently sold his services as a bridge designer only in St. Joseph and Elkhart Counties during the Twenties and Thirties, his contributions there were considerable.[18]

(South Bend Tribune, 1969)

William S. Moore with his slide-rule.

Whether the consequence of talent or of inclination, Moore was not as inventive a designer as Luten. He did not create, for example, a new reinforcing system or a novel form of design. Bridge design was only one of Moore's many professional activities. Still, bridges intrigued him. He stayed abreast of research and design practice, collected plans Luten and others submitted for bidding in St. Joseph and Elkhart Counties, and saved copies of ISHC drawings. While Moore incorporated ideas from others, his designs always retained a personal distinctiveness.

IX.2A Arches

Moore designed six, filled-spandrel arch bridges for the two-county area between 1926 and 1940 (Elkhart #371, #408, #3869, #6152, #6193; St. Joseph #21 1). He shared a lot with ISHC design. He uniformly used the Thacher style of two-tier longitudinal reinforcing with stirrups in three-centered rings as did the ISHC; he treated the piers elastically as did Luten and the ISHC by extending the ring reinforcing into them; and often he used the depressed, bush-hammered, rail panels that Titus made the ISHC's hallmark. At the same time, he designed rings with crowns thin enough that Luten would have applauded and Titus at the ISHC censured. Moore also had his spandrel walls bush-hammered as did Luten, and he went a step further to include the pier pilasters in the decorative treatment. Two of his structures used open-oval rail panels, a pattern Moore especially favored when decoration seemed appropriate.

In 1928 and 1929, Moore designed three open-spandrel arch bridges which shared some features with Luten and with state design while maintaining some distinctiveness (Elkhart #374, #6085; St. Joseph #207). All three designers used ribs rather than solid-barrel rings, although Luten and the ISHC preferred two ribs per span and Moore four. While all used two-tier reinforcing with stirrups in the rings and extended the reinforcing into the substructure, Luten integrated the ring-ribs with the substructure more fully, the ISHC integrated them least, and Moore compromised. The three differed over the relative thickness of their ribs at the crown. The ISHC designed the heaviest crown sections, Luten the lightest, and Moore once again favored the middle although on the state side of the center line.

Billy Moore did organize one of his decks in a novel way.[19] Both the ISHC and Luten had spandrel columns capped with transverse beams to support a concrete slab deck and transfer most loads directly to the arch ribs. Moore followed the same pattern in two of his open-spandrel arches and innovated with the third. In the County Line Bridge (Elkhart #6085), Moore introduced longitudinal beams adjacent to (rather than directly above) the ribs and turned the beams into T's by integrating their reinforcing with the deck slab above. With each of the beam's reinforcing integrated longitudinally as well as vertically and the beam supported by several sets of spandrel columns, Moore made his beams continuous, delivering deck loads to the rib-rings at several points away from midspan. This practice avoided loading the crown, an area of special vulnerability in ring-ribs.

Nothing made Moore's open-spandrel arches stand out more than the decoration he designed into them. Sidewalks curbed all the roadways, rails with open-oval or arch panels bordered the deck, and ornamental light standards highlighted the structure. Pier pilasters were usually accented, sometimes with depressed geometric paneling, at other times rising from the cutwaters like obelisks. The decor of the pier pilasters continued through the rail posts as well. Billy Moore never forgot his days as a designer of City Beautiful spans.

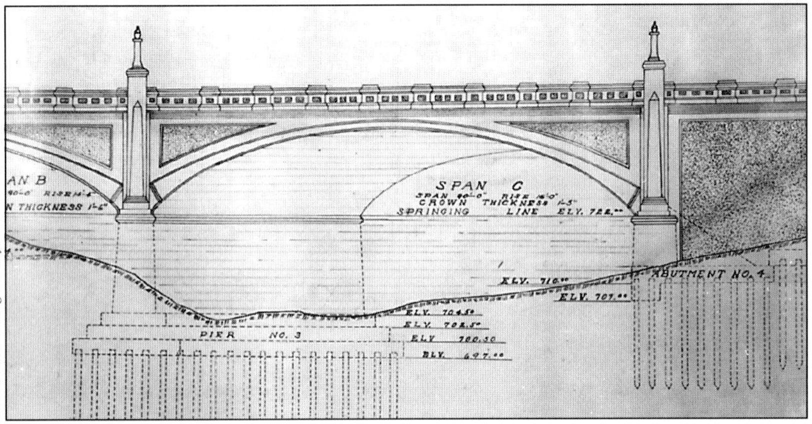

(Elkhart #3869; Lang, Feeney)

(Elkhart #3869)

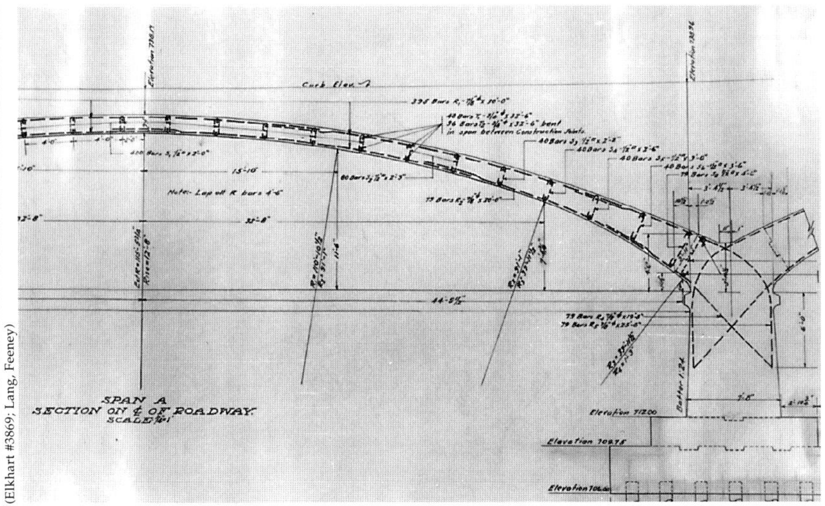

(Elkhart #3869; Lang, Feeney)

In the second phase of his career, William S. Moore practiced some of the style he learned in the first. For the Main Street Memorial Bridge over the St. Joseph River in Elkhart (above), he designed a structure with multiple functions. Commissioned in part to commemorate the sacrifices made by American soldiers in the First World War, Moore called on his City-Beautiful design experience. On the inside, Moore had moved beyond the Melan system of reinforcing of the earlier era to two-tier, Thacher-style ring reinforcing with stirrups, and he took a step toward elastic pier treatment, although he continued to shape his arches as nearly segmental (upper left). On the outside, he had his walls bush-hammered to set off the rings, piers, and rails (lower left). He extended his abutment and pier pilasters here into commemorative towers framed in part by open-column rail and multiply-coped posts. Metal replacement railings partly undermine the structure's intended message (right).

(Elkhart #3869)

Moore designed the Angela Avenue Bridge over the St. Joseph River in South Bend) towards the end of his career. Like other of his filled-spandrel arches built after the First World War, Moore used two-tier longitudinal reinforcing and stirrups with his nearly segmental rings, although in this later case he fully integrated the rods of the ring with those of the pier (below).

(St. Joseph #211)

(St. Joseph #211)

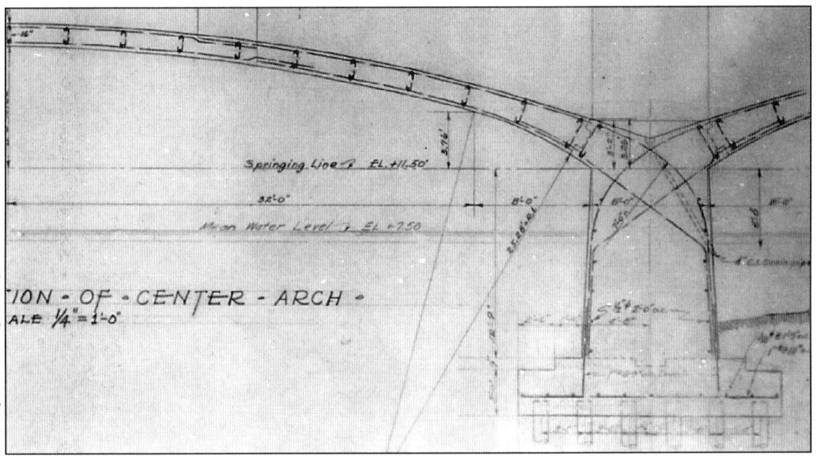

(St. Joseph #211; Lang, Feeney)

(St. Joseph #211)

Once again the decor stands out. His piers merge into receding, streamlined pilasters (right, lower) whose thrust is re-echoed in the rail posts (right, upper). The open-column rail panels also accent the vertical.

(Elkhart #6085)

(Elkhart #6085)

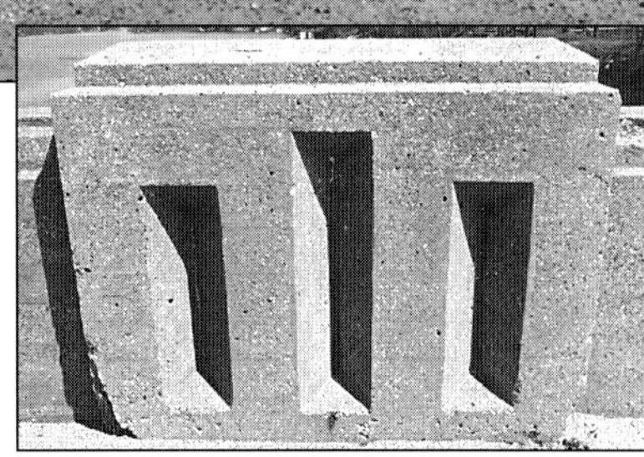

County Line Bridge over the St. Joseph River.

(Elkhart #6085)

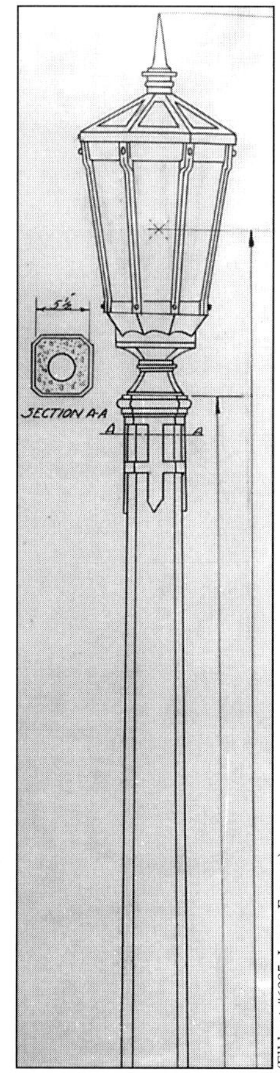

(Elkhart #6085: Lang, Feeney)

IX.2B McAnlis and Continuity

Local government and private designers did not have to be constrained by the dogma of scientific correctness which held sway in the federal-state highway alliance. The locals and the entrepreneurs could respond to the flurry of stories about continuous-beam or slab spans and rigid framing which the *ENR* and other engineering journals carried. But in Indiana this response was sparse and scattered during the Twenties. Only the engineers of the city of Indianapolis and of Marion County continued their earlier use of continuous and rigid design on small structures, and a designer occasionally experimented elsewhere on a rural span.[20] National discussion of continuity and framing continued in the Thirties, and by mid-decade Professor Hardy Cross had provided a relatively simple mathematical demonstration to certify the scientific correctness of continuous and rigid-frame design. Indiana's private designers finally responded with two of the three most fashionably designed bridges built in the state during the Thirties.[21]

Chauncey R. McAnlis, Fort Wayne's City Engineer, designed Indiana's longest and best publicized continuous T-beam structure in 1937. The Portland Cement Association presented the six-hundred-and-thirty-foot Lockport Bridge (Carroll #142) to the national engineering public with a full-page advertisement in the *ENR*.[22] The cement association promoted continuous design for longer spans (here up to ninety-five feet), more flexible pier locations, reduced pier width (four feet at top and seven at base), shallower deck (seven inches), and fewer expansion joints (one at each end).

For McAnlis, private consulting on bridge design seems to have been very much an occasional matter. The Lockport Bridge may be his only extant work. Probably because entrepreneurial design opportunities were so limited in the middle of the Depression, McAnlis neither maintained a business office nor advertised. The apparent absence of business promotion should not be read, though, as an absence of engineering credentials. Following graduation from Pennsylvania State College, McAnlis taught at Cornell University and the University of Illinois and "engaged in active engineering work." How this fully-registered, yet little-known professional engineer actually became the designer of the Lockport Bridge remains a mystery.[23]

(Carroll #142; *ENR*, 1940)

The Lockport Bridge over the Wabash River as advertised.

Despite McAnlis' limited visibility as an entrepreneurial engineer, he employed some current, although hardly avant-garde elements in design when he won the chance to plan a bridge. More particularly, the Lockport Bridge appears to owe much to continuous T-beam structures which Kansas and Illinois State Highway engineers had designed and built somewhat earlier in their respective states and then reported through the *ENR*.[24] With the Kansas design, McAnlis shared, for example, the size of variable-depth T-beams, flange-deck depth, and step-cut vertical construction joints in beams with some of their longitudinal reinforcing bars displaced from near

(Carroll #142)

(Carroll #142)

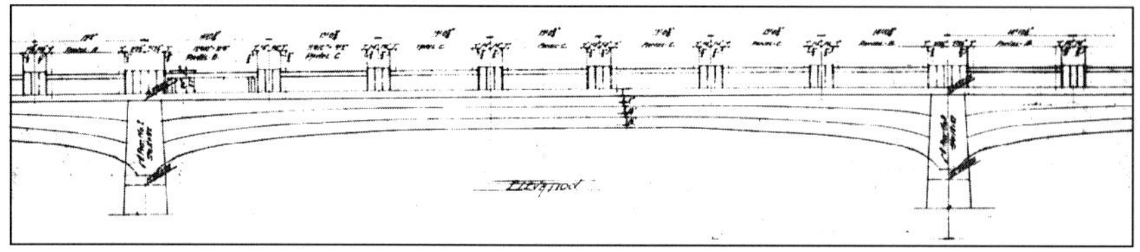

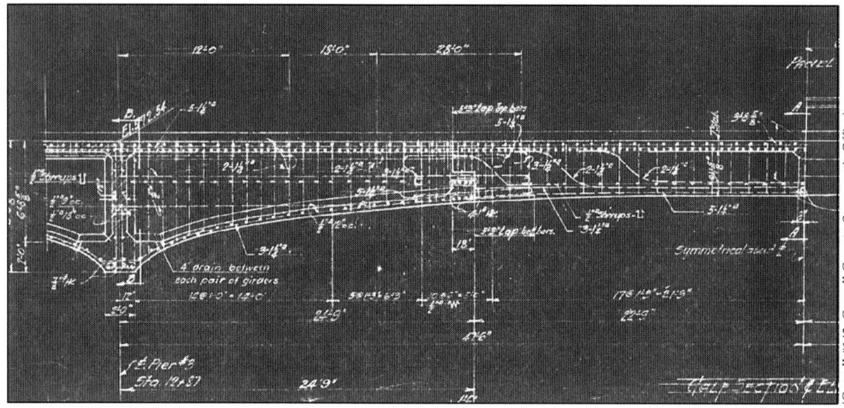

(Carroll #142; Carroll County Surveyor's Office)

Its outer beams streamlined in harmony with their depth, the Lockport Bridge projects graceful lines above and across the Wabash. The beams sit on metal rollers atop most parts of the substructure and are likewise detached from the pier pilasters to allow the superstructure to shift with temperature change and motion. The beams can expand and contract over the substructure and behind its pilasters. The "Half Section" (right, lower) shows where three of the lower lines of longitudinal reinforcing peel off towards the upper edge of the beam as they approach the construction joint at approximately the beginning of the last quarter of the span.

the lower to the upper edge towards the joints. McAnlis fixed his beams to the fourth pier and placed them on rollers or shoes over the other piers and at the abutments, allowing the superstructure to expand or contract from the center out to the ends. In all, he produced by far the most striking continuous-beam structure designed and built in Indiana before the Second World War. The ISHC, in contrast, never designed a continuous-beam structure at even half the length or with spans sixty percent as long as McAnlis' photogenic bridge.

IX.2C Moore's Frames and Boxes

Credit for the most adventurous, reinforced concrete design built in Indiana during the Thirties, though, belongs to William S. Moore. The Great Depression had brought a virtual halt to all bridge construction in his two-county area for much of the decade, allowing his design capabilities to lie fallow from 1931 to 1938. Federal aid gradually began to replace some of what local governments had spent on public works in more prosperous times. But not until the late Thirties, when the economy picked up enough to provide local governments with revenues sufficient to match federal funding on large bridge projects, did major construction resurface. Moore got the chance to make his most advanced contribution to bridge design only at the decade's end. The Bridge Street Bridge in Elkhart (#387) stands as his best.

(Elkhart #387)

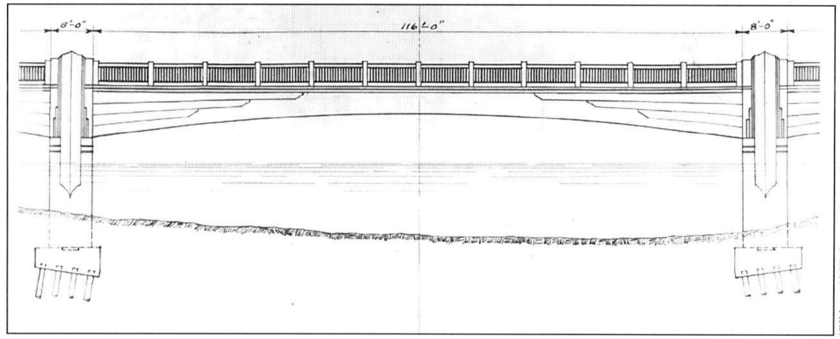

(Elkhart #387; Lang, Feeney)

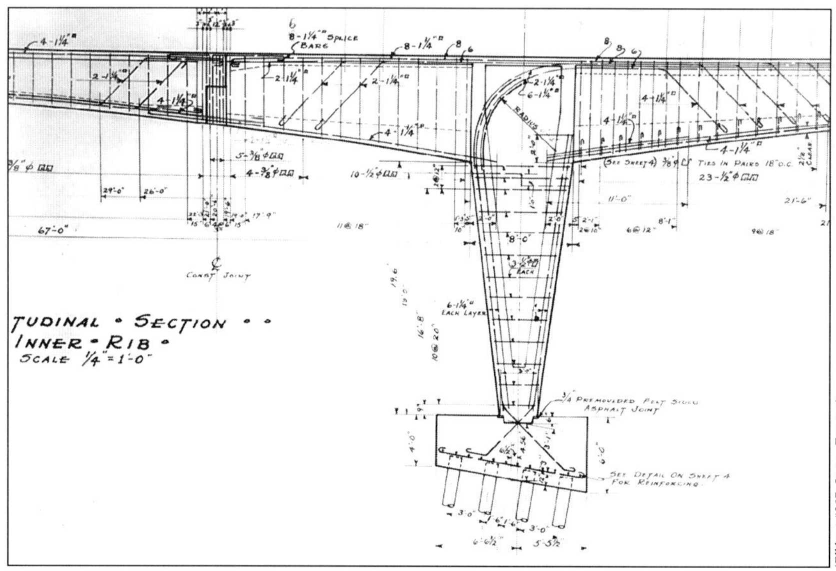

(Elkhart #387; Lang, Feeney)

Bridge Street Bridge over the St. Joseph River (upper, right). The reinforcing for this rigid-frame structure is much more complicated than for the ISHC's North Street Bridge in Delphi (Carroll #1788). The Bridge Street loads are heavier and its spans longer. The frame in both structures rests on a hinge from which the reinforcing extends up the sides to the knee. For Bridge Street, the frame is heavier and the reinforcing rods larger and more numerous. The lines of integrating reinforcing here (right, lower) rise out of the frame's side most distant from the beam to the right (center span), arc through the knee towards the top of the beam, and then one line after another shifts diagonally toward the lower region of the beam where the rods are hooked around longitudinal ones. A construction joint and reinforcing strategy similar to that used with the Lockport Bridge can be seen on the end-span plans (left beam) for Bridge Street.

Here Moore put together three spans totaling three hundred and eight feet of variable-depth T-beams, which can also be described as continuous and framed. At one hundred and sixteen feet, the center span is the longest of beams built in Indiana before the Second World War. The reinforcing of the center-span beam is fully extended and tapered into the piers as were Arthur Hayden's rigid frames. The reinforcing of each beam of the sixty-seven-foot end-spans is integrated and therefore continuous with the adjacent one in the center span. Moore provided for expansion and contraction joints only at the abutments, where the beams sit upon rollers. The variable depth of the beams underlines the continuity and the framing. At the piers, all are ninety-three inches deep. At the crown of the center span and at the abutments of the end-spans, they are forty inches deep.

Still more remarkable, Moore designed the only cellular or hollow superstructure known to have been built in Indiana before the Second World War. Cellular construction had long been used with bridge abutments and piers. Indeed, the open-spandrel arches which the Big Four Railroad's engineers designed for Hendricks and Putnam Counties in 1906 have hollow piers. Not until the mid-Thirties, though, was the cellular principle applied to bridge superstructures where its advantages were even greater than for substructures. A hollow or box beam eliminated concrete which did little or no carrying. The reduction in weight of the structure lessened stress, allowed for longer spans, and required fewer and sometimes lighter piers. In 1936, the State of Washington built and advertised through the *ENR* a number of box-beam spans, including two rigid-frame structures along with a number of continuous-beam bridges. The hollow rectangles which Moore designed inside

(deconstruction photo from Elkhart County Highway Department)

Moore designed two different sizes of beams and boxed voids for the Bridge Street Bridge (lower left). The designer kept his pier pilasters separated from the frame and beams so they could expand and contract above the lower hinge. A hole accidentally made in the structure's deck (left) documents Moore's boxing. Note that H. L Maddocks, the builder, had used a wooden frame to keep the concrete when poured from filling the planned void in the beam.

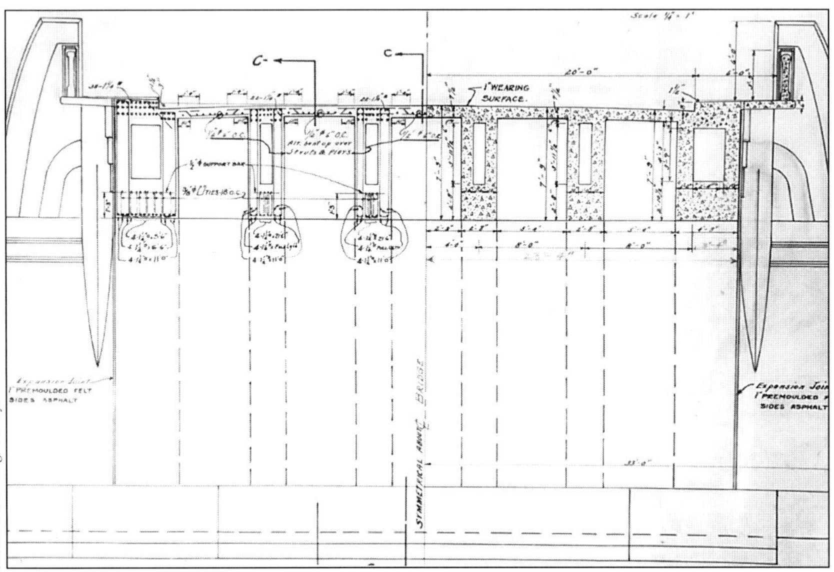

(Elkhart #387; Lang, Feeney)

(Elkhart #387)

each of his beams were about four feet deep and ranged from one to two feet wide.[25]

The arched lower edges of Moore's Bridge Street beams point towards the piers, which visually integrate the structural elements into streamlined vertical sweeps whose lines also wing across parts of the adjacent beams. The interior sides of the rail posts echo the vertical sweep, and their panels carry Moore's favorite open-ovals.

With their linear geometric shapes and vertical accents, Moore's original plans for Bridge Street suggest Art Deco, an abstract Twenties and Thirties substyle of the earlier Classical Revival so dear to the City Beautiful architects.[26] The revised plans, though, speak a bit more of Moderne, a Thirties revision of Deco in which the decidedly linear was rounded and the decidedly vertical became horizontally accented. Moore may have come – or been prodded to come – to what has sometimes been called "PWA Moderne," a combination of the substyles of Art Deco and Moderne which Moore's funding agency, the Public Works Administration, apparently favored. However their style is identified, the Bridge Street spans perch as assured and expansive as an American eagle ready for flight into another postwar era.

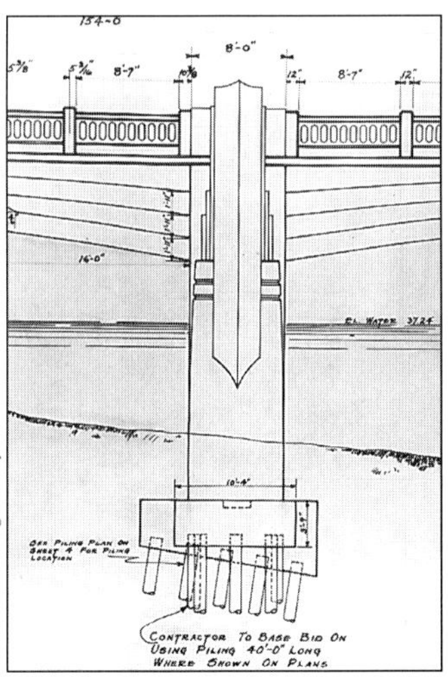
(Elkhart #387; Lang, Feeney)

(Elkhart #387)

From Art Deco (left) to Moderne (right) on Bridge Street.

WILLIAM S. MOORE'S DESIGN LEGACY[27]

	County	Number	Form	Date
1.	St. Joseph	#205	filled-spandrel arch	1913
2.	Elkhart	#6193	filled-spandrel arch	1926
3.	Elkhart	#6152	filled-spandrel arch	1926
4.	Elkhart	#408	filled-spandrel arch	1927
5.	Elkhart	#3869	filled-spandrel arch	1927
6.	Elkhart	#374	open-spandrel arch	1928
7.	Elkhart	#6085	open-spandrel arch	1929
8.	St. Joseph	#207	open-spandrel arch	1929
9.	Elkhart	#371	filled-spandrel arch	1931
10.	Elkhart	#387	framed box beams	1939
11.	St. Joseph	#211	filled-spandrel arch	1940

NOTES

[1] "Arch Study and Practice," *ENR* (21 March 1929) 102: 451.

[2] "Grand Slam," SCIENTECH Club of Indianapolis, *News Letter* (in Indiana State Library), 12 (2 March 1931): #9.

[3] See VI.2B, D.

[4] J. M. Henry, *A Short History of the Indiana State Highway Commission* (Indianapolis, 1926), chpt. II; Fountain County, "Commissioners Record," (September 1919-February 1920) 23: 431-32, 457-62, 475-76, 481, 502, 556, 572-77.

[5] William J. Titus to Franklin County Board of Commissioners (14 October 1922, 14 March 1923, 14 May 1923), W. J. T., "Office Memorandum" (8 November 1922), Titus to Earl Crawford (28 December 1922), Titus to Frank I. Siebert (12 June 1923), ISHC Correspondence on Roads, Indiana State Archives.

[6] See "Surviving Spans of Luten Design in Indiana," II: Appendix.

[7] Daniel B. Luten, *Hard Roads and Permanent Bridges* (Indianapolis, 1914), 3.

[8] In a letter to the author (3 June 1994), Daniel B. Luten, Jr., recounted what his father told him around 1925.

[9]Malverd A. Howe, *A Treatise on Arches* (New York, 1897), chpt. XI; District Court of the U. S. for New Jersey, *Daniel B. Luten vs. The Dover Construction Co.: Complainant's Record* (Trenton, New Jersey, 1913), 208.

[10]Fortunately the New Hampshire Department of Transportation kept both its own plan and the Luten design of the Main Street Bridge in Lancaster (1929). Special thanks go to three persons: (1) John W. Moore, Office Engineer (now retired) of the NHDOT's Bureau of Bridge Design, for figuring the amount of concrete required for the Luten design; (2) Marion Scott, Professor Emeritus of Civil Engineering, Purdue University, for calculating steel usage on the Luten design; (3) Lynne Emerson Monroe, Preservation Company, Kensington, N. H., for offering a warm and informed introduction to the bridge and its history. See also Lynne E. Monroe, "Main Street Bridge," HAER NH-17.

[11]The difference between the bids on the two plans is somewhat greater than these numbers suggest. Since the state paid all the expenses for developing its own plans from taxes, construction bids on it included no design costs. The bid on the spandrel-braced plan, on the other hand, included Luten's regular ten percent design fee.

[12]The BPR's approach to concrete bridge design, including arches, and the Bureau's research on those forms is discussed in more detail in VI.1D; VII.1-A.

[13]"Engineers Discuss Stresses in Multiple Arches," *ENR* (11 September 1924) 93: 440; Albin L. Gemeny and W. F. Hunter, "Loading Tests on a Reinforced Concrete Arch," *Public Roads* (December 1928) 9: 185-208; "Arch Study and Practice," *ENR* (21 March 1929) 102: 451-52.

[14]"New Knowledge of Arches," *ENR* (29 August 1929) 103: 318-19.

[15]"It Didn't Come," SCIENTECH, *News Letter* 12 (26 October 1931): #44.

[16]"SCIENTECHNICNACS," SCIENTECH, *News Letter* 12 (28 December 1931): #52; 14 (16 January 1933): #3; Luten Company President Dies," *Indianapolis Star* (11 July 1946).

[17]See I.3 and VI.2C. After leaving the ISHC, Moore served briefly as vice-president and general manager of the Bethlehem Spark Plug Company in Pennsylvania. Although one of the few civil or structural engineers in Indiana cross-licensed in other states, he was apparently not one to stay away from South Bend for long. Moore returned from Pennsylvania to join Charles W. Cole in an engineering partnership in 1922, an association that lasted for only a couple of years. He also served as Mishawaka's city civil engineer for a while in the mid-Twenties. "Engineering Partnership in South Bend," *ENR* (25 May 1922) 88: 889; *Indiana State Board of Registration for Professional Engineers and Land Surveyors* (Indianapolis, 1922-1930); South Bend city directories for 1904-1943.

[18]The following discussion of Moore's design work owes a great deal to several persons. First, William Lang and Bernard Feeney of Lang, Feeney and Associates, South Bend, the last of Moore's business associates and the holders of his bridge plans. I am deeply indebted to Messrs. Lang and Feeney for allowing me to see and to photocopy parts of those plans. They were gracious about my inevitable disruption of their business lives and weekend recreational time. Second, Maynard C. Schenk, Elkhart County Highway Engineer (now retired), helped me find my way to Lang, Feeney and Associates as well as through the county Engineering Division's records. His support was invaluable.

[19]I am especially grateful to Stephen Weintraut of Butler, Fairman & Seufert, Inc., for alerting me to this design feature in Elkhart #6085.

[20]County design is difficult to trace since the counties have not routinely retained their bridge plans. Clay and Hamilton Counties have a larger-than-average number of structures in their systems that may have been designed as continuous or rigidly framed.

[21]The ISHC provided the third example with its memorial bridge at Vincennes (Knox #690). See VII.2C.

[22]*ENR* (1 August 1940) 125: 91.

[23]McAnlis' appointment as the Fort Wayne City Engineer was reported in the *ENR*. But, while in the city, he did not list himself in the local directory among the civil, consulting, or mechanical engineers. He was always listed by residence and by employment as City Engineer. "Personals," *ENR* (10 January 1935) 114: 64; *Polk's Fort Wayne City Directory* for 1936, 1938, 1940.

[24]LaMotte Grover, "Long Continuous-Girder Bridge of Concrete in Kansas," *ENR* (25 January 1934) 112: 110-11; Alfred Benesch and Charles Morgan, "Designing Concrete Girder Bridges for Continuity," *ENR* (23 April 1936) 116: 604-6.

[25]The size of the hollow cells depended upon the size of the beam. The wider outer beams contained twenty-eight-inch wide cells. L. R. Need, "Hollow Concrete Piles Made by Centrifugal Process," *ENR* (10 March 1932) 108: 374-75; "Cellular Concrete Bridges Being Built in Washington," "Reducing Dead Weight," *ENR* (5 November 1936) 117: 637-38, 656; J. A. Dunford, "Record Rigid-Frame Bridge," *ENR* (24 June 1937) 118: 939-42; F. R. Easterday, "Concrete Box-Girders of Record Span," *ENR* (3 March 1938) 120: 339-42.

[26]Alan Gowans, *Styles and Types of North American Architecture: Social Function and Cultural Expression* (New York, 1992), 250-54.

[27]See the Elkhart and St. Joseph county sections of the inventory in Part Three for additional information on each bridge.

Epilogue

On the Road to Invisibility

The present-day demands of safety in highway traffic are of very recent growth, and it could hardly be expected that ... horse-and-buggy bridges which still survive would be altogether satisfactory in service under dense thirty-mile-an-hour motor-vehicle traffic. ...

It is the function of a bridge to carry a road and its traffic across a depression ... conveniently and without delay or slowdown. From the viewpoint of the passing traffic, the bridge does not exist, but is merely part of the road.
– *editors,* Engineering News-Record[1]

The often beautiful and sometimes serene St. Joseph River does not rank among the nation's mightiest waterways. At least in metaphor, though, it can become a significant divide, and William S. Moore's Bridge Street spans can carry visitors from one era to another. Travelers approaching the Elkhart bridge observe a functional monument with soaring columns. Strolling or motoring across its deck, one moves along a boulevard whose open-panel rails both contain the passage and frame the park beyond. Here is an updated and fitting encore for City-Beautiful design, a structural statement the landscape architect might laud and a contrast with what was yet to come.

A. From Eden to Elkhart: 1890-1942

Before the First World War, captivated community leaders encouraged architects to landscape the city with parks punctuated with naturalistically constructed landmarks. In Indiana, City-Beautiful experts like J. Clyde Power, George Kessler, Alonzo Hammond, William S. Moore, and Otto Clayton designed stone and concrete structures that complemented and enhanced a setting for travelers and neighbors. Travelers on horse-back or in carriages observed man-made monuments as they sauntered over them, and pedestrians absorbed their vistas from sidewalks and refuge-bays. From within the landscaped garden, neighbors and off-highway passersby might have been struck as much by the aesthetic statements the spans made as by their mundane function of passage from one river bank to another.

The City-Beautiful bridges were, by and large, pieces of functional sculpture in which at least as much attention went into their decor as into their structure. The technology of these concrete-steel spans took the country beyond timber, iron, and steel bridges by providing a workable metal skeleton upon which an architect could hang sheathing either in easily-sculpted concrete or in less flexible but decidedly natural cut stone. The metal frame probably mimicked the systems which Josef Melan or Fritz von Emperger had patented and which had appeared to take design an important step beyond the age of wood and iron. Appearances can deceive. Here the step did not carry structural design far beyond relying on iron or steel for most of the load-bearing.

THE RAIL TALE ... FIRST: THE CITY BEAUTIFUL WITH STONE AND BRICK.

(Marion #1705)

Thirtieth Street Bridge, Indianapolis (above); Main Street Bridge, Fort Wayne (below).

(Allen #527)

Technological change, especially the internal combustion revolution, began to undercut the sway of landscape architects over bridge design almost as soon as it became established. Besides dramatically altering how we transported goods and services between the two world wars, motor vehicles increasingly separated where we worked from where we resided, changed our associations, friendships, and activities, and transformed our leisure. By the end of the Twenties, Americans considered the automobile a necessity. Even the "Okies" who trekked from the Oklahoma dust-bowl to California in the midst of the Great Depression went by motor vehicle. In short, the city in which the poor walked, the affluent rode in horse-drawn buggies, and each used street, interurban, and steam rail service as supplementary transport was being transformed by the opening of the First World War and had virtually disappeared by the beginning of the Second. The landscape architect's planning, including his bridges, became artifacts of a lost culture.

Social change and political reform also shortened the landscape architect's reign. One group which enhanced its professional status in the early decades of the century – engineers – quickly and directly challenged the architect's vision of bridge design. As practitioners of applied science, engineers thought in terms of structural efficiency and focused on utility. They moved from concrete-steel, where the steel skeleton mostly carried and concrete mostly dressed a structure, to reinforced concrete, where steel and concrete were yoked together as a team in the carrying.

The engineers were not of one mind, however, about how science ought to be applied. Some worked out of the more entrepreneurial, empirical tradition inherited from the British which emphasized cost-effective invention sometimes intuitively discovered. Others preferred a science-centered tradition which the French had long promoted, that American colleges of engineering had incorporated, and which worked well in large-scale private and public organizations. The science-centered tradition insisted on systematic, step-by-step analysis which explained in quantitative terms how a structure worked, not just that it functioned. Whatever an engineer's professional persuasion might be, though, almost none would rank bridge decor equal to or before structural efficiency as an architect might do.

The formal education of scientists has not disparaged as much as it has ignored the aesthetic dimensions so central to the training of artists and humanists; engineers are typically less aesthetically lettered than architects. But, even without a course in "Elementary Aesthetics," the engineer who designs structures must make aesthetic judgments. Most have implicitly supported a structuralist aesthetics almost invariably in conflict with the preferences of landscape architects who took a hand at designing bridges in Indiana before 1920. Most engineers believed that the bridge's structural lines ought to govern its appearance: The line of duty became the line of beauty.

But persons from the two traditions of engineering and their accompanying employment patterns did not necessarily interpret structuralism in the same way. Daniel B. Luten best represented the entrepreneurial and empirical tradition in Indiana. To sell innovative designs that lacked scientific certification, Luten insisted on construction guarantees to the purchaser; and to keep paper guarantees from becoming financial and public relations liabilities, he favored experienced and skilled construction. A bridge should incorporate the most efficient structural technology, and it should present that technology with a style and craftsmanship that highlighted, even flattered, the structural streamlines. Antiquated technology should not be caked over with heavy makeup, as Luten thought many City-Beautiful architects had been wont to do. Instead the most innovative structural technology should be used and its features underlined to project a message of quality, of craftsmanship – and to advertise for its designer and builder.

THE RAIL TALE ...
SECOND: THE COUNTRY PRACTICAL WITH SCULPTED CONCRETE.

(Wayne #701)

Country Club Bridge, Indianapolis (left); South Side Bridge, Richmond (right).

The engineers out of the science-centered tradition who joined the federal-state highway alliance articulated a somewhat different structuralist aesthetics. Designers in the Indiana State Highway Commission [ISHC], for example, confronted some considerable constraints. First, the drive for detailed, standardized bridge planning assumed construction at the lowest common denominator. Second, design was a state monopoly and bridges, therefore, did not need a more styled appearance than the public minimally expected. Third, automobility set some limits on design and suggested others. Spiraling motor-vehicular speed, size, weight, and volume led logically to the construction of heavily built and easily widened deck bridges. Designers sought to pass motorists through an environment as unobtrusively as possible. The ISHC's key bridge designer in the Twenties and early Thirties, William J. Titus, responded to these constraints in an interesting and also predictable way. He successfully insisted on a dash of upper-level decor. He stylized the state's bridge rails, the only member that drivers and passengers likely noticed.

To return to Elkhart from our metaphorical journey takes us off the state highways and across William S. Moore's Bridge Street spans where the designer's aesthetic message resonates with the City-Beautiful movement even as the entrepreneurial engineer incorporated some of the latest structural innovations. Moore remained

THE RAIL TALE ... THIRD: THE COUNTRY PRACTICAL WITH DECORATIVE EFFECT IN CONCRETE.

(Franklin #35)

Inscribed and bush-hammered panels (above); ISHC pattern (below).

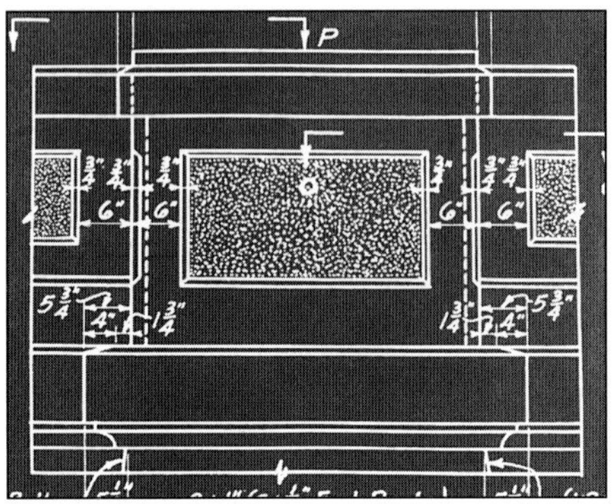

(Montgomery #184; INDOT)

committed to his profession and stayed ahead of the government designers within it. Indeed, he forecast a number of structural features of bridge design on postwar highways – spans of beams, continuous and sometimes framed, cellular or boxed.

B. The Trip to Erewhon: 1945 and After

When the Second World War ended, the federal-state highway alliance returned to its mission without the nagging presence of bridge-design challengers. Architects did not re-enter the fray, and private design engineers accommodated to the funding power and science-based practices of the government alliance. As automobility continued to turn the nation into a collection of ever larger, metropolitan, drive-by societies, the government engineers finally contributed what the automotive interests had begun to demand before and after the First World War – an interstate highway system of cross-country expressways.

Structural design of bridges also moved along. Around 1950, postwar designers added another significant feature to those Moore had included in his Bridge Street spans – prestressing. Here, instead of waiting for a load to generate stress, stressing was induced in structural members before or during construction. To prestress in concrete beams, manufacturers held high-strength steel rods in a stretched position while they poured concrete around the reinforcing and allowed the concrete to set. Then the externally induced tension was released. As the steel attempted to return to its original state, the reinforcing compressed the concrete.[2] The technique both improved the physical properties of the two materials and developed a closer working union between the concrete and the steel, thereby increasing the structural feasibility of longer-span beams and decreasing their cost. The precision, predictability, and economy prestressing introduced profoundly affected reinforced concrete construction without in most cases requiring any visible change in structural form.

Manufactured in large numbers to fairly precise specifications, factory-made prestressed beams suited federal-state bridge programs very well. Highly standardized beams could be shipped to a project site and erected rather speedily on abutments and piers prepared for them. Prestressing also made reinforced concrete beams competitive with steel ones and helped to reduce the range of skills required of bridge contractors. Turned back to function as they had in the modern iron age, contractors once again became builders of abutments and piers and erectors of manufactured superstructures.

Like their colleagues in other parts of the highway alliance, the ISHC's bridge engineers were technical persons trained at engineering institutions to do things the "correct" way. Along with structural efficiency, students had been taught to calculate cost-effectiveness, expedite traffic, value reliability, and consider administrative flow when making design decisions. Such factors controlled the engineer's preferences for the materials and forms proposed for particular bridge sites and generated off-the-shelf designs. With the advent of prestressing in the Fifties, the state started to order a large number of standardized, precast concrete beams just as it had begun by the late Thirties to buy large rolled steel I-beams for bridges.

As prestressing suited the structural preferences of the government's engineers, so automobility and the architectural fashions of the post-war era supported their aesthetic and political inclinations. Led by Walter Gropius, Ludwig Mies van der Rohe and others, the Thirties invasion of structuralism from Europe had evolved in the Forties and Fifties into Scientific Modernism, an architectural style that dominated many public works. The Scientific Modernists thought of theirs as an international style, beyond petty references to societal aspirations. Instead, they designed without extraneous motives and with the faith that whatever is scientific is good. The Modernists eschewed all ornament, emphasized twentieth-century materials and techniques, and, where possible, displayed working structure. Gropius had expressed the goal of the movement succinctly in 1922: "We want to create the clear organic bodies of buildings, naked and streaming forth from their inner laws, which affirm our world of machines, wires, and fast transport."[3]

With government and corporate leaders embracing Scientific Modernism in the post-Second World War period, utilitarian bridge-design fundamentalists operated in a quite favorable environment. All preferred easily assembled, simple shapes to complex and intricately organized ones. All preferred horizontal to vertical lines. All found virtue in unsheathed, simple, and clean surfaces rather than in coated, textured, or otherwise detailed ones. The government engineer had no difficulty pinning up the van der Rohe aphorism of "less is more" as a bridge-design slogan. In Indiana, the ISHC even dropped William J. Titus' token rail decorations shortly after the Second World War. It presented its bridges quite *au naturel*.

With automobility, government bureaucracy, and Scientific Modernism reinforcing each other, the bridge drifted gradually out of sight. By near mid-century, the traveler sped along a paved highway too fast to see a landmark safely and with a machine which, when out of control, could destroy parts of any landmark within reach. Where feasible, practical government engineers used standard bridge designs to build structures out of the motorist's reach and beyond view. The bridge melded into the federal-state highway by the outbreak of the Second World War, and Scientific Modernism helped postwar bridges stay tucked away. Undecorated and often even unsheathed, shorn of most vertical thrust and any suggestion of extra-utilitarian meaning, the bridge increasingly became aesthetically invisible. Even as the number of spans grew dramatically across the landscape, their low, repetitive, and non-descript profiles helped them to recede, even to disappear, from public view. Unless there is a substantial change in design consciousness, travelers over tomorrow's typical bridges will hardly notice them. Like Erewhon, these spans will seem to be "nowhere" even while they are under foot and wheel.

THE RAIL TALE ... LAST: ARRIVAL AT EREWHON.

(Howard #506)

Slab walls replace rails.

(Dubois #74)

NOTES

[1]"Bridges and Traffic Safety," *ENR* (2 July 1925) 95: 2-3.

[2]A second prestressing technique, post-tensioning, is also used with reinforced concrete beams. After box beams, for example, have been placed on a bridge, reinforcing can be passed through their hollows and tightened, stretching the reinforcing and compressing the already-set concrete. For a fuller introduction to prestressing, see Carl W. Condit, *American Building Art: The Twentieth Century* (New York, 1961), 187-94, 216-18.

[3]Alan Gowans, *Styles and Types of North American Architecture: Social Function and Cultural Expression* (New York, 1992), 260, 281-85; Stephen B. Goddard, *Getting There: The Epic Struggle between Road and Rail in the American Century* (New York, 1994), 197-98.

UNSOLICITED PUBLIC COMMENT

(Wayne #245)

A trip under Highway 66.

(Elkhart #387)

In memory of those who passed through the original rail at this spot.

(Clinton #138)

To rock like a Rolling Stone.

(Parke #208)

Virgil was here with his pencil in 1918.

PART THREE

An Inventory of Selected Concrete Spans

Hamilton #151

INVENTORY OF BRIDGES BY COUNTY

INRODUCTION

This inventory includes about a thousand concrete bridges built in Indiana before 1942 and still standing at the time of publication. All concrete structures known or believed to have been designed and constructed before 1919 – the era in which both design and construction underwent their greatest experimentation and development – are listed. For the Twenties and Thirties, new designers, new design forms or elements, and the structures of a major new builder – the Indiana State Highway Commission – are featured. Bridges built in these later decades which followed by then established design patterns are also included where there is something evidently noteworthy about them. At a minimum, a given structure may represent the scarce remnant of reinforced concrete bridge-building in the county or region before the Second World War. Where, on the other hand, a given design survives in a great many structures, mostly unaltered examples and those with secondary characteristics of special interest have often been singled out for inclusion.

The inventory remains a "work in progress" for several reasons. A number of bridges on roads re-routed or abandoned some years ago are now overgrown and officially forgotten in woods or between fields. Doubtless many bridges enclosed in the landscape, especially those situated on once-active Hoosier steam rail or interurban lines, have not yet been located. The cultural significance of many bridges, both within and beyond the inventory, is still to be determined. Until, for example, rail archives and the ninety-two sets of county records have been adequately consulted, some significant structures are likely unidentified or their importance under-recognized. Even after a decade and a half of research on Indiana's older bridges, results are, alas, incomplete.

☛ **537 Combs St Br** **NRC**
Tecumseh St, Ft. Wayne; Maumee R
Filled-spandrel arch (3) 1912
G Jaap, dsgr
H Tapp Const, bldr
<u>Note:</u> *100-ft center span; Melan system reinforcing; sandstone facing; high rounded cutwaters suggestive of refuge bays; neo-classically balustraded rails; designer criticized in ENR (24 June 1915); (I.4)*

The following line-by-line explanation should help to unpack a listing's necessarily compressed meaning:

☛ **First Line: Number, Name, And Significance**

A **bold** first line may provide three pieces of information about a bridge – its number, name, and rating of historical significance. County and state highway departments assign numbers to their bridges, numbers which are often stenciled on state bridges or their abutments and which are sometimes also found on county structures and road maps. Local officials and residents often gave the names of old nearby fords, mills, churches, roads and streets, or of significant property owners to a local bridge.

The historical assessment of a bridge is presented here in abbreviated form. First, inclusion within the inventory alone represents a minimal acknowledgement of significance. Only about two-thirds of the surviving concrete structures build before 1942 are listed. Second, the most significant structures – those considered eligible for listing on the National Register of Historic Places – have been designated as NRC. The ratings provided here are those of the author/surveyor. The Indiana Historic Bridge Committee, which advises the state Division of Historic Preservation and Archaeology (DHP&A), saw the proposed ratings in an earlier draft and approved the same with some suggested revisions. The following inventory incorporates those proposed revisions.

Undoubtedly more research could better connect some bridges with an important function or special event in community life or associate a span with a particular engineering or manufacturing achievement. Such findings would add structures to the inventory and elevate others to eligibility for listing on the National Register.

537 Combs St Br **NRC**
☛ Tecumseh St, Ft. Wayne; Maumee R
☛ Filled-spandrel arch (3) 1912
G Jaap, dsgr
H Tapp Const, bldr
<u>Note:</u> *100-ft center span; Melan system reinforcing; sandstone facing; high rounded cutwaters suggestive of refuge bays; neo-classically balustraded rails; designer criticized in ENR (24 June 1915); (I.4)*

☛ Second Line: On And Over

The second line (and its indented extensions) first indicates the facility which the bridge carries (usually a road, street, or railroad), often with a locational hint following a comma. After a semicolon, the listing identifies the river, road, or other feature the structure spans.

☛ Third Line(s): Form or Type, Number of Spans, And Date of Construction

The next line identifies the major structural form or type, notes the number of spans if more than one, and dates or estimates the date (c.19xx) of the structure. If the bridge contains other kinds of main or approach spans, they are identified in the same manner in a following line or two.

The occasional nameplate attached to a structure or information inscribed into the concrete or stone provide an easy means to date a structure. Written reports in the *Engineering News-Record* or local newspaper, engineering or construction company records, and county or state archives help to date bridges, too.

537 Combs St Br **NRC**
Tecumseh St, Ft. Wayne; Maumee R
Filled-spandrel arch (3) 1912
☛ G Jaap, dsgr
H Tapp Const, bldr
☛ <u>Note:</u> *100-ft center span; Melan system reinforcing; sandstone facing; high rounded cutwaters suggestive of refuge bays; neo-classically balustraded rails; designer criticized in ENR (24 June 1915); (I.4)*

☛ Next Line(s): Engineer/Designer and Builder

Additional lines are used where the engineer, designer, or builder are known or can be inferred (in brackets).

☛ <u>*Note:*</u> A Final Line

The <u>*Note*</u> line and its indented extensions call attention to unusual age, patented or inventive design, rare survivals, decoration, other special characteristics, as well as modifications of the original structure effected through repair and reconstruction. Of all structural elements, rails have been most subject to removal (rmvd), although a few have been replaced with copies or replicas (rpltd) of the originals. Textual discussion directly relevant to the structure is noted by chapter, section, and sometimes subsection (e.g., I.2B) within parentheses.

KEY AND ABBREVIATIONS

* = bypassed

Ave = avenue

Bh = branch

bldr = builder

blt = built

Blvd = boulevard

bm = beam

Br = bridge

bx = box

c19xx = about 19xx

Ck = creek

Co = county (in location line); company (in engineering, designing, and construction lines)

concr = concrete

Const = construction

Cont = contracting

CR = county route

Dh = ditch

Dn = drain

dsgr = designer

engnr = engineer

ENR = *Engineering News-Record*

Fk = fork

ft = foot/feet

in = inch(es)

nr(ly) = near(ly)

NRC = considered eligible for the National Register of Historic Places

Pd = pond

Pk = park

R = river

Rd = road

rmvd = original feature removed; replacement, if any, of different form or style

rpltd = original feature replicated or restored to original form or style

RR = railroad

SR = state route

St = saint or street

Twp = township

US = federal highway

w19xx = widened, or widening in, 19xx

Additional Information Available

The Historic American Building Survey and the Historic American Engineering Record have documented a few of Indiana's bridges, some of which have then found their way to the National Register of Historic Places. Thomas Slade summarized the national inventories in *Historic American Building Survey in Indiana* (Bloomington, Ind., 1983), although the book covers federal activities only through 1979. To consult or to order reproductions of HABS or HAER documents, contact the Division of Prints and Photographs, Library of Congress, Washington, DC 20540.

Since 1975, the state's DHP&A has funded inventories of historic sites and structures in most Indiana counties. These surveys, almost all of which the Historic Landmarks Foundation of Indiana (HLFI) conducted, include a number of bridges. Copies of many of the published county interim reports can be obtained from HLFI, 340 W. Michigan St., Indianapolis, IN 46202-3204. The DHP&A (W274, 402 W. Washington St., Indianapolis, IN 46204) retains the original county and bridge survey reports.

Additional Information Sought

Please correct, supplement, and update the information presented in this inventory. Send observations, written materials (including newspaper articles), and visual information (photographs, postcards) to J. L. Cooper, 629 E. Seminary St., Greencastle, IN 46135. Your help with improving the understanding of Indiana's engineering and industrial heritage is very much appreciated.

ADAMS [1]

42
CR 300W, nr Antioch Ch; Holthouse Dh
Girder/floor bms c1915

131
CR 500S, W of CR 200E; Gates Dh
Girder c1920
Note: skewed

1822
SR #218, N of Linn Grove; Wabash R
Filled-spandrel arch (4) 1941
ISHC design
Note: 270 ft long; w1986 with cantilevered slab deck rmvd rails

3414
SR #218, E of SR #1 16; Wabash R overflow
Continuous slab (3) 1941
ISHC design
J R Gates, bldr

ALLEN [2]

330
W Jefferson Blvd or Old US #24, Ft Wayne; St Marys R
Filled-spandrel arch (3) 1934
ISHC design
Note: skewed; 271 ft long; brackets carry sidewalks; spandrel pilaster lines extended into rail posts

525 Fourth St Br
Fourth St, Ft Wayne; Spy Run Ck
Filled-spandrel arch 1921
A W Grosvenor, dsgr
H W Tapp Const, bldr
Note: Melan system reinforcing; open-arch rail panels; lights rmvd

(Allen #527)

For Fort Wayne's Main Street Bridge, A. W. Grosvenor set the concrete of the spandrel walls off against the limestone-capped and posted brick rails. Renovation has obliterated this quite special element of the structure's decor.

527 W Main St Br NRC
Main St, Ft Wayne; St Mary R
Filled-spandrel arch (3) 1921
A W Grosvenor, dsgr
H W Tapp Const, bldr
Note: skewed; 290 ft long; Melan system reinforcement; rounded, raised & capped cutwaters; stone rail posts & copings; brick rail panels; (Epilogue.A)

529 Columbia St Br NRC
Columbia St, Ft Wayne; Maumee R
Filled-spandrel arch (4) 1926
A W Grosvenor, dsgr
I E Smith Const, bldr
Note: skewed; 416 ft long; Melan system reinforcing; paneled & emblazoned pier pilasters originally extended into rail posts; 1972 cantilevered deck rmvd posts, open-arch rails & light standards

537 Combs St Br NRC
Tecumseh St, Ft Wayne; Maumee R
Filled-spandrel arch (3) 1912
G Jaap, dsgr
Tapp Const, bldr
Note: 100-ft center span; Melan system reinforcing; sandstone facing; high rounded cutwaters suggestive of refuge bays; neoclassically balustraded rails; designer criticized in ENR *(24 June 1915); (I.4)*

538 Parnell Ave Br NRC
Parnell Ave, Ft Wayne; St Joseph R
Open-spandrel arch (2) 1927
A W Grosvenor, dsgr
I E Smith, bldr
Note: four ribs for each 105 ft span; two-tier ring reinforcing with stirrups; paneled pier pilaster; 1993 deck carries sidewalks on floor bms extended beyond spandrel columns; new open-arch rail panels & reproduction lights

(Allen #527)

539 Tennessee Ave Br NRC
Tennessee Ave, Ft Wayne; St Joseph R
Filled-spandrel arch (2) 1912
A W Grosvenor, dsgr
Lafayette Engrg, bldr
Note: 105-ft spans; Melan system reinforcing; flemish-bond brick facing on spandrel walls & rails; limestone trim on rings, copings & rail posts; (I.4)

546 Evard Br NRC
State Blvd, Ft Wayne; Spy Run Ck
T-beam 1927
A W Grosvenor & O Darling, dsgrs
H W Tapp, bldr
Note: "similar to" ISHC design; concr light standards rmvd

(Allen #538)

Not all of Allen County's bridge renovation, however, has such mixed results. In 1970, for example, the county aesthetically-decapitated Grosvenor's Parnell Avenue Bridge in Fort Wayne (above). In the 1990s, though, the county assigned another rehabilitation of the bridge to a consulting engineer who recaptured the spirit of Grosvenor's City-Beautiful style (below).

(Allen #538)

547 State Blvd Br
State Blvd, Ft Wayne, or CR 225N; St Joseph R
Filled-spandrel arch (3) 1919
A W Grosvenor, dsgr
H Tapp Const, bldr
Note: *293 ft long; Melan system reinforcing; w1982 with prestressed bx bms raised on extended triangular cutwaters; open-arch paneled rails & lights gone*

BARTHOLOMEW [3]

50
CR 850N or #218, nr Decatur Co line; Duck Ck
Girder/floor bms (2) c1915

55 Carmichael Ford **NRC**
CR 775E or #229; Haw Ck
Filled-spandrel arch (2) 1908
[W H Rights, dsgr]
Harmon Bros, bldrs
Note: *heavy, flat rings; triangular cutwaters*

56 **NRC**
CR 900N or #72, nr Haw Ck Ch; Haw Ck
Filled-spandrel arch 1909
Harmon bros, bldrs
Note: *heavy, flat ring*

165
CR 600W, S of Georgetown Rd; Wolf Ck, Bh
Slab c1915
Note: *skewed; long slab span (20 ft)*

193
CR 700S or #10, W of I-65; White Ck, E Fk
Filled-spandrel arch c1907
Note: *flat ring*

289 **NRC**
7th St, Columbus; Haw Ck
Filled-spandrel arch (3) 1913
E Hughes, dsgr
Hackedorn Cont, bldr
Note: *brackets carry sidewalks beyond spandrels; balustrated rails replaced with ISHC-design panels (1928)*

1777
US #31, SE of Azalia; Morton Hall Dh
T-beam c1936
ISHC design
Note: *skewed*

BENTON [4]

8
CR 300W, NE of Earl Park; Sugar Ck
T-beam c1915
Note: *long bm span (42 ft); outer bms wider than inner ones; rails extended over wingwalls*

9 **NRC**
CR 400W, NE of Earl Park; Sugar Ck
T-beam c1916
Note: *long bm span (42 ft); rails extended over wingwalls*

10 **NRC**
CR 500W, N of Earl Park; Sugar Ck
T-beam c1925
Note: *long bm span (42 ft); rails extended over wingwalls*

15
CR 200W; Sugar Ck
T-beam c1916
Note: *rails extended over wingwalls*

16
CR 100W; Sugar Ck, Bh
T-beam c1917
Note: *outer bms wider than inner ones; rails extended over wingwalls*

(Bartholomew #55)

H. W. Rights, who later served briefly as an engineer with the ISHC, probably designed the Carmichael Ford Bridge when Bartholomew County surveyor and engineer.

(Benton #10)

95
CR 200S; Big Pine Ck, Bh
T-beam c1926
Note: *skewed*

BLACKFORD [5]

5867
SR #18, nr Matamoras; Salamonie R
Filled-spandrel arch (3) 1924
ISHC design
F N Hoffman, bldr
Note: *226 ft long; spandrels, deck & rails rmvd when w1968 by 10 ft on each side*

BOONE [6]

NRC
CCC&StL (Big 4) RR, nr Zionsville; Eagle Ck
Filled-spandrel arch (5) 1919
Note: *heavy semicircular rings built for two-track, high-speed rail use; raised pier pilasters*

Vandalia RR; Camp St, Lebanon
Filled-spandrel arch 1916
Note: *link in Pa RR system plan for direct & high-speed line with grade separations between Indy & Chicago; elliptical ring*

NRC
Vandalia RR; Sugar Ck, nr Mechanicsburg
Filled-spandrel arch (3) 1916
Note: *skewed; elliptical rings; paneled pier pilasters; long wingwalls; (see note above)*

51
E Co Line Rd, N of SR #47; Big Eagle Ck
Slab c1920
Note: *skewed; long slab span (23 ft)*

139
CR 875W, SW of Dover; Walnut Fk
Slab c1915
Note: *skewed; long slab span (20 ft)*

164
Old Union Rd or CR 550S, NW of New Brunswick; Big Walnut Ck, W Fk
T-beam c1937
Note: *like state design, rails sit on outer bm flange; skewed*

219
High School Ave, Zionsville; Eagle Ck
Filled-spandrel arch (6) c1935
Note: *steel plates frame rings*

(Boone #309)

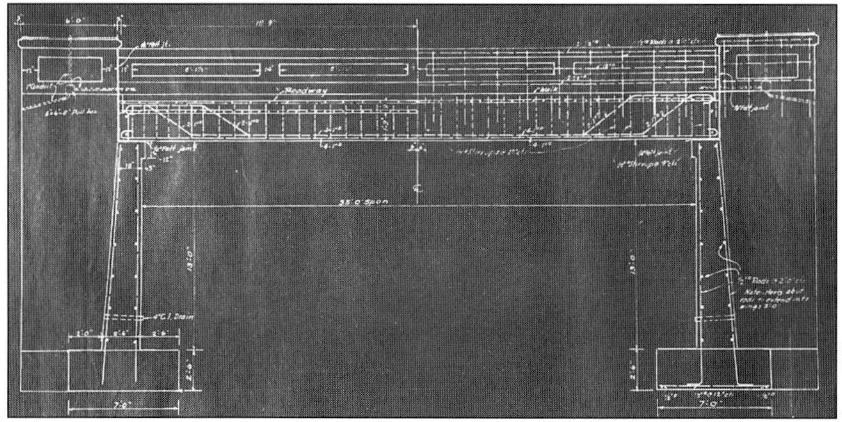

(Boone #309)

Boone County's plans for the East Fordice Street Bridge illustrate a rather conventional pattern for reinforcing T-beams. The longitudinal reinforcement runs near the bottom of the beam until some of the rods are selectively transposed to the upper areas as they approach the abutments (above). A transverse view through the beams (below) shows the longitudinal rods (presented as two rows of four dots per beam) bundled in stirrups which reach towards the tops of the beams where they are integrated into the reinforcing system of the flanges.

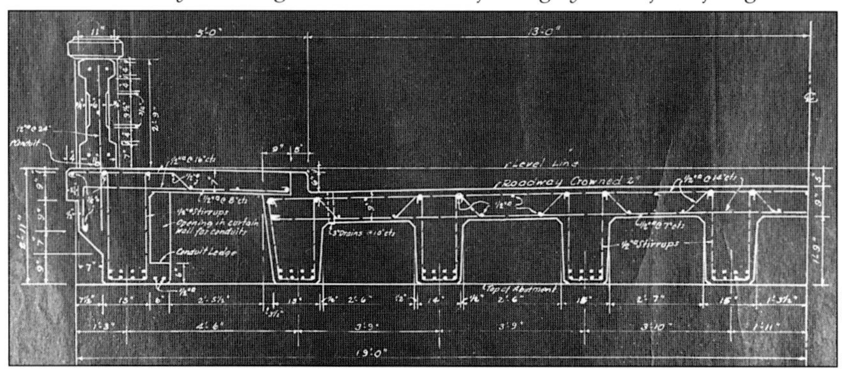

(Boone #309)

300
Patterson St, Lebanon; Prairie Ck
Filled-spandrel arch c1915
Note: *rails with open-diamond panels rmvd; new spandrel wall on E*

301
Lafayette Ave, Lebanon; Prairie Ck
Beam c1930
Note: *long bm span (44 ft); rails carried on outer bm flange*

303
N Jameson St, Lebanon; Prairie Ck
Filled-spandrel arch c1920
Note: *skewed; heavy semicircular ring strengthened with metal plate; spandrels coated & rails rmvd (1970)*

306 **NRC**
Royal St, Lebanon; Prairie Ck
Filled-spandrel arch c1920
Note: *skewed; flat ring; original neoclassically balustraded rail on N; house built over span; (IV)*

307
N Meridian St, Lebanon; Prairie Ck
Filled-spandrel arch c1920
Note: *w19xx with new arch differently skewed than original on each side*

308
N East St, Lebanon; Prairie Ck
Filled-spandrel arch c1930
Note: *flat ring; open rail panels now filled*

309 **NRC**
E Fordice St, Lebanon; Prairie Ck
T-beam c1930
Note: *rails sit on outer bm flange*

310
Park St, Lebanon; Prairie Ck
Beam c1930
Note: *skewed; rails sit on outer bm flange*

3139
US #52 [SE-bound lanes], NE of Thorntown; Sugar Ck
Filled-spandrel arch (4) c1941
ISHC design
Note: *249 ft long; flat rings; unsymmetrical outer spans; w1985 with cantilevered slab deck rmvd rails*

BROWN [7]

493
SR #46, Nashville; Salt Ck, N Fk
Open-spandrel arch 1929
T-beam (2)
ISHC design
J C Ellis, bldr
Note: *235 ft long; 150-ft center, 2-rib span; after w1974, bms carry approach span loads; spandrel columns, arched walls & rails with open-arch panels & bush-hammered posts all rmvd*

CARROLL [8]

19 S Moore Br **NRC**
CR 500W, S Co line; Wildcat Ck, Mid Fk
Filled-spandrel arch (2) c1916
Note: *spandrel walls decorated like rough-cut stone; keystone emblem*

132 Carrollton Br **NRC**
Carrollton Rd; Wabash R
Filled-spandrel arch (6) 1927
D B Luten, dsgr
Nat Concr, bldr
Note: *615 ft long; rings spring from high on piers; deck extended on brackets beyond spandrels; balustraded rails; (cover)*

(Carroll #19)

A folk, country version of City-Beautiful decor.

142 Lockport Br **NRC**
CR 275W, Lockport; Wabash R
Continuous T-beam (8) 1938
C McAnlis, dsgr
H L Maddocks, bldr
Note: *longest bm br (635 ft); among longest center spans (95 ft); variable-depth bms anchored at center pier & on rollers or shoes over other piers & at abutments; flared, triangular cutwaters & pier towers; gently arced streamlines inscribed on outer bms; (IX.2B)*

357
US #421, S Washington St, Delphi; Deer Ck
Filled-spandrel arch (2) 1936
ISHC design
H Tharp, bldr
Note: *skewed; unsymmetrical rings; sidewalks carried on brackets*

1788 **NRC**
N St, Delphi; S Washington St or US #421
Rigid frame slab 1935
ISHC design
H Tharp, bldr
Note: *premier example of state framed design; skewed; stairway between streets; (VIII.2B)*

CASS [9]

113* Zimmerman Br **NRC**
CR 600E; Wabash R
Filled-spandrel arch (4) 1918
D B Luten, dsgr
Nat Concr, bldr
Note: *linear rings; slim, high, cone-capped piers with raised pilasters; neoclassical rail balusters gone*

(Cass #123)

The Lewisburg Bridge as it looked at completion. Luten's bush-hammering or roughening of the spandrel walls created quite a contrast in color and texture to his polished arch rings.

(Cass #4178)

Before it was "renovated," the Market Street Bridge offered Logansport a taste of the City Beautiful.

123 Lewisburg Br **NRC**
CR 825E; Wabash R
Filled-spandrel arch (5) 1913
D B Luten, dsgr
Nat Concr, bldr
Note: *453 ft long; rings spring from high on piers; spandrel walls built separately from copings, allowing patented savings on wall construction; bush-hammered spandrels; neoclassical rail balusters rmvd; (II.4A)*

162
E Co Line Rd; Deer Ck, S Fk
Filled-spandrel arch c1917

3841
Michigan Ave, Logansport, or SR #25; Horney Ck
Filled-spandrel arch c1918
Note: *paneled abutment pilasters & balustraded rails*

4177 Third St Br **NRC**
Third St, Logansport, or SR #17; Eel R
Filled-spandrel arch (4) 1918
D B Luten, dsgr
Nat Concr, bldr
Note: *375 ft long; bush-hammered walls & pilasters; balustraded rail panels & inscribed posts; (II.4A)*

4178 Market St Br **NRC**
Market St, Logansport, or SR #24; Eel R
Filled-spandrel arch (4) 1930
J H Chase, dsgr
Nat Concr, bldr
Note: *441 ft long; long inner spans (110 ft); skewed; Melan system reinforcing; sidewalks carried on brackets; 1986 deck rmvd open-oval rail panels, inscribed posts & lampposts; vertical highlighting of pier towers partly hidden*

CLARK [10]

1343
US #31, Speed; Muddy Fk & St Joe Rd
Filled-spandrel arch (3) c1932
ISHC design
Note: *skewed; w1984 with cantilevered slab deck rmvd rails*

CLAY [11]

CR 6N, E of Perth; CCC&StL (Big 4) RR
Continuous T-beam (6) 1908
CCC&StL RR design
Note: *narrow bm stems & wide flanges*

CR 7N, SW of Lena; CCC&StL (Big 4) RR
Slab c1908
CCC&StL RR design
Note: *deep; under fill*

CR 13N, S of Lena; CCC&StL (Big 4) RR
Continuous T-beam (5) 1907
CCC&StL RR design
Note: *narrow bm stems & wide flanges*

1
CR 6N, NW of Brazil; Otter Ck
[Continuous] beam (2) c1920

12
CR 6W, W of Brazil; Waterworks Ck
[Continuous] beam (2) c1915

22 **NRC**
CR 4E, nr E Co line; Gun Ck
Girder 1915
G Sheehan, dsgr
McCullough & Londine, bldr

23
CR 16S or E Co Line Rd; Gun Ck
Girder c1915
Note: *deck tied into top of girders, leaving most load-bearing below rather than above roadway; slab deck; low rail rmvd on E*

26
CR 7N, N of US #40; Van Buren Ck
Filled-spandrel arch c1925

38
CR 9N, W of Shady Lane; Croys Ck
Girder c1915
Note: *slab deck*

(Clay #22)

50
CR 9S, SE of Staunton; Little Birch Ck
Girder c1915
Note: *slab deck tied into girders just below their middle*

55
CR 30N, N of Cloverland; Sulphur Ck
Girder c1915
Note: *slab deck*

84
CR 14S, N of Wesley Ch; Croys Ck
[Continuous] beam (2) c1920

96 **NRC**
CR 14E, SE of Hirt Cor; Tillman Ck
Girder 1915
G Sheehan, dsgr
McCullough & Londine, bldr
Note: *slab deck*

109
CR 51W, W of Zion Ch; Clear Bh
Girder c1910
[B F, C H, E B H, & K R, bldrs]
Note: *slab deck*

116
CR 27W, W of SR #59; Crooked Ck
[Continuous] beam (2) c1923

122 **NRC**
CR 51S, N of SR #46; Big Slough Dh
[Continuous] girder (2) c1910
Note: *few continuous girders; skewed; slab deck*

128
CR 47W, SW of Saline City; Birch Ck, Bh
[Continuous] beam (2) c1916

139 **NRC**
CR 76S, nr S Co line; Sulphur Ck
[Continuous] slab-girder (2) c1910
Note: *looks like small girders assist slab deck in carrying; spans angled up to pier; (VIII.2C)*

145
CR 34E, N of Ashboro; Birch Ck, Bh
Slab-girder c1916
Note: *looks like small girders assist slab deck in carrying*

157
CR 54S, S of SR #42; Knob Ck
Beam c1916
Note: *very modest waterway clearance*

171
CR 79S, NW of Old Hill; Big Slough Ck
Slab c1920
Note: *long slab span (20 ft)*

176
CR 61W, W of Old Hill; Big Slough Ck, Bh
Filled-spandrel arch c1915
Note: *skewed*

199
CR 97S, N of SR #246; Lafferty Dh, Bh
Slab c1913
Note: *skewed; long slab span (23 ft)*

211 **NRC**
CR 65W, N of Clay City; Lick Ck
Girder 1911
G Sheehan, dsgr
L Gibben & Son, bldr
Note: *oldest known girder structure; slab deck*

245
CR 129S, E of Buchanan Cor; Wabash & Erie Canal, Bh
Girder 1928
Hilbert Ol_nn, bldr
Note: *skewed; slab deck*

248
CR 125S, N of Howesville; Guirl Dh, Bh
Girder 1921
Note: *slab deck*

249
CR 133S, N of Buchanan Cor; Wabash & Erie Canal, Bh
Girder c1913

260
CR 106S, NE of Brunswick; White Oak Ck
[Continuous] beam (2) c1913

267
CR 118S, nr E & S Co lines; Pond Ck
Girder c1915

273
CR 121E, nr S Co line; Wabash & Erie Canal
Girder c1915
Note: *slab deck*

333 **NRC**
CR 84S, N of Clay City; Connelly Dh, Bh
Girder 1916
G Sheehan, dsgr
Lasell & Gibbons, bldr
Note: *slab deck*

634
CR 68S, S of Fatch Ch; Turkey Ck
Slab c1930
Note: *long slab span (20 ft)*

744
CR 113W, NW of Howesville; Lateral #4
Girder c1915
Note: *slab deck*

1315
SR #46, W of Bowling Green; Killion Dh
T-beam (4) 1937
ISHC design
Note: *157 ft long*

CLINTON [12]

19
CR 600W, S Co line; Campbells Run
Filled-spandrel arch (2) c1917
Note: *rough pebble finish on walls; rails largely gone*

21

CR 750N, SW of Rossville; Cripe Run
Filled-spandrel arch c1917
Note: *heavy, flat ring; pipe rail*

26
CR 400W, SW of Mattix Cor; Wildcat Ck, S Fk
Filled-spandrel arch c1915
Note: *100-ft span; heavy, flat ring; large entrance rail posts, fill & spandrels gone*

31
CR 580W, S of Hamilton; Kilmore Ck
Filled-spandrel arch (2) c1917
Note: *heavy, flat rings; rails rmvd*

32
CR 580W, N of CR 200N; Wildcat Ck, S Fk
Filled-spandrel arch (2) c1917
Note: *heavy, flat rings; low rails*

36 **NRC**
CR 950W, SW of Mulberry; Wildcat Ck, S Fk
Filled-spandrel arch (2) c1917
Note: *heavy, flat rings; low rails*

46
Gasline Rd or CR 500W, N of CR 200N; Wildcat Ck, S Fk
Filled-spandrel arch (2) c1916
Note: *heavy, flat rings; rails rmvd*

53
CR 600W, N of CR 200N; Wildcat Ck, S Fk
Filled-spandrel arch (2) c1917
Note: heavy, flat rings; low rails

55 **NRC**
CR 250W, S of CR 400N; Kilmore Ck
Filled-spandrel arch (2) c1917
Note: heavy, flat rings; triangular cutwater; low rails

66
CR 50E, S of CR 450N; Kilmore Ck
Filled-spandrel arch (2) c1917
Note: heavy, flat rings

70 Peters Br
CR 400E, N of CR 400N; Kilmore Ck
Filled-to-open spandrel arch (2) 1906
Kelleher & Slipher, bldr
Note: heavy, flat rings; fill & spandrels rmvd; transformed (1980) into open-spandrel arch with cantilevered slab deck

74
Michigantown Rd, nr Michigantown; Wildcat Ck, S Fk
Filled-spandrel arch c1928
Note: gentle rise to ring

124
CR 580S, W of US #421; Sugar Ck
Filled-spandrel arch c1924
Note: skewed; gentle rise to ring

(Clinton #70)

Clinton County's engineers have not wasted their arches. They have widened a number and transformed others. With Peters Bridge, for example, the county removed the walls and fill of a filled-spandrel structure and turned it into an open-spandrel arch.

(Clinton #70)

131 **NRC**
Scotland Rd or CR 680S, E of CR 400E; Davis Dh
Spandrel-braced arch c1928
[D B Luten, dsgr]
Note: two solid, semicircular ribs; cantilevered slab deck; (IX.1B)

138 **NRC**
CR 100W, S of CR 500S; Brush Ck
Slab c1931
Note: skewed; long slab span (27 ft); (Epilogue.B)

160
CR 150W, NW of Frankfort; Wildcat Ck, S Fk
Filled-spandrel arch (2) c1915
Note: heavy, flat rings; low rails

501
Kyger St, Frankfort; Prairie Ck
Filled-spandrel arch c1925
Note: heavy, flat ring; w19xx to carry walk on S; low rails

502
Green St, Frankfort; Prairie Ck
Filled-spandrel arch c1923
Note: heavy, flat ring

503
Clay St, Frankfort; Prairie Ck
Filled-spandrel arch c1924
Note: skewed; heavy, flat ring

505 **NRC**
E Washington St, Frankfort; Prairie Ck
Filled-spandrel arch c1915
Note: skewed; heavy, flat ring; spandrel walls decorated like stone; stone wings; rails probably not original

507
South St, Frankfort; Prairie Ck
Filled-to-open spandrel arch c1915
Note: heavy, flat ring; balustraded rails, fill & spandrels rmvd when transformed into open spandrel (1989)

508 **NRC**
Armstrong St, Frankfort; Prairie Ck
Filled-spandrel arch c1927
Note: heavy, flat ring; brick rails with diamond-shaped stone in center of panels

509 **NRC**
S Clay St, Frankfort; Prairie Ck
Filled-spandrel arch c1917
Note: skewed; flat ring; open-arch rail panels

1807
SR #26, W of Rossville; Campbells Run
Filled-spandrel arch c1937
ISHC design
Note: skewed

CRAWFORD [13]

1457
SR #37, English; Camp Fk Ck
Slab (3) 1935
ISHC design
Note: skewed; long slab span (24 ft); pipe rail

DAVIESS [14]

154 **NRC**
Old US #50 or CR 50S, SE of Montgomery; Prairie Ck, S Fk
T-beam 1923
ISHC design
J F Hopkins & Son, bldr
Note: early extant state-design bm br; skewed

1379
SR #57, N of Elnora; Vertrees Dh
T-beam c1933
ISHC design
Note: skewed

DEARBORN [15]

NRC
CCC&StL (Big 4) RR; Tanners Ck, W Fk, & Kondradi Rd or CR #136
Filled-spandrel arch (3) 1930
CCC&StL RR design
Moore-Mansfield, bldr
Note: *skewed; heavy, semicircular rings; heavy, rounded, low cutwaters; raised & paneled pier towers; tall & long wingwall*

Bonnell Rd; Tanners Ck, W Fk, Bh & Burtzelbach Rd [old CCC&StL (Big 4) RR spur]
Continuous T-beam (3) 1916
CCC&StL RR design
Note: *skewed*

CCC&StL (Big 4) RR; Tanners Ck, W Fk, & Bonnell Rd, 1 mi W of Guilford
Filled-spandrel arch (3) 1930
CCC&StL RR design
Moore-Mansfield, bldr
Note: *heavy, semicircular rings; heavy, rounded & low cutwaters; raised & paneled pier towers*

NRC
CCC&StL (Big 4) RR; Tanners Ck, W Fk, & Bonnell Rd, Guilford
Filled-spandrel arch (3) 1929
CCC&StL RR design
Moore-Mansfield, bldr
Note: *skewed; 228 ft long; heavy, semicircular rings; heavy, rounded & low cutwaters; raised & paneled pier towers*

7 Peter Ck Br
Spangler Rd or CR #8, W of Dillsboro; Peter Ck
Filled-spandrel arch c1920

11
Sangamaw Rd, nr Caesar Ck; Boyd Bh
Girder/floor bms c1915
Note: *skewed*

12
Cutter Rd, SW of Farmers Retreat; Laughery Ck, Bh
T-beam c1915

20 Chesterville Br NRC
Chesterville Rd, W of Stitts Hill; Allen Bh
Filled-spandrel arch c1910
J M & D B, bldrs
Note: *streambed floor; flat ring*

(Dearborn - CCC&StL RR, Bonnell Rd)

(Dearborn #20)

Pavement ties help to sustain the Chesterville Bridge's very flat arch ring.

24
Cold Spring Rd, nr Cold Springs; S Hogan Ck, Lee Bh
Filled-spandrel arch c1907
Note: *skewed; flat ring*

34 Short Ridge Br
Short Ridge Rd or CR #16, E of Mt Sinai; N Hogan Ck, Bh
T-beam c1918
Note: *skewed*

48
Bonnell or Konradi Rd, or CR #132, NW of Bonnell; Tanners Ck, W Fk, Bh
Girder c1930
Note: *slight arch to girder tops; sits on high stone abutments of previous br*

53 NRC
Kuebel Rd, NW of Guilford; Fox Run
Girder 1919
E O Kennedy, bldr
Note: *(II.3D)*

54
Greenbriar Rd, nr Guilford; Tanners Ck, W Fk, Bh
Filled-spandrel arch c1920
Note: *skewed*

62
Bonnell Rd, W of Guilford; Tanners Ck, W Fk
Girder c1905

77 Wolluing Rd Br NRC
Wolluing Rd or CR #154, E of Weisburg; Taylor Ck
Girder/floor bms 1917
[J W B C & C C P S, bldrs]
Note: *skewed; (II.3D)*

82
Hyland Lane, SE of St Leon; Ennis Ck
Filled-spandrel arch c1916
Note: *semicircular ring*

102 Taylor Ck Br
N Dearborn Rd, E of Hubbells Cor; Taylor Ck
Filled-spandrel arch c1910
Note: *skewed; w1972 on N with steel plates placed under ring*

135
Weisburg Rd or CR #154, Weisburg; Tanners Ck, W Fk
Continuous T-beam (2) c1910
Note: *skewed; rails rmvd*

3032 **NRC**
SR #46, E of Lawrenceville; Tanners Ck, E Fk
Filled-spandrel arch (3) 1937
ISHC design
Note: *261 ft long; close to semicircular ring in center span with three-centered rings for outer spans*

DECATUR [16]

CCC&StL (Big 4) RR; CR 850E, New Point
Filled-spandrel arch 1906
CCC&StL RR design
A J Yawger, bldr
Note: *blt to help improve rail profile; semicircular ring under considerable fill; old rails used for reinforcing in many overpasses; described in* ENR *(16 May 1907)*

2
Michigan Rd or Old US #421, NE of Adams; Clifty Ck
Filled-spandrel arch (2) c1935
ISHC design
Note: *204 ft long; unsymmetrical rings*

20 **NRC**
CR 650N or #222, N of St Paul; Mill Ck
Girder/floor bms c1920
Note: *long girder span (50 ft)*

25
CR 525W, nr Adams; Clifty Ck
Girder/floor bms (2) c1925

41 Clifty Ck Br **NRC**
CR 400N or #194, nr Adams; Clifty Ck
Filled-spandrel tied arch 1904
D B Luten, dsgr
Nat Br Co, bldr
Note: *among Luten's oldest; linear ring; cambered copings; design & construction discussed in* ENR *(28 June 1906)*

(Decatur #41)

(Decatur #41)

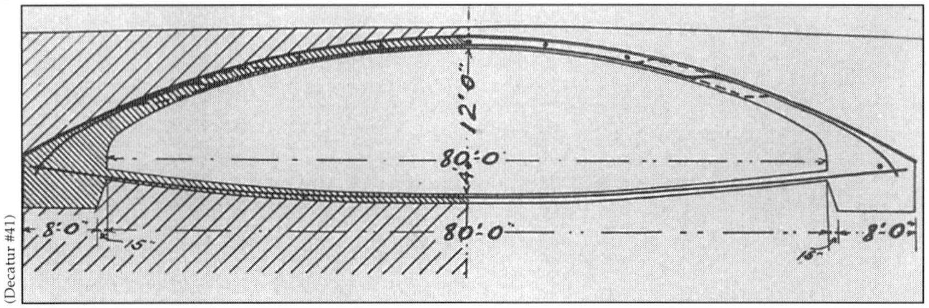

The Clifty Creek Bridge shortly after completion. A Luten-design tied arch.

53 **NRC**
CR 150N or #174, NE of Greensburg; Sand Ck
Slab (2) c1913
[W H L, bldr]
Note: *nicely-appointed country structure*

73
CR 150N or #180, SE of St Maurice; Salt Ck, Fk
Filled-spandrel arch c1910
Note: *spandrel wall & rails stepped up with ring to crown*

95
CR 300S or #42, S of Greensburg; Sand Ck
Girder/floor bms c1925
Note: *long girder span (49 ft)*

119 **NRC**
CR 700W or #17, N of SR #46; Clifty Ck, Fall Fk
Filled-spandrel arch (3) 1904
Kessing-Thompson, bldr
Note: *short spans with flat rings*

120
CR 550W or #29, N of where SR #3 & #46 join; Clifty Ck, Fall Fk
Slab c1920
Note: *long slab span (21 ft)*

129
CR 850W or #15, S of SR #46; Clifty Ck, Fall Fk
Filled-spandrel arch c1920
Note: *heavy, flat ring*

141
CR 1000W or #7, S of SR #46; Clifty Ck, Fall Fk
Filled-spandrel arch c1923
Note: *heavy, flat ring; deck rides over ring at crown*

943 **NRC**
SR #46, E of Hartsville; Clifty Ck, Mid Fk
Filled-spandrel arch (2) 1930
ISHC design
Watt Constr, bldr
Note: *designed for fill above spandrel walls; few of such state multi-spans; skewed; low, flat rings*

DEKALB [17]

NRC
Old Ft Wayne & NW RR, W of Altona, [under B&O RR]; Little Cedar Ck
Slab c1912
W J Watson, dsgr
Ind Concr, bldr
Note: *inverted siphon culvert blt to thread interurban Ry between stream below & steam RR above*

2 **NRC**
S Co Line Rd, E of SR #327; Cedar Ck, Bh
Beam (3) c1930
Note: *113 ft long*

DELAWARE [18]

503 Washington St Br **NRC**
W Washington St, Muncie; White R, W Fk
Filled-spandrel arch (4) 1930
C A Armintraut, dsgr
Burk Constr, bldr
Note: *skewed; 281 ft long; gentle rise to rings*

514
Broadway Ave, Muncie; White R, W Fk
Filled-spandrel arch (4) 1936
ISHC design
Note: *305 ft long; sidewalks carried on brackets; walks & rails rmvd*

742
SR #28 or #67; Mississinewa R
Filled-spandrel arch (4) c1928
ISHC design
Note: *270 ft long; gentle rise to rings; w1957 rmvd deck & rails; substitute rails have open-arc rectangular panels*

(Dekalb - FtW& NR RR)

Sandwiching an interburan line between the stream below and the steam railroad above.

(Delaware #503)

The Washington Street Bridge, Muncie.

DUBOIS [19]

S RR; CR 175E
Slab 1907
S RR design
Note: skewed

74 **NRC**
CR 350W; Patoka R
Filled-spandrel arch (2) c1920
Note: fairly flat rings; sections of neoclassical rail balusters in river; (Epilogue.B)

3262 **NRC**
SR #64, E of CR 75W; Hunley Ck overflow
Continuous slab (2) 1939
ISHC design
R P Olinger, bldr
Note: oldest unaltered continuous state-design slab; long slab spans (20 ft); depth of slab reduced under walks & rails; (VIII.2B)

5825 **NRC**
SR #56, E of Dubois Crossroads; Drainage Dh
T-beam (3) 1922
ISHC design
M & P Contr, bldr
Note: oldest extant state-design T-bm; skewed; w1968 with precast channel bms

5828
SR #56 at CR 960E; Davis Ck, Bh
Slab (3) 1923
ISHC design
R N Sullivan bldr
Note: among oldest ISHC-designed slabs; w1968 with precast channel bms

5829
SR #56, nr Hillham; Davis Ck, Bh
Slab (3) 1923
ISHC design
R N Sullivan, bldr
Note: among oldest ISHC-designed slabs; skewed; w1968 with precast channel bms

ELKHART [20]

371 Middlebury St Br
Middlebury St, Elkhart; Elkhart R
Filled-spandrel arch (3) 1931
W S Moore, dsgr
Bontrager Constr, bldr
Note: 160 ft long; unsymmetrical outer spans; sidewalks carried on counterforted brackets; deck, walks & lampposts rmvd (1985); only entrance rails with open-oval panels retained; (IX.2A)

374 **NRC**
Elkhart Ave, Elkhart; Elkhart R
Open-spandrel arch 1928
W S Moore, dsgr
Note: skewed; ribbed ring; 117-ft span; sidewalks carried on brackets; deck, walks, lamp standards & rails with open-oval panels rmvd; (IX.2A)

387 Bridge St Br **NRC**
Bridge St, Elkhart; St Joseph R
Continuous framed box beams (3) 1939
W S Moore, dsgr
H L Maddocks, bldr
Note: Federal Works Agency funded; long bm br (272 ft); long bm spans (123 ft); center span bm reinforcing extended through piers; earliest known box bms; cantilevered slab deck; open-oval paneled rails; Moderne accents on piers & outer bms; (before VI;IX.2C; Epilogue.B)

408 Lincoln Ave Br **NRC**
Lincoln Ave, Goshen; Elkhart R
Filled-spandrel arch (3) 1927
W S Moore, dsgr
Rieth-Riley Const, bldr
Note: skewed; 237 ft long; light standards rmvd; (IX.2A)

(Elkhart #408)

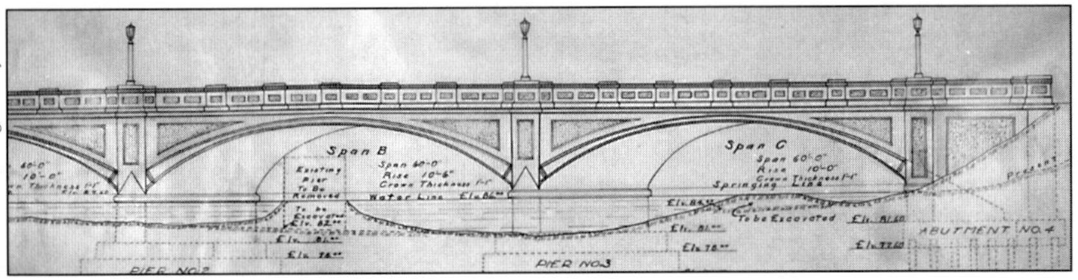

(Elkhart #408; Lang, Feeney)

William S. Moore's Lincoln Avenue Bridge, Goshen, as planned (above) and executed (top and right, detail).

(Elkhart #408)

3869 Main St Memorial Br **NRC**
N Main St, Elkhart, or SR #19; St Joseph R
Filled-spandrel arch (3) 1927
W S Moore, dsgr
Miller-Taylor Const, bldr
Note: *skewed; 375 ft long; "a monumental structure"; art-moderne decor for high entrance abutment, pier towers & rail posts; light standards & rails rmvd; (IX.2A)*

6085 County Line Br **NRC**
SR #219; St Joseph R
Open-spandrel arch (3) 1929
W S Moore, dsgr
Rieth-Riley Const & E F Smith, bldrs
Note: *c300 ft long plus 185-ft filled-wall approach; ribbed rings; spandrel piers; deck carried above longitudinal bms over crown of span; sidewalks extended beyond spandrel columns; deck, walks, open-arch rail panels, abutment dies with cut-triangular columns, & lampposts with cast bronze fixtures rmvd (1994); renovation sensitive to much of original decor; (IX.2A)*

(Elkhart #6193; Lang, Feeney)

Moore planned the Pike Street Bridge in Goshen to look like this (above). Unfortunately the designer for the structure's rehabilitation (right) lacked Moore's sense of proportion and symmetry. The disproportionally tall reproduction lights are also not appropriately centered with the structure's lines.

(Elkhart #6193)

6152 Lexington Ave Br **NRC**
W Lexington Ave, Elkhart, or SR #20; St Joseph R
Filled-spandrel arch (3) 1926
W S Moore, dsgr
M E White, bldr
Note: *skewed; 311 ft long; bush-hammered spandrels, pilasters & rail panels; lights rmvd; (IX.2A)*

6193 Pike St Br **NRC**
Pike St, Goshen, or SR #33; Elkhart R
Filled-spandrel arch (3) 1926
W S Moore, dsgr
H H Stoddard, bldr
Note: *258 ft long; bush-hammered spandrels & pilasters; deck, paneled rails & light standards rmvd (1995); new rail design & light placement eccentric; (IX.2A)*

FAYETTE [21]

1
CR 400N, NE of Harrisburg; Lick Ck
Beam (2) c1915

2 **NRC**
CR 400N, NW of Harrisburg; Williams Ck
Girder/floor bms (2) 1924
Slab
[R P, bldr]
Note: *metal caissons shore up slab*

30
CR 150W, NE of Harrisburg; Lick Ck
Beam (2) c1916

35
CR 450W, NW of Columbia; Fall Ck
Beam c1913
Note: *skewed*

37
CR 350W, SW of Connersville; Fall Ck
Beam c1915

38
CR #100N, NW of Tyner Crossing; Little Williams Ck
Beam c1916

54
CR 600S, SE of Orange; Sanes Ck, Bh
Beam c1917
Note: *skewed*

61
CR 350E, SW of Alquina; Village Ck, Bh
Beam c1923
Note: *low rails*

86 **NRC**
E 5th St, Connersville, or SR #44; Whitewater R, W Fk
Filled-spandrel arch (7) 1936
ISHC design
I E Smith, bldr
Note: *532 ft long; skewed; brackets carry sidewalks*

93
CR 610S, W of Garrison Ck Church; Sanes Ck, Bh
Beam c1916

FLOYD [22]

3
Chapel Hill Rd or CR #125W, Scottsville; Thompson Ck
[Continuous] slab (2) c1916
Note: *skewed*

24
Georgetown-Greenville Rd or CR 550W, NW of Indian Ck Church; Richland Ck
[Continuous] slab (2) c1929
Note: *skewed*

(Floyd #70)

(Floyd #70)

The ISHC widely publicized its design and construction of the Spring Street Bridge in New Albany.

61
Budd Rd or CR 30E, SW of New Albany; Knob Ck
Girder/floor bms c1915
Note: *skewed*

70 **NRC**
Spring St, New Albany; Silver Ck
Open-spandrel arch (4) 1926
ISHC design
C I & G Walter, bldr
Note: *old state design; 358 ft long; walled approaches; long center span (150 ft); 2-rib rings; open-arch spandrels; extended floorbms carry sidewalks; half of open-arch rail paneling rmvd*

(Fountain #62)

The Fountain County engineer designed the Leightle Bridge.

(Fountain #122)

(Fountain #122)

The Galloway Bridge is one of six of Luten's spandrel-braced arches to survive in Fountain County.

89
Chapel Hill Rd or CR 650N, NW of Navilleton; Jersey Park Ck, Campbell Bh
[Continuous] slab (2) c1914
Note: skewed; long slab spans (20 ft); pipe rail

FOUNTAIN [23]

62 Leightle Br **NRC**
CR 270E, SE of Veedersburg; Coal Ck, E Fk
Beam (3) 1917
Col W Sentman, dsgr
W E Ireland, bldr
Note: 101 ft long

95
CR 500E, S of Mellott; Dry Run Ck
Filled-spandrel arch c1912
Note: fairly flat ring

96
CR 500E, W of Mellott; Turkey Run
T-beam (2) c1921
Note: wide bm flanges

108
CR 325E, N of CR 300N; Turkey Run Ck
Filled-spandrel arch c1915
Note: fairly flat ring; rails rmvd

110
CR 425E, S of Stephens Crossing; Dry Run
Slab (3) c1927

122 Galloway Br **NRC**
CR 450N, E of CR 400W; Rattlesnake Ck
Spandrel-braced arch 1930
D B Luten, dsgr
Luten Engrg, bldr
Note: deck cantilevered over solid ribs; bush-hammered spandrels & rail panels; (IX.1B)

124 Loeb Br
Portland Arch Rd or CR 700W, nr Covington; Mallory Bh
Spandrel-braced arch 1930
D B Luten, dsgr
Luten Engrg, bldr
Note: deck cantilevered over solid ribs; w1982 with prestressed bx bms; bush-hammered spandrels; coped & paneled rails in streambed; (IX.1B)

126 Prescott/ Massey Br
9th St, Covington, or CR 670W; Mallory Bh
Spandrel-braced arch 1930
D B Luten, dsgr
Luten Engrg, bldr
Note: deck cantilevered over solid ribs; bush-hammered spandrels & rail panels; (IX.1B)

129 Smith Br **NRC**
CR 150N, nr Covington; Mallory Bh
Spandrel-braced arch 1931
D B Luten, dsgr
Luten Engrg, bldr
Note: skewed; flat spandrel ribs; deck cantilevered over solid ribs; bush-hammered spandrels & rail panels; (IX.1B)

130 Geo McDonald Br **NRC**
Union St, Covington, or CR 730W; Hamilton Bh or Conrail Ck
Spandrel-braced arch 1929
D B Luten, dsgr
Luten Engrg, bldr
Note: deck cantilevered over solid ribs; 50-ft span; bush-hammered spandrels & rail panels; (IX.1B)

140
CR 140E, nr Rob Roy; Big Shawnee Ck
Filled-spandrel arch (2) c1915
Note: three-centered rings with sharp descent over haunches

151 Possum Hollow Br **NRC**
CR 1300N, S of Riverside; Opossum Hollow
Spandrel-braced arch 1928
D B Luten, dsgr
C Dunning Supply, bldr
Note: *deck cantilevered over solid ribs; bush-hammered spandrels & rail panels; (IX.1B)*

169 **NRC**
Old US #41, N of Veedersburg; Dry Run Ck
T-beam (3) 1926
ISHC design
W H Grammer & Son, bldr
Note: *old unaltered state design; long (124 ft); skewed*

3885
US #41, nr Aylesworth; Coal Ck
Filled-spandrel arch (2) 1925
ISHC design
Ellis & Scamahorn, bldrs
Note: *skewed; w1967 covered spandrels & rmvd rails*

3887
US #41, nr Rob Roy; Big Shawnee Ck
Filled-spandrel arch (2) c1924
ISHC design
Note: *unsymmetrical; flat rings; w1960 covered spandrels & rmvd rails*

(Franklin #5)

These T-beams sit atop the substructure of the previous bridge on this site. Dry weather surfaced the timber grillage (left) on which the old stone abutments had originally been erected.

FRANKLIN [24]

NRC
CCC&StL (Big 4) RR; Co Line or Underpass Rd, Batesville
Filled-spandrel arch 1905
Note: *skewed; heavy, semicircular ring; old rails used for reinforcing in many overpasses*

2
Little Salt Ck Rd, CR 600N or #202, NE of Andersonville; Little Salt Ck
T-beam c1907
Note: *skewed; old stone abutments behind new ones*

5 **NRC**
Chapel Rd or CR 275N, SE of Andersonville; Little Salt Ck, S Fk
T-beam c1915
[H C, bldr]
Note: *timber grillage in streambed*

17 **NRC**
Laurel Rd or CR #19, Laurel; Whitewater R, W Fk
Filled-spandrel arch (5) c1925
[D B Luten, dsgr]
[Nat Concr, bldr]
Note: *482 ft long; deck extended on brackets; (IX.1A)*

27
Bath Rd or CR #74, W of SR #101; Templeton Ck
Filled-spandrel arch c1910
Note: *pipe rail*

35 **NRC**
Harvey Bh Rd or CR #15, S of Rail Fence Rd; Salt Ck, Brown Bh
Filled-spandrel arch c1912
Note: *bush-hammered rail panels; (Epilogue.A)*

39 **NRC**
US #52, S of Stavetown; Little Cedar Ck
Filled-spandrel arch 1922
ISHC design
J H Thornburgh, bldr
Note: *w1954 rmvd spandrels & rails; open-arch paneled replacement rails*

41
Snail Ck Rd, CR 250 or #160, S of Coffee Rd; Snail Ck
T-beam c1925
Note: *24-41, 42, 43, 44, 45 are all on Snail Ck Rd & skewed; all appear to have been cut from the same patterns at about the same time; those spanning about 25 ft are slabs & those around 30 ft are T-bms*

41
SR #1, N of Blooming Grove; Duck Ck
Filled-spandrel arch 1925
ISHC design
D Y Stout, bldr
Note: *skewed; modest rise to ring; w1960 rmvd spandrels & rails*

43
Snail Ck Rd or CR #160, NE of Shop Rd; Snail Ck
Slab c1925
Note: *skewed; long slab span (25 ft)*

44
Snail Ck Rd or CR #160, NE of Shop Rd; Snail Ck
Slab c1923
Note: *skewed; long slab span (25 ft)*

45
Snail Ck Rd or CR #160, NE of Shop Rd; Snail Ck
Slab c1925
Note: *skewed; long slab span (25 ft)*

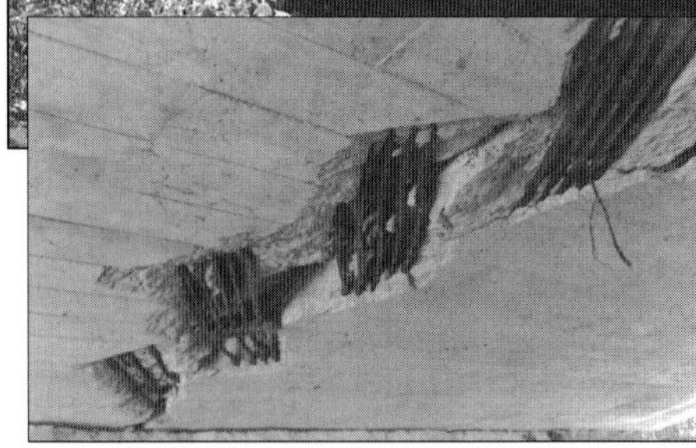

(Franklin #43)

The cracking of this unusually long slab span has revealed what appear to be Luten "trusses" (left).

(Franklin #100)

A nicely crafted, very traditionally designed country arch.

47 **NRC**
Pipe Ck Rd or CR #159, S of Haytown Rd; Pipe Ck, Trace Bh
Filled-spandrel arch 1911
T J Wilson, bldr
Note: *nrly semicircular ring*

49
Levee Rd or CR #46, W of Brookville; Snail Ck
Filled-spandrel arch c1910
Note: *semicircular ring; w19xx with I-bms & steel rail on N*

60
Reily Rd, CR 150N or #234, E of Darrell Rd; Big Cedar Ck
T-beam c1930
Note: *skewed*

61
Liberty Pike, CR #51 or 625E, N of SR #252; Big Cedar Ck
Filled-spandrel arch c1910
Note: *fairly flat ring*

62
Merril Rd or CR #44, NE of Mt Carmel; Whitewater R, Dry Fk
Girder/floor bms c1920
Note: *skewed*

67
Urban Rd, CR 760E or #57, N of Mt Carmel; Big Cedar Ck
T-beam c1915

89
Highland Center Rd, CR 100E or #35, S of Highland Center; Blue Ck, E Fk
Slab c1923
Note: *long (23 ft) & deep (2 ft) slab span*

90 **NRC**
Camptown Rd, CR #18 or 400E, W of Cedar Grove; Ramsey Ck
Filled-spandrel arch 1908
W Wilhelm, bldr
Note: *flat ring; pipe rail*

94
Blue Ck Rd or CR #33, SW of Stavetown; Blue Ck, E Fk
Filled-spandrel arch c1915
Note: *skewed*

95
Big Cedar Rd, CR 420E or #47, N of US #52; Big Cedar Ck, Bh
Slab c1920

98
Johnson Fk Rd, CR 850E or #59 at CR 500S; Whitewater R, Johnson Fk
T-beam c1925

99
Johnson Fk Rd, CR 850E or #59 between CR 500S & 600S; Whitewater R, Johnson Fk
Filled-spandrel arch 1909
Note: *three-centered ring*

100 **NRC**
Johnson Fk Rd or CR #59, N of Rockdale; Whitewater R, Johnson Fk, Bh
Filled-spandrel arch 1909
T A Wilson, bldr
Note: *semicircular ring; spandrel walls, rails, & wings woven together*

117
Columbus Ave, Huntersville, CR 840S or old SR #46; Little Laughery Ck
T-beam (2) c1935
[ISHC design]
Note: *rails sit on outer bm flange*

163 **NRC**
Wasserstrasse, Oldenburg; Harvey Bh
T-beam 1929
J L Stewart, dsgr
New Point Stone, bldr
Note: skewed

225
US #52, W of Metamora; Whitewater Canal
T-beam 1932
ISHC design
W A Gutzwiler, bldr
Note: skewed

3124 **NRC**
SR #46, N of Huntersville; Laughery Ck
T-beam c1938
ISHC design
Note: long state-design bm span (45 ft); skewed; (VII.3A)

3181
SR #229, SW of Metamora; Salt Ck
Filled-spandrel arch (4) 1940
ISHC design
Note: 255 ft long; rings raised high on slim piers

FULTON [25]

40 **NRC**
CR 375E, N of CR 550N; Reister Dh
Girder 1917
[J L, bldr]

GIBSON [26]

NYC RR; Cedar Grove Rd or CR 1000S, W of SR #65
Filled-spandrel arch 1910
NYC RR design
Note: skewed; semicircular ring

29
J D Thompson Rd, NE of Johnson; Higginbotham Dh
Filled-spandrel arch c1910
Note: w1950 with I-bms rmvd rails

85
Mt Tabor Rd, NE of Wheeling; Yellow Ck
Filled-spandrel arch c1905
Note: low ring; w1968 on W with I-bms

313
SR #65, N of Mounts Rd; Black R
T-beam 1924
ISHC design
W H Grammer & Son, bldr
Note: w1966 with precast channel bms rmvd rails

3195 **NRC**
SR #57, S of Buckskin; Pigeon Ck overflow
Slab (4) 1939
ISHC design
Note: slab depth reduced over rails; long spans (20 ft) & structure (87 ft) for state design; (VII.3B)

GRANT [27]

139
CR 650S, between SRs #37 & #9; Deer Ck
Filled-spandrel arch c1912
Note: skewed; heavy, flat ring

165 **NRC**
CR 825E, SW of Upland; Lake Bh Ck
T-beam c1916
Note: long span (46 ft)

1442
SR #5 or #221, N of Upland; Walnut Ck
T-beam (2) c1932
ISHC design
Note: skewed

1982
SR #15, N of Shady Hills; Mississinewa R
Filled-spandrel arch (4) c1936
ISHC design
Note: 347 ft long; w1984 with cantilevered slab deck; rails rmvd

GREENE [28]

15 **NRC**
CR 490N, N of Solsberry; Dry Bh
[Continuous] slab (3) c1920
Note: long slab spans (20 ft)

HAMILTON [29]

9
W 276th St, nr W Co line; Pearce Dh
Slab c1915
Note: skewed; decorated like stone

17 **NRC**
W 246th St, SE of Boxley; Little Cicero Ck
[Continuous] slab (2) c1913
Note: decorated like stone; streambed floor

23
Joliet Rd, NW of Jolietville; Little Eagle Ck, Bh
Slab c1930
Note: skewed; long slab span (23 ft)

26
Lamong Rd, N of Lamong; Finley Ck
Slab c1926
Note: skewed; long slab span (21 ft)

27
Centennial Rd, SW of Lamong; Finley Ck
Slab c1913
Note: skewed; long slab span (24 ft)

44
161th St, N of Eagletown; Little Eagle Ck, Bh
Filled-spandrel arch c1912
Note: skewed; heavy ring for short span

45
Six Points Rd, SW of W 136th St; Henley Ck
Slab c1919
Note: skewed; long slab span (21 ft)

61
Websters Pit Rd, N of E 276th St; Little Cicero Ck
[Continuous] slab (2) c1915
Plate girder
Note: slabs decorated like stone

62
Anthony Rd, N of E 236th St; Hinkle Ck
[Continuous] slab (2) c1913
Note: modest waterway clearance

73
E 256th St, W of Walnut Grove; Long Bh
Slab c1928
Note: long slab span (21 ft)

79
E 256th St, W of Cornell Rd; Duck Ck
[Continuous] beam (2) c1915

82 **NRC**
Whistler Rd, NE of Arcadia; Cicero Ck
Slab (8) c1916
Note: longest slab br (183 ft); most slab spans; long slab spans (23 ft)

83
Nightenhelser Rd, nr N Co line; Forkner Dh
Slab c1928
Note: skewed

(Hamilton #82)

Probably the longest extant old slab structure in Indiana.

84
E 286th St, E of Fall Rd; Weasel Ck
Slab c1928
Note: skewed

87
Startsman Rd; Weasel Ck
Filled-spandrel arch c1917
Note: fairly heavy semicircular ring

88 **NRC**
Lacy Rd, NW of Walnut Grove; Long Bh
Filled-spandrel arch c1925
Note: skewed; ring more linear than arced

100 **NRC**
E 225th St, SE of Deming; Hinkle Ck
[Continuous] slab (3) c1916
Note: decorated like stone

102
E 216th St, SE of Deming; Hinkle Ck
Filled-spandrel arch c1930
Note: fairly flat ring

103
E 216th St, SW of Deming; Jones Dh
Filled-spandrel arch c1917

120
191st St, W of Fishersburg; Stony Ck
[Continuous] slab (2) c1920
Note: long slab spans (21 ft); decorated like stone; rails rmvd

121
E 196th St, E of Mystic Rd; Lock Dh
[Continuous] slab (2) c1915
Note: skewed; long slab span (22 ft)

123
E 211th St, W of Cyntheanne Rd; Lock Dh
Slab c1913
Note: skewed; long slab span (26 ft)

128
Victory Chapel Rd, N of E 196th St; Dry Run Ck
Slab c1920
Note: skewed; long slab span (23 ft)

132 **NRC**
Pennington Rd, SW of Durbin; Weaver Dh
Filled-spandrel arch c1920
Note: skewed; heavy, very flat ring

133
E 186th St, W of Pennington Rd; Stony Ck
[Continuous] slab (3) c1916
Note: triangular cutwaters; decorated like stone

136
E 146th St, W of White R; Kirkendall Ck
Filled-spandrel arch c1923
Slab
Note: built as though continuous; heavy, unusually flat arch ring; w19xx with prestressed bx bms rmvd rails

142 **NRC**
Cherry Tree Ave, E edge of Carmel; Mitchener Dh
Filled-spandrel arch c1930
Note: flat ring; spandrel walls built outside ring; balustraded rails

145
Oak Rd, N of Carmel; Cool Ck
Filled-spandrel arch c1923
Note: almost semicircular ring

151 **NRC**
Cumberland Rd, S of Noblesville; Stony Ck
[Continuous] slab (3) c1916
Note: decorated like stone; cut-stone wingwalls; long slab br (73 ft); long slab spans (23 ft)

(Hamilton #88)

At the crown, the ring seems more like the apex of a triangle than a gentle arc. Did the builder choose for his convenience to frame with linear segments rather than with arcs, or does the external suggestion of linearity mirror some internal structural arrangement?

161
E 116th St, E of White R; Delight Ck
Slab c1940
Note: end rails curve outward over wingwalls

172
Cyntheanne Rd, S of SR #38; Mud Ck
Filled-spandrel arch c1917
Note: heavy, flat ring; inscribed spandrels

173
Cyntheanne Rd, nr Cyntheanne Church; Thorpe Ck
Slab c1919

175
E 146th St, W of Cyntheanne Rd; Mud Ck
[Continuous] slab (2) c1913
Note: skewed; decorated like stone; compact abutments

176
E 136th St, E of SR #238; Mud Ck
[Continuous] slab (3) c1919

177
Prairie Rd (closed), N of SR #238; Mud Ck
T-beam (2) c1912
Note: triangular cutwater

181
Promise or Klepfer School Rd, nr I-69 overpass; Sand Ck
Filled-spandrel arch c1930
Note: skewed; heavy, flat ring

185
E 136th St; Sand Ck
[Continuous] slab (2) c1916
Note: triangular cutwater

188 **NRC**
Klepfer School Rd, S of E 116th St; Mud Ck
Filled-spandrel arch c1913
Note: flat ring; imposing abutment & approach rail columns

190
Connecticut Ave, off SR #238 & nr E Co line; Flatfork Ck
Filled-spandrel arch c1915
Note: semicircular ring; deck extended on brackets

193
E 113th St, E of Olio Rd; Thorpe Ck
[Continuous] slab (3) c1919
Note: long slab spans (22 ft); triangular cutwaters

230 John Owen Br **NRC**
E 146th St; White R, W Fk
Filled-spandrel arch (5) 1920
D B Luten, dsgr
Nat Concr, bldr
Note: 516 ft long; spans of 101 ft; w1972 with prestressed bx bms rmvd rails

236
E 101st St, W of Fortville; Flatfork Ck
Filled-spandrel arch c1935
Note: semicircular ring; low rails

HANCOCK [30]

108 **NRC**
CR 900E or Co Line Rd, N of CR 300S; Sixmile Ck
Filled-spandrel arch c1925
Note: flat ring

1709
US #40, Charlottesville; Sixmile Ck
Filled-spandrel arch c1921
Note: probably not blt to ISHC design; w1942 replaced spandrels & rails with state design

3505 **NRC**
US #40, Greenfield; Brandywine Ck
Filled-spandrel arch c1921
Note: probably not of ISHC design; skewed; sidewalks carried on brackets; open-arch rail panels & post motifs

HARRISON [31]

1444
SR #64, W of Falkerson; Alstott Bh
T-beam c1934
ISHC design
Note: skewed

(Hancock #3505)

(Hancock #3505)

Decorative accents for an urban setting.

THE HAUNTED CATACOMBS OF AVON AND DANVILLE

As today's Conrail trains pass over certain bridges in the dead of night on their way to and from the large switching yard in Avon, they evoke the past. The Big Four Railroad's engineers who designed the open spandrels of the Haunted Bridge and Twin Bridges early in the century were probably unaware of how those structures would speak and what future generations would tell about them.

Little did the Big Four's designers and builders realize that the story of the African-American construction worker who was allegedly buried alive in the wet concrete of the bridge at Avon or at Danville would outlive their corporate existence. The worker's pick and the outline of his face allegedly show on the side of the bridge at Avon. His saw had reportedly extended above the hardened concrete at Danville until other workers cut it off. Visit the bridges at midnight and reportedly you can hear the worker's screams particularly when a train passes over his entombed body.[1]

Cries in the night, however, do not all come from the buried African-American, especially at the Avon Bridge. Some come from a mother, a baby, a pregnant woman, a boyfriend, and a Chinese man. Partway across the bridge, a mother wails for her baby who died under the wheels of a passing train; or a baby who slipped from mother's arms and fell over the bridge cries from the stream below. Two who jumped to their death from the bridge shriek – one, a pregnant woman whose husband had abandoned her; the other, a boyfriend caught halfway across the structure by a passing train as he headed home after a late date. A Chinese man screams as he slips under a passing train.

The evidence of haunt are not all limited to cries. Some hear the rattling – resonating upwards from the bridge's foundations as a train passes overhead in the dark – of the leg irons of the convict who fell and drowned in the wet concrete. Some see an apparition, the

(continued on next page)

Haunted Bridge, Avon (left). An old postcard of Twin Bridges, Danville, (right) shows the open-spandrel concrete arches of the double-tracked railroad behind the stone abutments for the iron span (now removed) on the old single-track line. The twin, a wrought iron highway bridge, is seated below the central railroad span.

woman in white. Her ghostly figure purportedly floats in the foggy moonlight within the chamber opening over the crown of the central span.

These widely reported folk tales have been woven into legends that physical contact with the structures themselves recharge. By designing high and wide spans to accommodate two tracks at minimum grade adjustment, the Big Four Railroad's engineers erected structures which are imposing to the viewer. The engineers also used cellular design where feasible to eliminate structural redundancy. As a consequence the pier towers are hollow and the spandrel columns contain inner openings which act as passageways between windowless rooms at span ends.

Traveling across an arch ring through the spandrel columns into the tower enclosures offers a mildly athletic challenge. Making the trip in the dark of night adds to the adventure, and a train passing overhead sends reverberations careening eerily through the structure's caverns. What sport for the teenage male striving to establish his manhood! Carloads of young men periodically make the initiatory passage and confirm their accomplishment by reporting on the haunt to others, especially to young women. The legends surrounding the Haunted and Twin Bridges are not limited to adolescent conversation, however. Adults reinforce the legends swirling around the Haunted and Twin Bridges by periodically repeating them in print.[2]

Negative evidence also confirms the power of legend. No one has bothered to recount another of these structures' claims to fame. In an attempt to define and illustrate *Artistic Bridge Design*, Henry Tyrrell listed structures in the United States which he considered artistic. The Big Four Railroad's open-spandrels ranked among the two dozen, reinforced concrete bridges that Tyrrell selected for praise in his widely acclaimed 1912 treatise.[3] While many Hendricks County residents know of at least some of the haunts of the Avon and Danville bridges, none is probably aware that Tyrrell once included these same spans in his national gallery of artistic structures.

Haunted Br **NRC**
CCC&StL (Big 4) RR; White Lick Ck & CR 625E, nr Avon
Open-spandrel arch (3) c1906
W M Dunne, engnr
Note: *among oldest open spandrels; 75-ft spans; heavy, coped rings; arched spandrel columns; pilasters curtain off fairly low piers; (Prologue)*

Twin Brs **NRC**
CCC&StL (Big 4) RR; White Lick Ck, W Fk & CR 150E, nr Danville
Open-spandrel arch (3) 1906
W M Dunne, engnr
Note: *among oldest open spandrels; heavy, coped rings; arched spandrel columns; pilasters curtain off fairly low piers; (Prologue)*

HENDRICKS [32]

(See box on pages 222-223)

CCC&StL (Big 4) RR; CR 450W, Hadley
Slab 1907
CCC&StL RR design
Note: *deep slab (c3 ft); under fill*

CCC&StL (Big 4) RR; CR 200W
Slab 1907
CCC&StL RR design
Note: *deep slab (c3 ft); under fill*

Blake St, Danville, or Cartersburg Rd; CCC&StL (Big 4) RR
Continuous T-beam (5) 1908
CCC&StL RR design
Note: *skewed; wider center bms; rails rmvd*

CR 550W; CCC&StL (Big 4) RR
Continuous T-beam (5) 1907
CCC&StL RR design
Note: *slight skew; wide flanges; rails rmvd*

NRC
Terre Haute, Indy & E Ry, S of Martin Rd, Cartersburg; White Lick Ck, W Fk
Filled-spandrel tied arch 1907
D B Luten, dsgr
Nat Concr, bldr
Note: *designed for 100-ton interurban cars*

Terre Haute, Indy & E Ry, off US #136, Jamestown; Big Walnut Ck
T-beam (3) 1908
THI&E RR design

1
CR 900N, NE of N Salem; Big Walnut Ck
Filled-spandrel arch c1910
Note: *fairly flat ring; w1977 with prestressed bx bms rmvd rails*

10
Ross Rd, S of Jamestown; Big Walnut Ck, W Fk
Filled-spandrel arch c1917
Note: *gentle rise to ring*

106 **NRC**
CR 550N, SW of N Salem; Big Walnut Ck, W Fk
Filled-spandrel arch c1915
Note: *heavy, flat ring; inscribed spandrels*

107
CR 675N, W of N Salem; Big Walnut Ck, W Fk
Filled-spandrel arch c1925
Note: *skewed; rather flat ring*

124
CR 300S, S of Hadley; Mill Ck
Filled-spandrel arch c1910
Note: *gentle rise to ring; minimal spandrel walling & fill; rails & roadway arch over ring*

168
CR 300S, NW of Clayton; Mill Ck, E Fk
Filled-spandrel arch c1917
Note: *fairly flat ring; inscribed spandrels*

175
CR 400S, NW of Clayton; Mill Ck, E Fk
Filled-spandrel arch c1925
Note: *skewed; rather flat ring*

192
Keller Hill, S Co Line Rd or CR 1100S, S of Center Valley; McCracken Ck
Filled-spandrel arch c1917
Note: *gentle rise to ring*

214
Old SR #267, S of I-70; Clarks Ck
Filled-spandrel arch c1925
Note: *skewed; fairly flat ring; w1960 with prestressed bx bms on timber bents rmvd rails*

220
CR 600S, W of SR #267; Clarks Ck
Filled-spandrel arch c1920
Note: *slight skew; flat ring; w1981 on S with prestressed bx bms rmvd half of rails*

227 **NRC**
CR 600S, off CR 1050E; White Lick Ck, E Fk
Filled-spandrel arch c1930
[O Y, bldr]
Note: *flat ring; inscribed spandrels*

[1842] **NRC**
off US #40, W of Plainfield; White Lick Ck, W Fk
Filled-spandrel arch (2) 1922
ISHC design
R H King, bldr
Note: *old unaltered state design; bypassed in World War II rearmament campaign; (VII.2C)*

1842
US #40, W of Plainfield; White Lick Ck, W Fk
Filled-spandrel arch (2) 1941
ISHC design
Note: *skewed; flat rings; w1986 rmvd rails*

HENRY [33]

3596 **NRC**
US #40, nr Cambridge City; Simmons Ck
Filled-spandrel arch 1921
ISHC design
H F Burk, bldr
Note: *old state design partly obscured & altered when w1944 to N*

HOWARD [34]

11
CR 1150W, at Division Rd; Wildcat Ck
Filled-spandrel arch (2) c1925
Note: *low, flat ring*

64
La Fountain St, Kokomo; Kokomo Ck
Filled-spandrel arch c1935
Note: *low, flat ring; w1970 with prestressed bx bms hid original spandrels & piers & rmvd rails*

108
CR 1100E, S of Phlox; Mud Ck
Filled-spandrel arch c1920
Note: *unsymmetrical, flat rings; rails rmvd*

506 LaFountain St Br **NRC**
Apperson Way, Kokomo; Wildcat Ck
Open-spandrel arch (2) 1929
[W A Ehrman], dsgr
I E Smith Const, bldr
Note: *1987 deck rmvd rails; (Epilogue.B)*

HUNTINGTON [35]

76 **NRC**
CR 600E, E Co line & S of I-69; Flat Ck
Filled-spandrel arch c1920
Note: *flat ring; inscribed spandrels; paneled & coped entrance rail posts*

103
CR 200E, N of Zion Church; Bull Ck
Girder/floor bms c1913
Note: *skewed*

144
CR 750W, SE of Andrews; Loon Ck
Girder/floor bms c1910

502 **NRC**
Briant St, Huntington; Little Wabash R
Filled-spandrel arch (3) 1912
I J Walker, bldr
Note: *215 ft long; inscribed spandrels; paneled pilasters; rails, entrance posts, lights rmvd*

JACKSON [36]

1340
US #50; Guthrie Ck
Filled-spandrel arch 1933
ISHC design
R L Schutt, bldr
Note: *designed for fill above spandrel walls; nr semicircular ring; skewed*

(Huntington #76)

A country arch meant to attract attention in Huntington County.

(Jay #62)

One of Luten's spandrel-braced arches for Jay County's Mike Mertz Bridge.

JASPER [37]

191
CR 900N or Old SR #10, E of CR 100W; Wheatfield Dh
T-beam 1940
ISHC design
Note: *skewed*

JAY [38]

62 Mike Mertz Br **NRC**
CR 225E, N of CR 70S; Limberlost Ck
Spandrel-braced arch 1927
D B Luten, dsgr
Carl Dunning Supply Co, bldr
Note: *50-ft span; deck cantilevered over solid ribs; skewed; (IX.1B)*

184
CR 207E, E of Salamonia; Salamonie R
Slab c1915
Note: *slab modestly curved at abutments; long slab span (30 ft)*

6182 **NRC**
S Meridian St, Portland, or SR #27; Salamonie R
Bowstring arch 1914
[E A Gast &] O O Clayton, dsgr
I E Smith, bldr
Note: *112-ft span; only concrete thru & tied arch in Indiana; (I.5; II.3E)*

(Jefferson #22JPG)

Which is more unusual: a solid-barrel ring on an open-spandrel arch or a railing defaced by shrapnel from artillery shells?

(Jefferson #2602)

Luten's Madison Hill Bridge with tell-tale signs of steam locomotives having passed below.

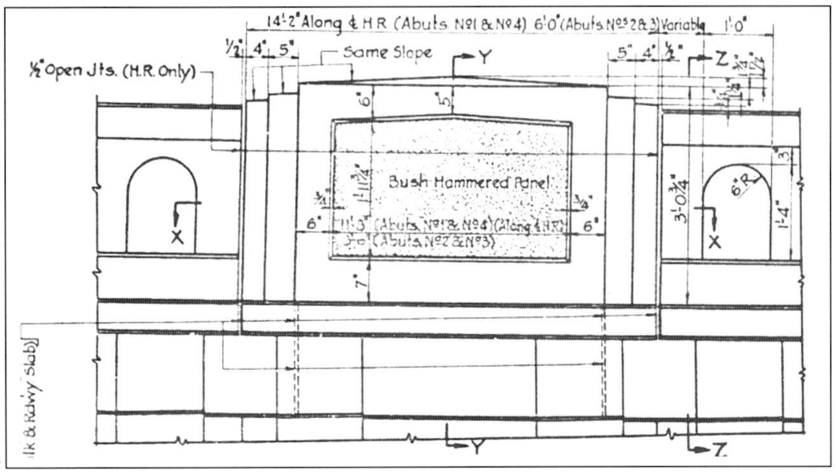

(Jefferson #3051)

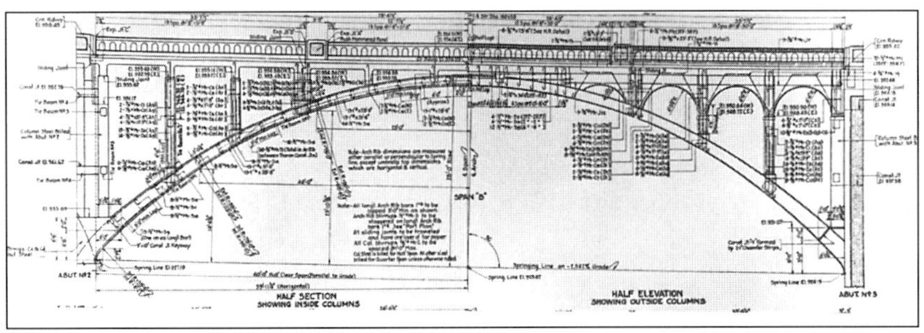

(Jefferson #3051)

(Jefferson #3051)

Left – A classic ISHC-designed open-spandrel arch with conventional reinforcing and the state's usual concession of open-rail panels for its soaring ribs.

JEFFERSON [39]

22 JPG NRC
D Rd, Jefferson Prov Grnds; Big Ck
Open-spandrel arch c1920
Slab (6)
Note: *among older open spandrels; semicircular ring; light piers; limited use of spandrel columns*

39
CR 300N, SE of Deputy; Lewis Ck
Girder/floor bms c1915
Note: *raised entrance posts*

92 NRC
SR #56, E of Madison; Eagle Hollow Ck
Filled-spandrel arch 1931
ISHC design
Dryden Bros, bldr
Note: *semicircular ring; fill above spandrel walls; long, heavily filled span (60 ft)*

2602 Madison Hill Br NRC
off SR #7, Madison State Hospital; Pa RR
Filled-spandrel arch 1921
D B Luten, dsgr
Note: *carries traffic over Madison incline, once the steepest grade for steam locomotives in the US; nrly semicircular ring; neoclassical rail balusters*

3051 NRC
Clifty Dr or SR #62, W of Madison; Big Clifty Ck
Open-spandrel arch (3) 1938
ISHC design
Note: *243 ft long; long center span (120 ft); 2-rib rings; extended floorbms carry deck beyond ribs & spandrels; bush-hammered pier pilasters; arched spandrel columns, open-arch paneled rails*

(Jennings #30)

A noteworthy example of a country arch probably designed and constructed by a mason. The ring is solid and close to segmental, the spandrel walls were not tied through reinforcing to the ring, the railings are undifferentiated extensions of the spandrels, and the concrete lacks a craftsman's finishing touches. Luten called this "inexpert" design and construction. Nonetheless, the structure has been tested by decades of use and continues to provide service.

JENNINGS [40]

2
CR 900W or #199, N of CR 800N; Rock Ck
T-beam c1910
Note: *rails rmvd*

7
CR 900N or #260, E of CR 400W; Bear Ck
T-beam c1907

9 NRC
CR 500W or #211, NE of Scipio; Bear Ck
Girder c1905
[A N T & W L, bldrs]
Note: *skewed; long girder span (50 ft)*

13
CR 300N or #44, W of CR 880W; Mutton Ck
Filled-spandrel arch c1915
Note: *fairly heavy ring*

14
CR 300N or #44, nr W Co line; Mutton Ck, Fk
Filled-spandrel arch c1915

15
CR 400N, Oak Grove Church W of CR 700W; Mutton Ck
Slab (2) c1906
Note: *skewed; rails largely gone*

16
CR 400W or #193, W of N Vernon; Twomile Ck
Filled-spandrel arch c1913
Note: *skewed; flat ring; spandrels & rails crumbling*

20
CR 900W or #47, N of CR 200N; Mutton Ck
T-beam c1911
Note: *reused stone abutment on N from previous br*

24
CR 75W, E of SR #3; Fish Ck
[Continuous] Slab (2) c1903
Note: *very old slab; rails gone*

26
CR #64, NE of Scipio; Rat Tail or Wyaloosing Ck
Filled-spandrel arch c1917
Note: *step-up spandrels/rails over ring's crown*

30
CR 850N or #66, W of SR #3; Wyaloosing Ck
Filled-spandrel arch c1917
Note: *spandrels/rails angled up to ring's crown*

35
CR 675N or #64, W of SR #3; Sand Ck, Bh
Filled-spandrel arch c1917
[J A D, bldr]
Note: *semicircular ring; SE spandrel wall rmvd*

37
CR 450N or #50, W of CR 150E; Fish Ck
Slab (2) c1912
Note: *reused stone abutment on E from previous br; rails mostly gone*

42
CR 310E or #247, N of US #50; Pleasant Run
Filled-spandrel arch c1915
Note: *flat ring*

43
CR 450E, N of Muscatatuck State School; Brush Ck
Filled-spandrel arch c1919
Note: rails rmvd

46
CR 400E, S of CR 850N; Bear Ck
[Continuous] slab (3) c1915
Note: W rail gone

48 **NRC**
CR 675E or #171, W of Nebraska; Brush Ck
Continuous slab (2) c1909
Note: long slab span (23 ft)

49
CR 860N or #288, nr E Co line; Muscatatuck R, Finch Bh
Filled-spandrel arch c1916

51 **NRC**
CR 1220N or #286, nr N & E Co line; Muscatatuck R, Vernon Fk
T-beam (3) c1915
Note: long bm br (117 ft)

53
CR 500E or #155, N of CR 150S; Crooked Ck
Filled-spandrel arch c1916
Note: nrly semicircular ring

58
CR 980N or #276, E of CR 500E; Sand Ck, Bh
Filled-spandrel arch c1910

59
CR 1000N or #80, W of CR 500E; Sand Ck, Bh
Slab c1913
Note: sits atop stone abutments of previous br; long slab span (21 ft)

63
CR 780E or #309, W edge Jefferson Prov Grnds; Graham Ck, Rush Bh
Slab c1909
Note: skewed; rails rmvd

71
CR 500S or #18, E of CR 600W; Polly Bh
Filled-spandrel arch c1915
Note: fairly heavy, flat ring

72
CR 600S or #14, SW of Lovett; Tea Ck
Filled-spandrel arch 1926

75
CR 75W, N of Paris; Bear Ck
Filled-spandrel arch c1910
Note: close to semicircular ring; w19xx to W with I-bms; step-up spandrel/rail over ring's crown remains to E

76 **NRC**
CR 800S or #10, E of SR #3; Big Graham Ck
Filled-spandrel arch (3) c1915
Note: 220 ft long; heavy rings; pipe rails

77
CR 800S or #10, nr E Co line; Bear Ck
Filled-spandrel arch 1910
[H C, bldr]
Note: close to semicircular ring

78
CR 200E or #15, E Co line; Bear Ck, Fk
Filled-spandrel arch c1920

80
CR 400S, SE of Vernon; Turkey Run
Filled-spandrel arch c1916
Note: close to semicircular ring; rails rmvd

A long, somewhat crudely constructed, and serviceable T-beam bridge.

81
CR 130E or #277, SE of Vernon; Turkey Run
Filled-spandrel arch c1913

82
CR 600S or #16, nr SE Co line; Bear Ck
T-beam c1915

84 **NRC**
CR 700S, S Co line; Bear Ck
Filled-spandrel arch c1915
J F Lewis, bldr

87
CR 650W, N of CR 700S; Tea Ck
Filled-spandrel arch c1915

88
CR 650W, N of CR 700S; Tea Ck
Filled-spandrel arch c1912
[J B J, bldr]

94
CR 595W, N of CR 500S; Polly Bh
Filled-spandrel arch c1915
[W M, bldr]
Note: flat ring

97
CR 550W, N of CR 750S; Tea Ck
Filled-spandrel arch c1917
Note: heavy, flat ring; L Lee w1975 arch to the W; E spandrels & rails rmvd

101
CR 500S or #18, W of CR 700W; Sixmile Ck, Bh
Filled-spandrel arch c1910
Note: flat ring; so little fill, roadway rides over ring

107
CR 925W or #2, S of CR 1100S; Cana Ck
Slab c1905
Note: long slab span (25 ft); no rails

(Johnson #501)

An early ISHC-designed arch for an urban setting on South Street in Franklin.

109
CR 1000S or #2, W of CR 550W; Slate Ck
Filled-spandrel arch c1910
Note: flat ring

117
CR 610W or #275, N of US #50; Sixmile Ck
T-beam (2) c1910
Note: most of low rail gone

145
CR 780W, S of CR 750N; Nettle Ck
Slab c1914

147
CR 150N or #38, W of CR 600W; Storm Ck
Filled-spandrel arch c1910

159
S Co Line Rd, CR 500S or #20, W of Jefferson Prov Grnds; Camp Ck
Slab c1913

854
US #50, W of N Vernon; Indian Ck
Filled-spandrel arch 1932
ISHC design
Ryan & Ryan, bldr
Note: ring more circular than average for considerable fill above spandrel walls; skewed

917 **NRC**
US #50 [E-bound lanes], Vernon; Muscatatuck R, Vernon Fk
Open-spandrel arch (3) 1930
ISHC design
A E Button, bldr
Note: 300 ft long; long state-design center span (180 ft); 2-rib rings; approach-span rings unsymmetrical; w1974 with new slab rmvd rails & arched spandrel walls

JOHNSON [41]

501 **NRC**
South St, Franklin; Youngs Ck
Filled-spandrel arch (3) 1934
ISHC design
R L Schutt, bldr
Note: first yr of state design for urban areas; skewed; unsymmetrical; 48 ft wide

KNOX [42]

235 Hazelton Br **NRC**
Old US #41, N of Hazelton; White R
T-beam (21) 1923
Parker thru truss (4)
Warren pony truss (4)
ISHC design
Stein Constr, bldr
Note: longest state highway br in Midwest at the time (2,002 ft); 841 ft of T-bm; (VII.3A; VIII.2A)

690 Geo Rogers Clark/ Lincoln Memorial Br **NRC**
Vigo St, Vincennes, or SR #441; Wabash R
Open-spandrel arch (7) 1932
ISHC design
Ferguson Constr, bldr
Bennett, Parsons & Frost, approach plaza dsgrs
Note: longest state-design arch structure (1,020 ft); long center span (180 ft); 2-rib rings; heavy pier towers; bush-hammered pilasters & rail posts; open-arch rail panels. Approach plaza: stone rails, carved phylons & pedestrian park entrances; lampposts with cast bronze fixtures; (VII.2C)

KOSCIUSKO [43]

(See box on pages 230-231)

LAKE [45]

149 **NRC**
Ridge Rd, Highland, or Old US #6; Pa RR
T-beam (5) 1934
ISHC design
Rieth-Riley Constr, bldr
Note: 199 ft long; skewed; (VII.3A)

244 **NRC**
Hohman Ave, Hammond; Little Calumet R
Filled-spandrel arch (3) 1913
Engnrg Const, bldr
Note: heavy, flat rings; piers decorated like stone; rails rmvd; end posts remain

1935
SR #53, E of Crown Pt; Deep R or Main Beaver Dam Dh
T-beam c1937
ISHC design
Note: skewed

LAPORTE [46]

73 **NRC**
Long Lane or CR 300W, SE of Union Mills; Mill Ck
[Continuous] slab (2) c1936
Note: long slab spans (21 ft)

3213
SR #8, at CR 400W; D S Myres or Tuesburg Dh, Hanna Arm
T-beam c1941
ISHC design

A TWIN FRAMED?

Located within a two-hour train ride of Chicago, Winona Lake with its ten-mile shoreline and deep, clear waters provided an ideal site for summer renewal assemblies for urban Protestants. The religious camps grew in number, length, and size from their late nineteenth-century beginnings. Billy Sunday, the noted early twentieth-century evangelist, helped to seal the lake's national religious prominence when he chose Winona as his home retreat.

Prominence spelled development. As cottages and some stately houses replaced cabins, as auditoriums supplanted tents and supplemented open-air amphitheaters, and as fields became parks, the president of the Standard Oil Company of Kentucky, Alexander McDonald, determined to transform the old military parade park from a swampy peninsula into McDonald island. He had the shoreline dredged, the land raised with the silt excavated from the lake, and a canal cut across the base of the peninsula.

The canal required bridging, and the spans needed to be appropriate for the setting. Who designed the bridges for the two main crossings – at Auditorium and Administration Boulevards, the former having been replaced – remains a mystery since Kosciusko County has not retained its older bridge plans. But the Winona Lake spans look like marginally-adjusted copies of a rigid-frame structure built in a Freeport, Illinois, park and promoted in the *ENR* in 1935.[4]

The Illinois and the Indiana bridges shared much. While at seventy feet the Illinois span was more than twenty feet longer than Indiana's, both used four beams of variable depth and width. Beam depth and width increased towards span-end. Both carried a slab deck for an approximately twenty-foot roadway with a sidewalk cantilevered out to one side.

(continued on next page)

Freeport Bridge.

(Kosciusko #280)

Administration Boulevard Bridge.

A slight camber gave the Illini and Hoosier spans "a graceful line," the curb provided "a shadow line" to partition the side elevation, streamlines on the outer beams' external side emphasized the horizontal, and the light metal rail above continued without overpowering the roadway's visual statement. In both cases, vertical off-sets in the wingwalls enclosed the structures, although the Indiana bridges did not follow the Illinois model's additional use of terminating concrete fins in the rails.

The Indiana spans were quite fit for their Thirties setting at Winona Lake, "one of the most beautiful towns in the Hoosier State. Its six-hundred-acre park, Venice-like canal spanned by four bridges, shady concrete walks, artificial ponds and flower beds all combine to form a garden spot on the lake shore for its visitors and residents."[5] If indeed a twin, the survivor remains significant as one of the handful of rigid frames built in the state in the interwar years.

(Kosciusko #280)

Streamlines and vertically off-set wingwalls on the Administration Boulevard Bridge (above).

280 **NRC**
Administration Blvd, Winona Lake; canal
[Rigid frame] c1935
<u>Note</u>: *long (48 ft), variable-depth & width bms braced at midspan; horizontal streamlines; sidewalk cantilevered beyond bms on N; fluted & painted metal rails & rail columns; (VIII.1B)*

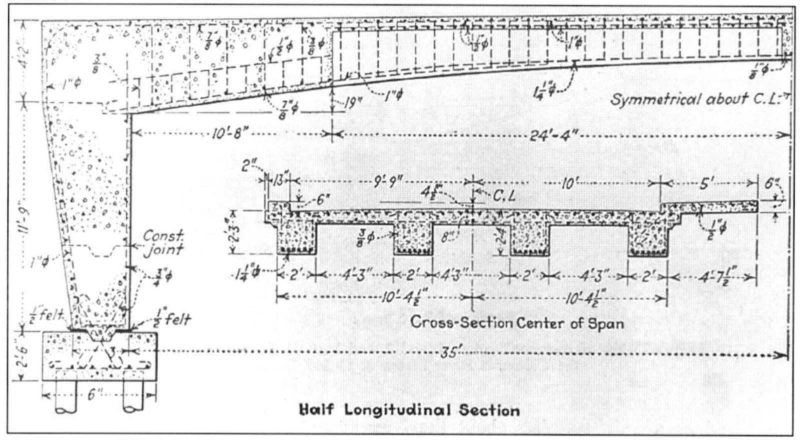

The Freeport Bridge used a solid slab for nearly eleven feet out from the frame and knee (above) and then converted the slab into beams. The designer of the Administration Boulevard Bridge, on the other hand, used beams across the span (below), although the beams visibly widen as they approach the knee and frame.

(Kosciusko #280)

(Lawrence #20)

Graceful ISHC-designed spans. The line of open-arch spandrel columns extends horizontally across the main span and into the short approaches, and their theme harmonizes with the line of open-arch rail panels above.

LAWRENCE [47]

CR 820S, N of Spring Mill State Pk; B&O RR
[Continuous] T-beam (3) c1930
B&O RR design
Note: skewed; pipe rail

20 NRC
CR 280W or Old SR #37, N of Needmore; Gulletts Ck
Open-spandrel arch (3) 1928
ISHC design
Ryan & Ryan, bldr
Note: 160 ft long; 2-rib rings; approach rings unsymmetrical; arched spandrel column walls; bush-hammered pier pilasters; open-arch rail panels

21
Oolitic Rd or Old SR #37, S of Oolitic; Salt Ck
Open-spandrel arch (3) 1935
ISHC design
H Hicks & Son, bldr
Note: 288 ft long; long center span (150 ft); 2-rib rings; arched spandrel column walls, open-arch paneled rails & bush-hammered pier pilasters rmvd (1990)

36
CR 675E, SE of Heltonville; Leatherwood Ck, S Fk
Filled-spandrel arch c1910
Note: inscribed spandrels

45
CR #59, S of Bartlettsville; Little Salt Ck, Brewer Bh
Girder/floor bms c1915
Note: roughly crafted

69
CR 600E, N of Heltonville; Henderson Ck
Filled-spandrel arch c1915
Note: little fill; roadway rides over ring

77
CR 850E, S of Bono; Sugar Ck
Filled-spandrel arch c1910
Note: rather flat ring; w1989 with cantilevered slab deck rmvd rails

78
CR 650S or #134, W of Rivervale; Mill Ck
Filled-spandrel arch c1910
Note: w1988 with cantilevered slab deck rmvd rails

160
Oolitic Rd, Bedford, or old SR #37; [CMStP&P RR]
T-beam (5) 1935
ISHC design
Note: 186 ft long; skewed

172 NRC
Cement Plant Rd, Bedford; Leatherwood Ck
[Continuous] girder (2) 1909
Note: early example, few continuous girders; (VIII.2C)

197 NRC
CR 575E, Heltonville; Leatherwood Ck
Filled-spandrel arch c1910
Note: with fill mostly gone, traffic rides directly on ring; (IV.2)

1092
SR #54, nr W Co line; Indian Ck
Filled-spandrel arch (3) c1931
ISHC design

1348
16th St, Bedford, or US #50; [CMStP&P RR]
Slab (3) 1933
ISHC design
A G Ryan & Sons, bldrs
Note: skewed; cut filled on S & rail rmvd

MADISON [48]

104
CR 400N, NE of Anderson; Big Killbuck Ck
Filled-spandrel arch c1920
Note: w1987 with cantilevered slab deck rmvd rails

(Madison #509)

This ISHC-designed arch on Jackson Street in Anderson is notable both for its length and the third abutment inserted between the second and third spans from the right.

122 NRC
Alford Rd or CR 580W, E of Ingalls; Lick Ck
Filled-spandrel arch (2) c1916

136 NRC
CR 750W, W of Ingalls; Lick Ck
Filled-spandrel arch c1927
Note: rather flat ring; entrance rails flared away from roadway

181
Alexandria Pike, Anderson; Big Killbuck Ck
Filled-spandrel arch (3) c1925
Note: 150 ft long; skewed

509 NRC
Jackson St, Anderson, or SR Business #9; White R & Killbuck Ck
Filled-spandrel arch (8) 1937
ISHC design
Note: 725 ft long; skewed; 3rd abutment inserted between 2nd & 3rd spans from N; w1984 with cantilevered slab deck rmvd rails

607
S A St, Elwood; Big Duck Ck
Filled-spandrel arch c1930
Note: skewed; rails rmvd; posts remain

1646
SR #67 or US #36, S of Ingalls; Lick Ck
Filled-spandrel arch c1925
ISHC design
Note: w1935; spandrels & rails rpltd

MARION [49]

BV600 NRC
Kent Ave, Ft Ben Harrison; Schoen Ck, E Fk
Filled-spandrel arch c1906
Note: possible Melan system reinforcing; neoclassical rail balusters; period light standards on rail posts

BV800 NRC
off Shafter Rd, Ft Ben Harrison; Lawrence Ck
Filled-spandrel arch c1900
Note: if unreinforced, only known extant example; flat ring; spandrel walls decorated

Country Club Br NRC
Woodstock Dr, Indy Museum of Art
Filled-spandrel arch 1910
D B Luten, dsgr
Nat Concr, bldr
Note: blt on H M Landon (of Indy Water Co) estate; to retain fill in place, spandrel walls extended over abutments, but with patented reduction in wall size to limit additional concrete needed from 120 to 30 cu yds; straight wingwalls cut DBL's savings more than would his typically curved ones; open rectangular rail panels; br exquisitely rpltd (1991); (II.2D; Epilogue.A)

Holcomb Estate Br NRC
nr Dunaway Ct, Spring Hill, off Cold Spring Rd
Double-drum, filled-spandrel arch 1919
D B Luten, dsgr
Note: two 4-in rings; bushhammered abutment pilasters & spandrel walls; neoclassically balustraded rails with planter urns on posts; (II.3G)

Ladywood Academy/ Fletcher Estate Br
Cathedral H S/Hudson Inst, off 56th St
Open-spandrel arch (3) c1920
Note: among oldest open spandrels; half-arch approach rings; flat center span ring with extra pier support added later; rough-coat finish on pilasters & rings; low rails arched over center span & decorated like stone; crosses inscribed in outer approach rails; heavy rail posts carry lampposts

Maple Rd Subway NRC
38th St; Crown Hill Cemetery
Filled-spandrel arch 1926
D A Bohlen, dsgr
E C Strathman, bldr
Note: stone-faced spandrel walls; keystone emblem on ring & bushhammered inner surface; neoclassical abutment pilaster columns & rail balusters; (page 275)

Spades Place Ftbr
Spades Pk; Pogues Run
Filled-spandrel arch 1907
Note: Pk Board design; spandrel walls decorated like stone; entrance rails curved; walk, walls & rails resonate with curve of ring; (I.2A-B)

NRC
Vandalia RR; US #40, Bridgeport
Filled-spandrel arch 1917
J E Starbuck, dsgr
Dunn & McCarthy, bldrs
Note: grade separations gave Pa rail system high-speed, direct link between Indy & Chicago for heavy traffic; skewed; elliptical ring; open-arch paneled rails cast in place; rectangular recessed panels on entrance rail posts & recessed triangular ones on spandrels; discussed in ENR (26 Oct, 14 Dec 1916)

(Marion - Country Club Br)

A modest Luten arch with appropriately decorated but not exclusive railings for a private estate in Indianapolis.

(Marion - Maple Rd Subway)

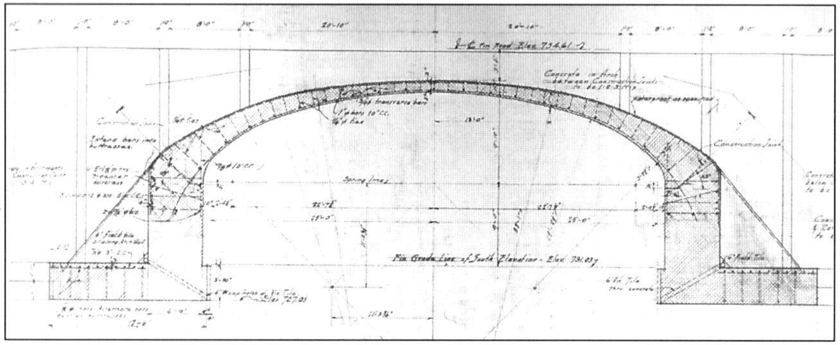

(Crown Hill Cemetery)

To maintain a respectful serenity in memory of those citizens buried therein, Crown Hill Cemetery in Indianapolis erected a grade separation between Maple Road (now 38th Street) and the cemetery grounds. Viewing the grounds as akin to a park, the Crown Hill leaders wanted a structural separation appropriate for the setting and therefore ordered the construction of a City-Beautiful arch with classical motifs. Covered on its sides with stone facing, the concrete ring contains conventional two-tier reinforcing with stirrups. Travelers along 38th Street can get a glimpse of only the neoclassical rail balusters behind the cemetery fence. In contrast, visitors to the grounds may take in the whole arch in classical garb.

(Marion - Maple Rd Subway)

310
W 76th St, Augusta; Crooked Ck
Filled-spandrel arch 1911
J E Sullivan, bldr

401
College Ave; Williams Ck
Filled-spandrel arch 1924
Nat Concr, bldr
Note: *rather flat, somewhat heavy ring*

403
Pennsylvania St; Williams Ck
Filled-spandrel arch 1911
Amer Constr, bldr
Note: *w1976 enclosed stonefacing of spandrel walls; rails rmvd*

405
Spring Mill Rd; Williams Ck
Filled-spandrel arch 1924
Meredith Hwy Constr, bldr
Note: *seated N on cut-stone abutment of previous br*

802
Wilson Rd; Fishback Ck
Filled-spandrel arch c1905
Note: *skewed; Melan system reinforcing; spandrel walls decorated like stone*

909 **NRC**
W 59th St, N Westway Pk; Little Eagle Ck
Filled-spandrel arch 1902
A W Moore, bldr
Note: *flat ring; spandrels decorated like stone; open-diamond rail panels*

910 **NRC**
Hollingsworth Rd, N Westway Pk; Little Eagle Ck
Filled-spandrel arch 1902
A W Moore, bldr
Note: *inscribed spandrels; open-diamond rail panels*

1007 **NRC**
Kessler Blvd; Crooked Ck
Filled-spandrel arch 1924
D B Luten, dsgr
Nat Concr, bldr
Note: *flat ring; lights rmvd*

1087 **NRC**
N Meridian St or US #31; White R & Riverview Dr
Filled-spandrel arch (8) 1933
ISHC design
Cunningham & Johnson, bldrs
Note: *long state-design, filled-spandrel arch (739 ft); long center spans (114, 120 ft); skewed; sidewalks carried on brackets; paneled pilasters; w1961 rmvd W rails; (VII.2A)*

1109
Guilford Ave; I W C Canal
Filled-spandrel arch 1906
[H W Klausmann, dsgr]
Volpp & Fritz, bldr
Note: *may use Melan system reinforcing; open-oval rail panels*

1111 **NRC**
Illinois St; I W C Canal
Filled-spandrel arch (2) 1920
D B Luten, dsgr
Nat Concr, bldr
Note: *Royce-formula tapered abutments; skewed; neoclassical rail balusters rmvd*

1407
Pendleton Pike, off SR #67; Indian Ck
Filled-spandrel arch (2) 1921
[J J Griffith, dsgr]
E F Smith, bldr
Note: *fairly flat, unsymmetrical rings; inscribed spandrels; rail balusters rmvd*

1410
E 52nd St, Lawrence; Indian Ck
[Continuous] slab (3) c1907
W C Smith, bldr
Note: *skewed; triangular cutwaters; decorated like stone; original rails gone & newer ones going*

1705 **NRC**
W 30th St, Taggart Riverside Pk; White R
Filled-spandrel arch (3) 1905
H W Klausmann, dsgr
Marion Co Constr, bldr
Note: 326 ft long; long center span (104 ft); Melan system reinforcing likely; faced with Indiana oolitic limestone; refuge-bays; stone rail balusters & posts; plaza entrances to park; (I.2A; Epilogue.A)

1707
W 30th St; I W C Canal
Filled-spandrel arch 1909
Note: skewed; stone-faced; deck, rails & lights rmvd

1712
W Roache Ave; I W C Canal
Filled-spandrel arch 1911
Amer Const, bldr
Note: skewed; open rail panels now filled

1715 **NRC**
Cold Springs Rd; Crooked Ck
Filled-spandrel arch 1914
[G E Kessler, dsgr]
Cleary-Kuert, bldr
Note: Pk Board design; flat ring; stone-faced with neoclassical detail; keystone emblem on ring; paneled spandrels & abutment pilasters; deck & rails rmvd; (I.2B)

1801
E 38th St; Fall Ck
Filled-spandrel arch (3) c1936
Note: 311 ft long; skewed; sidewalks carried on brackets; open-arch rail panels

1805
Delaware St; Fall Ck
Filled-spandrel arch (3) 1924
M E White, bldr
Note: 262 ft long; skewed

1808 **NRC**
E 39th St; Fall Ck
Filled-spandrel arch (3) 1905
Marion Co Const, bldr
Note: 242 ft long, raised high on piers; Melan system reinforcing likely; high triangular cutwaters with limestone caps shaped as seashells; lightlydressed stone facing; stone rails with small open-diamond panels

1814 **NRC**
W Washington St or US #40; Big Eagle Ck
Filled-spandrel arch (3) 1924
ISHC design
R H King, bldr
Note: skewed; unsymmetrical approach spans; w1938 by extending side-walks on brackets as part of World War II rearmament campaign

1912 Glen Ethel Br
Millersville Rd; Devon Ck
Filled-spandrel arch 1920
E F Smith, bldr

2305
W Michigan St; Little Eagle Ck
T-beam (2) c1941
Note: like state design; rails sit on outer bm flange

2401
Burdsal Pkwy; I W C Canal
[Continuous] T-beam (2) 1925
F C Lingenfelter, dsgr
E F Smith, bldr
Note: skewed; open-oval rail panels

2407
W 16th St; Fall Ck
Filled-spandrel arch (4) 1934
ISHC design
E F Smith, bldr
Note: 308 ft long; skewed; sidewalks carried on brackets

2501 Capitol Ave Br **NRC**
N Capitol Ave; Fall Ck
Filled-spandrel arch (3) 1912
G E Kessler, dsgr
Cleary-Kuert Constr, bldr
Note: skewed; 228 ft long; stone-faced with neoclassical detail; cartouches on ring crown, pier & abutment pilasters; open-arch stone rails; lampposts rmvd; (I.2B; III.2A; IV.2)

2502 Illinois St Br **NRC**
Illinois St; Fall Ck
Filled-spandrel arch (3) 1901
Melan Arch Const, dsgrs
Hoosier Br, bldrs
Note: 256 ft long; skewed; Melan system reinforcing with 10-in I-bms placed 3 ft apart; facing & rails of Bedford limestone; yellow pine creosoted paving blocks long gone; (I.2A)

2510
W Michigan St; Pogues Run
Continuous box (2) 1917
B J T Jeup, dsgr
J R Mead, bldr
Note: wide concrete abutment ahead of stone from prior br on W; balustraded rails rmvd; (VIII.1A)

2511 **NRC**
Dorman St; Pogues Run
Continuous framed slab (2) 1910
H W Klausmann, dsgr
Hackedorn Constr, bldr
Note: early integration of slab, pier, & abutment reinforcement; Klausmann known outside Indiana for his integrated design; skewed; open-circle paneled rails rmvd; (VIII.1A)

2514
N Rural St; Pogues Run
Filled-spandrel arch c1909
Note: w1962 with prestressed bx bms hid most spandrel decoration & rmvd rails

2516 Nowland Ave Br **NRC**
Nowland Ave, Brookside Pk; Pogues Run
Filled-spandrel, tied arch 1903
D B Luten, dsgr
Nat Br, bldr
Note: oldest Luten structure in Indy; patented design; balustraded rails; (II.3B; III.1Bbx)

2517
Commerce Ave; Pogues Run
Filled-spandrel arch 1911
B H Miller, dsgr
Note: stone-faced; w1960 rmvd deck & neoclassical rails & encased half of facing

2521
N St; Pogues Run
[Continuous framed] slab (2) 1911
Note: similar structure to #2510, although with heavier pier; rails extend around corner; (VIII.1A)

2527
Senate & Boulevard Aves; Fall Ck
Filled-spandrel arch (3) 1920
C E Cheney, dsgr
Amer Const, bldr
Note: not consulted on Co plans, Indy Pk Board prevented constr until parties agreed about embankment slopes, electric line placement, curbing, walks, & lighting; 295 ft long; 101-ft center span; skewed; paneled & emblazoned pier pilasters

2604
Emerson Ave; Pleasant Run
Filled-spandrel arch 1909
Volpp & Fritz, bldr

2615
Brookside Pk Dr; Pogues Run
Filled-spandrel arch 1908
Note: Pk Board design; spandrels decorated like stone; balustraded rails rmvd

CITY-BEAUTIFUL BRIDGES IN THE CAPITAL CITY'S BUSHES

As lost empires in temperate climates have their monuments recaptured by the forest, so a number of structures commemorating Indianapolis' prowess have been neglected and sometimes even decapitated. The celebration of Indianapolis' previous civic achievement is being naturally reclaimed as has, for example, many a Mayan or other temple.

(Marion #1808)

East 39th Street over Fall Creek.

(Marion #1808)

(Marion #1808)

(Marion #3311)

(Marion #3311)

Kentucky Avenue over Big Eagle Creek.

(Marion #1715)

Cold Spring Road over Crooked Creek.

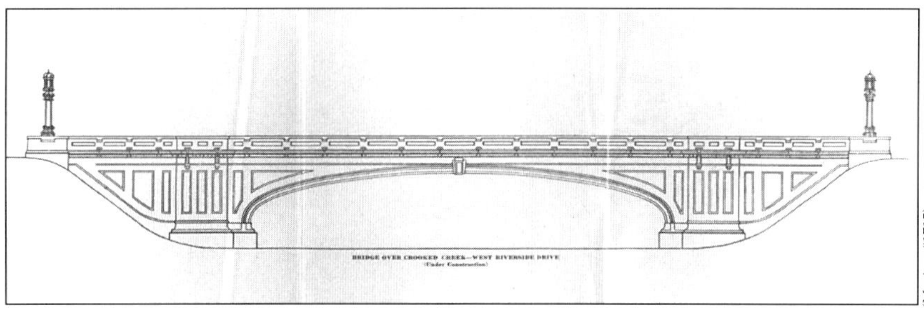

(Marion #1715)

2703
English Ave; Lick Ck
Filled-spandrel arch 1911
C Hoover, bldr
Note: *skewed; w1964 with precast channel bms rmvd rails*

2803
E Co Line Rd, S of Cumberland; Buck Ck
Filled-spandrel arch 1908
Volpp & Fritz, bldr
Note: *Melan system reinforcing*

3001
High School Rd, S of US #40; Pa RR
T-beam (5) c1939
Note: *like state design; 193 ft long; rails sit on outer bm flange; skewed*

3008
Farnsworth St, Mars Hill; State Dh
Filled-spandrel arch c1935
Note: *bush-hammered spandrel panels; rail balusters gone*

3012
Westbrook Ave; Neeld Dh
Slab 1925
H C Garland, bldr
Note: *highly skewed; deep slab with heavy reinforcinforcing bars*

3103
Bluff Rd, N of Troy Ave; Pleasant Run
Filled-spandrel arch c1935
Note: *skewed*

3104
Oliver Ave; White R
Filled-spandrel arch (5) 1925
Note: *680 ft long; long spans (120-130 ft); skewed; bush-hammered spandrel & pier pilaster panels; plate girder approach added (1960); slab deck extended on brackets; deck, paneled rails & lampposts rmvd (1964)*

3108 **NRC**
Morris St; White R
Open-spandrel arch (5) 1929
D B Luten, dsgr
Nat Concr, bldr
Note: *655 ft long; 120-130 ft spans; 3 ribs per ring on spandrel piers; bushhammered pier towers; deck cantilevered beyond ribs to carry sidewalks; deck, floor bms, spandrel columns, open-column rails, & lampposts rmvd (1963)*

3110 **NRC**
Kentucky Ave; White R
Filled-spandrel arch (7) 1925
Mead-Balch Constr, bldr
Note: *989 ft long; 121-146 ft spans; skewed; rails & light standards rmvd*

3111
Kentucky Ave; Big Eagle Ck
Filled-spandrel arch (3) c1920
Note: *210 ft long; skewed; brackets carry sidewalks*

3203 Shelby St Br
Shelby St; Pleasant Run
Filled-spandrel arch 1938
[ISHC design]
Note: *WPA-funded; skewed; brackets carry sidewalks*

3204
S State Ave; Pleasant Run
Filled-spandrel arch 1911
Note: *spandrel walls decorated like stone; open-panel rails now filled*

(Marion #3108)

In the pursuit of building with less material, Luten designed spandrel piers for the Morris Street, Indianapolis, Bridge. Luten used piers here only where ribs carried load to the substructure, and he eliminated pier sections not put to structural work.

3207
Shelby St, S of Garfield Pk; Bean Ck
Filled-spandrel arch c1926
Note: *Co paid half for constr*

3209
Churchman Ave; Pleasant Run
Filled-spandrel arch 1923
Note: *skewed; Pk Board design; double-line ring reinforcing with inclined stirrups*

3210
Villa Ave; Pleasant Run
Filled-spandrel arch 1916
Amer Const, bldr
Note: *skewed; paneled pier pilasters*

3213
Linden St; Pleasant Run
Filled-spandrel arch c1922

3215 **NRC**
Garfield Pk, E Rd; Pleasant Run
Filled-spandrel arch 1908
H F Hackedorn, bldr
Note: *Pk Board design; spandrel walls decorated like stone; deck rides on ring at crown; rails rmvd; (I.2A-B; III.1A)*

3216 **NRC**
Garfield Pk Rd (S East St); Pleasant Run
Filled-spandrel arch 1927
J E Perry, dsgr
E F Smith, bldr
Note: *Pk Board design; skewed; paneled pier pilasters; rails rmvd*

3219
Garfield Pk Rd; Bean Ck
Filled-spandrel arch 1923
J L Elliott, dsgr
Todd & Reid, bldr
Note: *Pk Board design; stone rails rmvd*

3220
Garfield Pk Rd; Bean Ck
Filled-spandrel arch 1923
J L Elliott, dsgr
Todd & Reid, bldr
Note: *Pk Board design; skewed; rails rmvd*

3221 **NRC**
Southern Ave; Bean Ck
Filled-spandrel arch 1907
T Frazier, bldr
Note: *skewed; stone-faced; stone rails with open-diamond panels*

3301
English Ave; Pleasant Run
Filled-spandrel arch (2) 1922
[D B Luten, dsgr]
Nat Concr, bldr
Note: *unsymmetrical rings; sidewalks carried on brackets; deck rmvd; rails rmvd except at abutments (1982)*

3311 **NRC**
Southeastern Ave; Pleasant Run
Filled-spandrel arch c1915
Note: *likely Pk Board design; neoclassical details; ring crown & abutment pilasters emblazoned; spandrels inscribed with bush-hammered panels; large entrance & midspan rail posts*

3313 **NRC**
Hobart Ave; Bean Ck
[Continuous] slab (2) 1906
Marion Constr, bldr
Note: *oldest known slab br; long 20 ft spans; decorated like stone*

3318
Main St, Beech Grove; Lick Ck
Filled-spandrel arch 1908
G W Fife, bldr
Note: *skewed; walls decorated like stone; deck & rails rmvd*

3802
Bluff Rd; Lick Ck
Filled-spandrel arch 1933
ISHC design
Mead-Balch Constr, bldr
Note: *skewed*

4005
9th Ave, Beech Grove, N of I-465; Beech Ck
[Continuous] slab (3) 1908
H F Hackedorn, bldr
Note: *91 ft long; w1985 with precast channel bms & substructure almost encases spans & stone-like decor*

4205
E Co Line Rd, S of Julietta; Doe Ck
[Continuous] slab (2) c1910
Note: *skewed; decorated like stone; rails in streambed*

4302
Mills Rd; Milhouse Ck
Filled-spandrel arch 1904
Moore-Mansfield Constr, bldr
Note: *spandrel walls decorated like stone*

4395 Joseph W Summers Mem/ Meridian St Br **NRC**
Meridian St; Fall Ck
Filled-spandrel arch (3) 1917
G E Kessler, dsgr
Dunn-McCarthy Const, bldrs
Note: *300 ft long; skewed; stone-faced with neoclassical detail; cartouches on ring crowns, pier & abutment pilasters, & rail posts; balustraded stone rails; ornate caps on entrance & central rail posts support period lampposts; flower boxes rmvd; (I.2B)*

4396 **NRC**
W Washington St or US #40; White R
Filled-spandrel arch (7) 1916
D B Luten, dsgr
Nat Concr, bldr
Note: *844 ft long; 106-120 ft spans; skewed; bush-hammered walls & pilasters; emblazoned pier towers; rails with Pettijohn balusters & light standards rmvd; (III.1B)*

4506
S Belmont Ave; Little Buck Ck
Filled-spandrel arch c1935
[ISHC design]

4507
S Harding St; Little Buck Ck
Beam (2) c1924
Note: *long; wide bm stems*

4508
S Belmont Ave; Orme Dh
Filled-spandrel arch c1915
Note: *low, flat ring; much-deteriorated walls once fully decorated like stone; rails rmvd*

4708 **NRC**
Shelbyville Rd; Little Buck Ck
Filled-spandrel arch 1903
Fritz & Frazier, bldr
Note: *stone-faced walls; rails rmvd*

6190 Meridian St Br
Meridian St or US #31; I W C Canal
Filled-spandrel arch (3) 1926
D B Luten, dsgr
Nat Concr, bldr
Note: *skewed; lampposts mostly gone*

6215
W Washington St, or US #40; Little Eagle Ck
Filled-spandrel arch 1913
Reliable Constr, bldr
Note: *stone-faced walls; stone rails remain on N*

MARSHALL [50]

226 **NRC**
S Michigan St, Plymouth; Yellow R
Filled-spandrel arch (2) c1917
D B Luten, dsgr
Note: *brackets carry sidewalks; deck & balustaded rails rmvd (1979)*

(Marshall #226)

Luten arches on South Michigan Street, Plymouth, shortly after construction.

(Marshall #227)

Garro Street, Plymouth.

227 **NRC**
Garro St, Plymouth; Yellow R
Filled-spandrel arch (2) c1930
Note: *skewed; triangular cutwater; paneled pier pilaster*

228
E Jefferson St, Plymouth; Yellow R
Filled-spandrel arch (2) c1927
Note: *skewed; w1987 with cantilevered deck rmvd rails & lampposts*

MARTIN [51]

862
US #50; Beech Ck
Filled-spandrel arch 1930
ISHC design
R P Olinger, bldr
Note: *designed for fill above spandrel walls; skewed*

6344
US #150; Natchez Bh
T-beam 1925
ISHC design
Scott Constr, bldr
Note: *skewed; w1956 with precast channel bms rmvd rails*

MIAMI [52]

132 **NRC**
CR 450E, S of Santa Fe, off SR #21; Pipe Ck
Filled-spandrel arch (2) 1907
D B Luten, dsgr
Nat Br, bldr
Note: *paneled rails mostly gone into ck; br closed to traffic*

MONROE [53]

3
Old SR #37, S of Colpert Hill Rd, Bloomington; Muddy Ck, Bh
T-beam 1926
ISHC design
Hancock & Kieffer, bldr
Note: *skewed*

6
Old SR #37, N of Anderson Rd; Buck Ck, Bh
T-beam 1926
ISHC design
E C Wright, bldr

(Monroe #182)

Vintage ISHC-designed slabs.

127 **NRC**
Harbison Rd or CR 350N, SE of Ellettsville; Jacks Defeat Ck
Continuous beam (2) c1915
Note: *open-panel rails; (VIII.2C)*

182 **NRC**
CR 360N at W Brook Downs, off SR #46; Owens Bh
Slab (2) 1927
ISHC design
Wright & Gray, bldrs
Note: *oldest unaltered standard state-design slab; skewed*

905
Allen St, Bloomington, just W of Walnut St; Jordan R
Continuous beam (2) c1916
Note: *skewed*

MONTGOMERY [54]

Peoria E RR, nr Waynestown; Coal Ck, E Fk
Filled-spandrel arch (3) c1906

139
CR 200W; Offield Ck
Beam (2) c1915

184
US #136 or Lafayette Ave, Crawfordsville; Sugar Ck
Filled-spandrel arch (4) 1933
ISHC design
E F Smith, bldr
Note: *325 ft long; w1989 rmvd some spandrel walls & all of deck & rails; (Epilogue.A)*

281
US #231, S of Crawfordsville; Dry Bh Ck, Fk
T-beam 1926
ISHC design
D L Reichard, bldr
Note: *skewed; w1960 with prestressed bx bms on timber bents*

501 **NRC**
Chestnut & Danville Sts, Crawfordsville; Dry Bh Ck
Filled-spandrel arch c1915
Note: spandrels decorated like stone; open-diamond rail panels

1199
SR #47, SW of Crawfordsville; Offield Ck
Filled-spandrel arch c1932
ISHC design

3046
SR #47, NE of CR 875E; Hazel Ck
T-beam c1941
ISHC design
Note: skewed; long state-design T-bm span (50 ft); bms braced; (VII.3A)

MORGAN [55]

Thomas Wheeler Br **NRC**
Awbrey Rd or CR 500N, W of Plano; Rhodes Ck
Slab 1913
H A Blunk, dsgr
L P Guthrie, bldr
Note: "flat-top" span includes 13 patented Luten trusses; rails rmvd; (II.3C)

9 Bryants Ck Br **NRC**
Bryants Ck Rd, CR 600W or #54; Bryants Ck
Filled-spandrel arch 1909
E O Gilbert, dsgr
N W Gilbert, bldr
Note: the Gilberts out-flanked Nat Concr Co with its Luten design

(Morgan #9)

(Montgomery #501)

A small touch of City-Beautiful bridge decor in Crawfordsville.

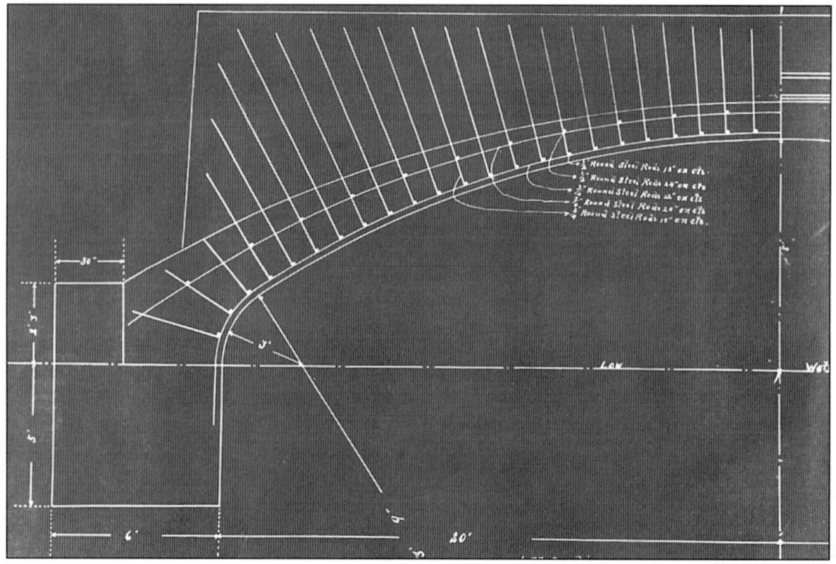

(Morgan #9)

E. O. Gilbert, Morgan County Surveyor, designed two-tier ring reinforcement tied into the spandrel walls for Bryant's Creek Arch.

30 Miller Br NRC
Mahalasville Rd, SW of Morgantown; Pike Ck
Girder/floor bms c1916
Note: *includes patented Luten trusses; skewed; (II.3C)*

44 Dean Arch NRC
Peavine Rd, S of Chetwynd; S Prong Ck
Filled-spandrel arch 1911
H A Blunk, dsgr
E O Gilbert, bldr
Note: *Luten plans bypassed*

49 Flake Arch
Abraham Rd, NE of Cope; N Prong, Stotts Ck
Filled-spandrel arch 1911
E O Gilbert, bldr
Note: *skewed; Luten plans bypassed*

56 Stockwell Br NRC
Teeters Rd, nr SR #37; Clear Ck, W Fk
Girder/floor bms 1916
Note: *uses patented Luten trusses; (II.3C)*

64 Crone Br
Bain Rd; Lambs Ck, Bh
Girder/floor bms c1916
Note: *uses patented Luten trusses; (II.3C)*

67
Big Hurricane Rd, W of Moster Rd; Lambs Ck, Bh
Girder/floor bms c1928
Note: *skewed*

103
Briarhopper Rd, NW of Wilbur; Lambs Ck
Girder/floor bms 1916
C Haase, dsgr
Note: *uses patented Luten trusses; (II.3C)*

129 Copeland Arch
Rinker Rd, SE of Brooklyn; Ck
Filled-spandrel arch 1911
H A Blunk, dsgr
E O Gilbert, bldr
Note: *skewed; Luten plans bypassed*

143 NRC
Hammer Rd or CR 1350N, N of Gasburg; McCracken Ck
Filled-spandrel arch 1907
[E O Gilbert, dsgr]
Kelleher & Slipher, bldrs
Note: *high rails & posts*

161 NRC
Old SR #37, S of Liberty Ch Rd; Little Indian Ck
Filled-spandrel arch 1922
ISHC design
E C Wright, bldr
Note: *among oldest unaltered state-design arches; skewed; (VII.2B)*

162
Old SR #37, Stines Mill Cor; Clear Ck
T-beam (3) 1925
ISHC design
R H King, bldr

163 Stotts Ck/ Exchange Br
New Harmony Rd or CR 475N, S of Exchange; Stotts Ck
Filled-spandrel arch (2) 1911
E O Gilbert, bldr
Note: *deteriorated & closed; useful as example of how filled-spandrels are organized (& sometimes come apart); (IV.3)*

(Morgan #30)

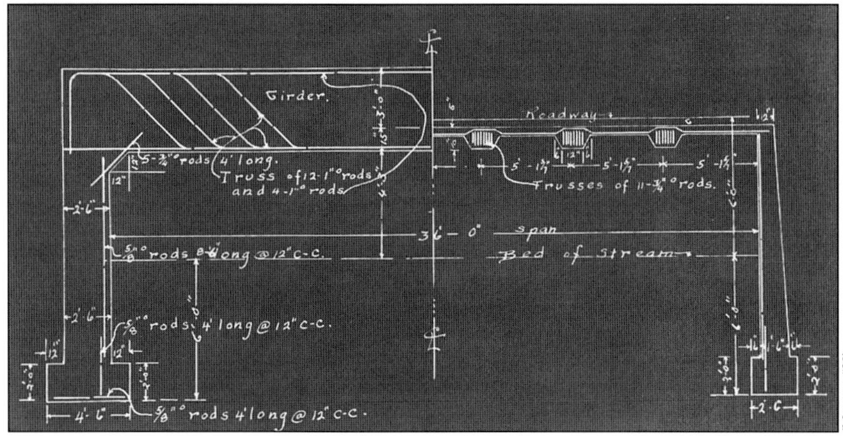

(Morgan #30)

Morgan County's Surveyors called for Luten "trusses" in many of their small beam and slab spans. Miller Bridge illustrates Luten trussed girders and floor beams.

166 **NRC**
Old SR #37, Waverly; Bluff Ck
Slab (2) 1925
ISHC design
W H Grammer & Son, bldr
Note: *oldest unaltered state-design slab; rare example of slab designed to carry fill over deck; no rails*

173 **NRC**
Country Club Rd, Brooklyn; Monical Bh
Filled-spandrel arch 1908
N W Gilbert, bldr
Note: *heavily skewed; rails extend beyond spandrel walls*

189 Patrick Arch
Paragon Rd or CR 300S, S of Paragon; White R, W Fk, Ck
Filled-spandrel arch 1910
N W Gilbert, bldr
Note: *skewed*

1560
SR #67, SW of Paragon; Fall Ck
T-beam c1939
ISHC design
Note: *skewed; rather long state T-bm span (40 ft)*

1966
SR #252, W of Morgantown; Crooked Ck, W Fk
T-beam c1937
ISHC design

NEWTON [56]

142
CR 125E, N of CR 1100S; Hunter Dh
Girder/floor bms c1930

NOBLE [57]

CR 200W; B&O RR
[Continuous] T-beam (3) c1921
B&O RR design

250 **NRC**
SR #9, N of US #6; Tamarack Ck or Gretzinger Dh
T-beam 1925
ISHC design
D Jenkins, bldr
Note: *early, unaltered state-design T-bm br; w1936 rpltd decor*

ORANGE [59]

50
CR 620W or #115, SE of French Lick; Lick Ck, Bh
[Continuous] slab (2) 1913
Note: *concr post & pipe rails*

55 Livengood Fd **NRC**
Old SR #154, S of French Lick; Lick Ck
Filled-spandrel arch 1910
D B Luten, dsgr
Nat Concr, bldr
Note: *linear ring; slab deck extended to carry paneled rails*

99 **NRC**
CR #18, E of Valeene; Patoka R
Slab (6) c1913
Note: *very long slab br (133 ft); long slab spans (21-22 ft)*

1962
SR #337, E of Orleans; Lost R
Filled-spandrel arch (2) c1940
ISHC design
Note: *skewed; N span unsymmetrical*

(Orange #55)

(Orange #55)

Luten offered a number of designs to builders in Orange County. The National Concrete Company, Luten's main construction arm in Indiana, secured the contract to build a small arch at Livengood Ford.

3710
SR #56, W Baden Springs; Lost R
Filled-spandrel arch (2) 1923
ISHC design
R H King, bldr
Note: *skewed; w1986 rmvd spandrels & rails*

4869
SR #37, N of Paoli; Lost R
Filled-spandrel arch (3) 1922
ISHC design
E C Wright, bldr
Note: *w1964 rmvd spandrels & rails*

5937
US #150, NW of Abydel; Lick Ck
Filled-spandrel arch (2) 1924
ISHC design
Cleary, Hailey & Barry, bldrs
Note: *w1968 with prestressed bx bms covered piers & most of spandrels & rmvd rails*

OWEN [60]

7
Turkey Neck Rd or CR #77; Mill Ck, Bh or Rhodes Ck, Bh
Filled-spandrel arch (2) c1917

91
CR #86, E of SR #157; Beach Ck
Girder/floor bms 1923
[R C & W M, bldrs]

137
CR #7, E of SR #157; Beach Ck, Bh
[Continuous] slab (2) c1905

143 NRC
CR #235, S of New Hope; Lick Ck
Girder\floor bms c1907
Note: long girder span (52 ft); (II.3D)

144
CR #235, between New Hope & Vilas; Raccoon Ck, Dh
Girder/floor bms c1905

148 **NRC**
CR #98, N of Freeman; Raccoon Ck
Girder/floor bms c1907
Note: long girder span (51 ft); (II.3D)

3102
SR #42, NE of Cunot; Doe Ck
Filled-spandrel arch (3) c1939
ISHC design
Note: symmetrical center span flanked by unsymmetrical outer ones

(Owen #143)
A long girder span.

PARKE [61]

(See box on pages 245-247)

CR 1200E, E Co line; CCC&StL (Big 4) RR
Continuous T-beam (5) 1907
CCC&StL RR design
Note: narrow bm stems & wide flanges; pipe rails rmvd

1
CR 1100S, S Co line, W of Lena; Croys Ck
[Continuous] T-beam (2) c1915

5
CR 1000S or #5, E of CR 850E; Otter Ck, N Bh
Filled-spandrel arch 1933
Note: ring rests on steel plate

13
CR 500E or #133, N of CR 700S; Little Raccoon Ck, Strangers Bh
Filled-spandrel arch c1914

18 Wayne Day Br
CR 720S, W of SR #59; Big Raccoon Ck, Bh
Girder 1928

29
CR 900S, W of SR #59; Raccoon Ck, N Bh
Slab c1925
Note: long slab span (21 ft)

33
CR 500S, E of CR 80E; Little Raccoon Ck, Bh
Slab c1913
Note: long slab span (23 ft)

43
CR 890S, NE of Rosedale; Raccoon Ck, Bh
Slab c1930

76
CR 40E or #126, S of Rockville; Little Raccoon Ck, Sunderland Bh
Filled-spandrel arch c1910
Note: w on E with 2nd arch; nrly semicircular rings

91
CR 275S or #105, off CR 40E; Little Raccoon Ck, Sunderland Bh
Girder c1915

92
CR 35E, off CR 40E; Little Raccoon Ck, Sunderland Bh
Girder c1917

107
CR 700E or #135, nr S Co line; Ebenezer Ck
T-beam c1919
[W H __mure & Dawell, bldrs]

148
CR 1050E or #235, S of Milligan; Little Raccoon Ck, S Fk
T-beam (2) c1920

156 Dunn's Ford Br **NRC**
Waveland Rd or CR #325, N of Guion; Little Raccoon Ck
Filled-spandrel tied arch 1905
D B Luten, dsgr
C C King & Son, bldr
Note: patented design; (I.1A: "Arch Terms"; II.3B)

157
CR 720E, N of SR #236; Little Raccoon Ck
Girder (2) c1925
Note: triangular cutwater

160
CR 1000E or #205, S of CR 350N; Little Raccoon Ck, S Fk
T-beam (2) 1911
[H A B & M R, Jr, bldrs]

HOOSIER CAMELBACKS?

Little, indeed, is known about the design of many small reinforced concrete bridges still in place on Indiana's rural roads. County surveyors and engineers designed some; builders planned others. While in most cases the designs had once been committed to paper, the counties have typically discarded their bridge plans and bidding files, and small firms have disposed of their business records. The surviving structures that surveyor-engineers and entrepreneurs designed years ago often contain the only remaining evidence of their designers' intentions. But with reinforced concrete structures, where at least half the scheme for load-bearing is encased in cement, the designer's voice lies entombed.

The girders with arched tops spanning Parke County streams, for example, reveal a shape not found anywhere else in the state, and, with seven structures extant, their number suggests that the shape was not altogether casual. Although the dimensions of the seven survivors vary, some possibly suggestive patterns emerge. At fifty or more feet, most carry quite long clear spans for girders. All the girders are two feet wide, are typically five feet high at their ends, and all are arced, making them about two feet higher at midspan. Their width and height at span-end is consistent with rectangular girders. The quantity of concrete used differs from regular girders in only two

(continued on next page)

Seventy-five-foot Michigan camelback at L'Anse (above). Standard Michigan plans, denoting the vertical reinforcing (right, top) and the longitudinal reinforcing tied at span-end (right, bottom).

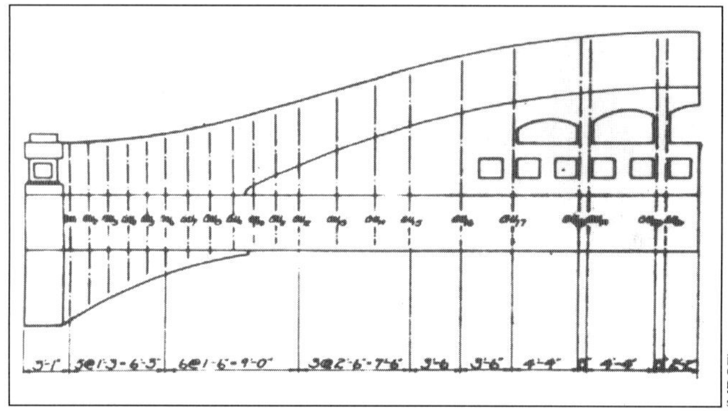

(MDOT)

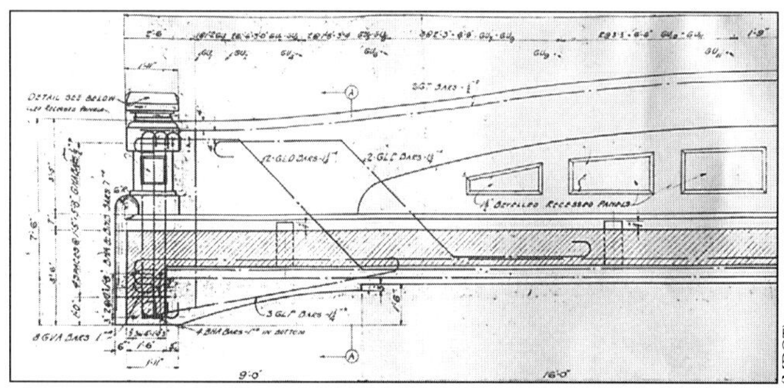

(MDOT)

ways. With their curved tops, Parke County's special girders require extra materials under the arc toward midspan, although they partly offset the addition with indented panels, which made the central regions of the girder slimmer.

Without either written evidence or an adequate number of physical clues, the detective may resort to analogy with better documented structures to hypothesize the patterns of reinforcement used in the more mysterious spans. The Parke County structures, for example, might be considered as small versions of concrete camelback girders. Michigan's state highway commission designed a camelback model, started building it around 1915, and constructed others at least into the mid-Twenties.[6] Engineers in neighboring Indiana knew of the Michigan designers' longer and wider girder spans.

Michigan camelbacks operate a lot like tied arches or bowstrings, similar in ways to the Meridian Street Bridge in Portland, Indiana (Jay #6182).[7] Longitudinal reinforcing runs near the upper edge of the arched girder top and along the horizontal lower edge. The upper and lower runs are tied together towards their ends, causing the significantly lateral thrust of the arced upper run to be resisted and offset by the horizontal lower run, thereby delivering largely vertical pressure to the abutments. To support the deck and to prevent the uppermost and lowermost rods from whipping out of the concrete, stirrups periodically tie rods from the two extremities together.

Collision, flooding damage, settling, and weather provide a glimpse here and there into the reinforcing pattern(s) of Parke County's seven arced girders. The clues, even when taken as a composite, remain disconnected shards: heavy longitudinal bars near the lower edge of the girder, somewhat lighter rods near the upper edge, and stirrups running vertically about every eighteen inches between the upper and lower rods. Of these clues, only the placement of the stirrups differs from the standard rectangular girder. If the designer thought the arced tops of his girders generated lateral thrust that needed to be resisted, then the upper and lower bar reinforcing systems are probably tied together at the girders' ends. The answer to this query, however, lies buried in concrete. Until one of these girders is knocked apart for replacement – and unless an interested observer witnesses that deconstruction even then – the Parke County designer's exact structural strategy will remain a mystery.

(continued on next page)

(Parke #34)

(Parke #106)

(Parke #220)

CAMELBACKS?
continued

21
CR 1025S or #84, W of SR #59, nr S Co line; Otter Ck, N Bh
[Bowstring] girder c1917
Note: *51 ft long*

34 Perrin Br NRC
CR 500S, Catlin; Sunderland Bh
[Bowstring] girder 1921
[C Ott, dsgr]
W J Nowling, bldr

98 NRC
CR 160S or #156, SE of Rockville; Williams Ck
[Bowstring] girder c1920
Note: *51 ft long*

106
CR 700E or #135, nr S Co line; Otter Ck, N Bh
[Bowstring] girder 1921
[C Ott, dsgr]
R J Teague, bldr

135 NRC
CR 475S or #128, SE of Ferndale; Rocky Fk Ck
[Bowstring] girder c1915
Note: *51 ft long*

220
CR 830N or #275, W of US #41; Roaring Ck
[Bowstring] girder c1915
Note: *50 ft long*

237
CR 400S, off CR 40E; Little Raccoon Ck, Sunderland Bh
[Bowstring] girder c1915

165
CR 160E, S of Nyesville; Sand Ck
T-beam c1910

197
CR 250W, W of Bloomingdale; Cat Ck
Continuous slab (2) c1910
Note: *long slab spans (24 ft); heavily flared pier*

203
CR 550W or #333, SE of Howard; Mill Ck, Jim Bh
Girder c1919
Note: *skewed*

208 NRC
CR 1050N or #345, E of Tangier; Rush Ck, E Fk
Tied beam 1918
[V Osborn, bldr]
Note: *rare example of bms tied through streambed pavement; bldr/worker's name penciled on bm; (Epilogue.B)*

217
CR 325W, NE of Tangier; Rush Ck, E Fk
Slab c1930
Note: *long slab span (25 ft)*

218
CR 175W, N of Annapolis; Sugar Ck, Bh
Slab c1930
Note: *long slab span (21 ft)*

229 Wilkins Mill Conc Br
CR 110E or #293, E of US #41; Sugar Mill Ck
T-beam (2) c1915

278 NRC
Greencastle Rd, W of CR 1000E; Rocky Fk Ck, Straight Bh
Slab 1941
Note: *WPA-funded; long slab span (22 ft); skewed; streambed pavement, possibly from previous br*

5864 NRC
US #41, S of SR #47; Roaring Ck
Open-spandrel arch 1926
ISHC design
Steele Bros, bldrs
Note: *old state-design open spandrel; 4-rib, nrly semicircular, ring; abutments atop stone walls on ck banks (probably from previous br); w1968 from 20 to 38 ft*

5865
US #41, N of SR #47; Sugar Ck
Filled-spandrel arch (3) 1926
ISHC design
R McCalman, bldr
Note: *319 ft long; 105-ft center span; w1969 rmvd spandrels & rails*

7056 NRC
Turkey Run State Pk, nr Inn; Turkey Run
Filled-spandrel arch 1914
H L Davies, dsgr
W E Ireland, bldr
Note: *hike the ravine below & view the span against the trees and sky; (before I)*

PERRY [62]

6024
SR #62, E of Uniontown; Anderson R, Mid Fk
Filled-spandrel arch 1925
ISHC design
B Seufert, bldr
Note: *w1984 rmvd spandrels & rails*

POSEY [65]

143 NRC
Upper Poseyville & Cynthiana Rd, W of SR #165; Black R, Bh
Slab c1905
Note: *long slab span (22 ft); low rails*

PULASKI [66]

211 NRC
CR 200W, NE of Ripley; Quigley Marsh Dh
Slab c1915
Note: *skewed; long slab span (25 ft); low rails*

238
CR 325E, NE of Star City; Prather Dh
Slab c1921
Note: *long slab span (24 ft); low rails*

1245
SR #14, SW of Ripley; Hathaway or Agnew Bh Dh
Slab 1932
ISHC design
Continental Engrg, bldr
Note: *12-ft span directly across ditch; 32 ft on skew; rails do not overlap; (VII.3B)*

3065
US #35, Winamac; Quigley Marsh Dh
T-beam c1937
ISHC design

(Pulaski #1245)

This ISHC-design slab is so skewed that the rails do not overlap.

PUTNAM [67]

CCC&StL (Big 4) RR; CR 800W & Snake Ck
Filled-spandrel arch c1906
CCC&StL RR design
Note: *semicircular ring under considerable fill*

CR 600W; CCC&StL (Big 4) RR
Continuous T-beam (4) 1909
Timber beam (3)
CCC&StL RR design
Note: *timber spans added to N; 187 ft long; skewed; narrow bm stems & wide flanges; latticed rails*

NRC
CR 275S; CCC&StL (Big 4) RR
Continuous T-beam (5) 1910
CCC&StL RR design
Note: *narrow bm stems & wide flanges; latticed rails; (VIII.1A)*

Big Four Arches **NRC**
CCC&StL (Big 4) RR; Hurley Rd or CR 500W & Little Walnut Ck
Open-spandrel arch (3) 1906
W M Dunne, engnr
Note: *among oldest open spandrels; skewed; heavy, coped rings; arched spandrel columns; pilasters curtain off low piers; (Prologue; Hendricks: "Haunted Catacombs")*

CR 400W; CCC&StL (Big 4) RR
Continuous T-beam (6) 1908
CCC&StL RR design
Note: *skewed; 166 ft long; narrow bm stems & wide flanges; pipe rails*

CCC&StL (Big 4) RR; Big Walnut Ck
Open-spandrel arch (3) 1906
W M Dunne, engnr
Note: *among oldest open spandrels; heavy, coped rings; arched spandrel columns; pilasters curtain off low piers; (Prologue; Hendricks: "Haunted Catacombs")*

CCC&StL (Big 4) RR; US #231, Greencastle
Filled-spandrel arch 1906
CCC&StL RR design
Note: *skewed; semicircular ring under considerable fill*

CR 100E; CCC&StL (Big 4) RR
Continuous T-beam (5) 1907
CCC&StL RR design
Note: *skewed; narrow bm stems & wide flanges; pipe rails rmvd*

Arnold Rd, NW of Fillmore; CCC&StL (Big 4) RR
Continuous T-beam (5) 1908
CCC&StL RR design
Note: *narrow bm stems & wide flanges; pipe rails*

42
CR 25W, SE of Fincastle; S Ramp Ck
Girder/floor bms c1920
Note: *skewed*

62 Keene Br **NRC**
CR 500N, W of Heritage Lake; Clear Ck
Spandrel-braced arch 1927
D B Luten, dsgr
C Dunning Supply Co, bldr
Note: *deck cantilevered over solid ribs; rails rmvd; (IX.1B)*

86 Grimes Arch
Morton Rd, SE of US #36; Owl Ck, Bh
Filled-spandrel arch 1922
Hamilton & Crawley, bldr

111 Dills Ford Br **NRC**
CR 50N, W of CR 450W; Little Walnut Ck
Filled-spandrel arch (2) 1909
D B Luten, dsgr
S H Gibson, bldr
Note: *linear rings with Luten-design reinforcing; rail balusters in stream*

148
CR 50N, E of Fillmore; Deer Ck, Bh
Slab c1925

159 Reelsville Br **NRC**
CR 650W, Reelsville; Big Walnut Ck
Open-spandrel arch 1929
Beam (5)
D B Luten, dsgr
Luten Engrg, bldr
Note: *2-rib, 120-ft, parabolic center span; floor bms extend deck beyond spandrel columns & bents; reuse of cut-stone S abutment of prior br allowed for slim concr anchoring wall; (I.1A: "Arch Terms")*

193 Allee Br
CR 600S, E of Belle Union; Belle Union Bh Ck
Spandrel-braced arch 1927
D B Luten, dsgr
C Dunning Supply Co, bldr
Note: *deck cantilevered over solid ribs; w1976 with prestressed bx bms to N rmvd rails; (IX.1B)*

206
CR 300S, nr E Co line; Mud or Sallust Bh Ck
Girder/floor bms c1920
Note: *skewed*

221
CR 875S, nr CR 600E; Higgins Bh Ck
Filled-spandrel arch c1915

229 Staley Br **NRC**
CR 925E, S of I-70 & nr E Co line; Mill Ck
Open-spandrel arch 1929
Beam (2)
D B Luten, dsgr
Luten Engrg, bldr
Note: *skewed; light, 2-rib ring; bm approaches anchor ribs; deck extended on floor bms; paneled pier pilasters & rails*

233
CR 925E, NW of Broad Pk; Cotton Bh Ck
Girder c1911

237 **NRC**
CR 550S or old US #40, E of Putnamville; Deer Ck
Open-spandrel arch (4) 1923
ISHC design
E F Smith, bldr
Note: *oldest state-design open spandrel; 276 ft long; 2-rib rings; arched spandrel columns; openarch paneled rails; bypassed in "Highway Rearmament" for World War II; (VII.2C)*

248
CR 650E, S of SR #236; Big Walnut Ck
Filled-spandrel arch (2) c1910
Note: *skewed; unsymmetrical rings; w1982 with cantilevered slab deck rmvd rails*

301 **NRC**
CR 750S or old US #40, W of Pleasant Gardens; Big Walnut Ck
Filled-spandrel arch (3) 1923
ISHC design
E C Wright, bldr
Note: *skewed; 234 ft long; bypassed in preparation for World War II*

1077
US #231, N of Greencastle; Big Walnut Ck
Filled-spandrel arch (3) c1931
ISHC design
Note: *skewed; 204 ft long; w1969 rmvd spandrels & rails*

1101
US #36, E of Bainbridge; Big Walnut Ck
Filled-spandrel arch (2) c1932
ISHC design
Note: *skewed; unsymmetrical rings*

(Putnam #229)

The Staley Ford Bridge, shown just after completion (above), is one of two sleek open-spandrel arches that Luten designed and built in Putnam County in 1929. Luten braced his light ribs (below, left) and secured their ends with a box-like slab approach. The ribs, columns, and braces were all chamfered.

(Putnam #229)

(Putnam #229)

1833
US #40, W of CR 800W; Johnson Bh
T-beam c1940
ISHC design
Note: skewed

1834
US #40, W of Pleasant Gardens; Big Walnut Ck
Filled-spandrel arch (3) 1941
ISHC design
Note: wartime replacement for 67-301; skewed; 234 ft long; 120-ft center span; w1986 rmvd deck, spandrels & rails; new slab deck cantilevered

1835 **NRC**
US #40, E of Putnamville; Deer Ck
Filled-spandrel arch (3) 1937
ISHC design
Note: wartime replacement for 67-237; skewed; w1992 rmvd deck & rails; new slab deck cantilevered

RANDOLPH [68]

29 **NRC**
CR 200W, SE of Ridgeville; Fetid Ck, Bh
Slab c1920
Note: skewed; long slab span (30 ft); low rails

RIPLEY [69]

NRC
CCC&StL (Big 4) RR; CR 300E, Batesville
Filled-spandrel arch 1905
CCC&StL RR design
W J Oliver, bldr
Note: Tennesseans blt as part of "Batesville cutoff" or bypass; fairly flat for ring under considerable fill; skewed; old rails used for reinforcing in many overpasses; described in ENR (16 May 1907)

NRC
CCC&StL (Big 4) RR;
CR 1400N, Morris
Filled-spandrel arch 1906
CCC&StL RR design
Southern Contr, bldr
Note: Kentucky firm blt to help improve rail profile; semicircular ring under considerable fill; old rails used for reinforcing in many overpasses; described in ENR (16 May 1907)

9 **NRC**
Cavehill Rd or CR 400E, NE of Olean; Laughery Ck
Filled-spandrel arch (2) c1910
Note: 202 ft long; unsymmetrical rings

18 **NRC**
CR 900S, W of CR 75W; Indian-Kentuck Ck
Girder/floor bms (2) c1907
Note: among longest girder structures (77 ft)

23
CR 925S, nr Haney Cor; Indian Ck, Vestal Bh
Girder/floor bms c1915

24
CR 200S, SW of Versailles; Big Graham Ck
Girder/floor bms c1919

35 **NRC**
US #421, SW of Rexville; Big Camp Ck, Bh
Filled-spandrel arch 1922
ISHC design
P S Hauk, bldr
Note: oldest state-design, highly-filled arch; w1958 rmvd spandrels

56
Old Michigan Rd, W of Osgood; Long Bh
Girder/floor bms c1916

(Ripley #9)

Highly unsymmetrical arch rings springing from high on the pier when compared with the abutments.

(Ripley #76)

76 **NRC**
Hopewell Rd, N of Holton; Long Bh
[Continuous] T-beam (3) c1920

1234
SR #101, S of Milan; Hogan Ck
Filled-spandrel arch 1933
ISHC design
O W Pegree & F F Wildman, bldrs
Note: *skewed; semicircular ring designed to carry considerable fill; (before VI)*

RUSH [70]

3
CR 1000N, nr W Co line; Sixmile Ck
Filled-spandrel arch (2) c1920
Note: *fairly flat rings; triangular cutwater; raised pier pilaster*

12
CR 1030N, SW of Carthage; Threemile Ck
T-beam c1940
Note: *skewed; like state design, rails sit on outer bm flange*

71
CR 600E, NW of Almouth; Shawnee Ck, Bh
Filled-spandrel arch (2) c1920
Note: *flat rings*

89 **NRC**
CR 450E, N of SR #44; Ben Davis Ck, Bh
Filled-spandrel arch c1910
Note: *skewed; very flat segmental ring*

90
CR 150N, off CR 450E; Ben Davis Ck, Bh
Filled-spandrel arch c1915
Note: *very flat segmental ring*

110 **NRC**
CR 550W, W of Henderson; Farmers Stream
Filled-spandrel arch c1920
Note: *skewed; very flat segmental arch; concr post & pipe rails*

115
CR 300N, NE of Arlington; Little Blue R
Filled-spandrel arch (2) c1920
Note: *flat rings; rails & much of spandrel walls rmvd*

118 **NRC**
CR 400N, W of Henderson; Little Blue R
Filled-spandrel arch c1915
Note: *heavy, flat, segmental ring*

141
CR 200W, N of CR 500S; Flatrock R, Bh
Filled-spandrel arch c1915
Note: *rail panels with open arcs above deck*

142
CR 650W, SE of Homer; Deer Ck
Filled-spandrel arch c1912
Note: *rather flat ring; pipe rails*

163
CR 800E, E Co line & S of Orange; Sanes Ck
Filled-spandrel arch c1915
Note: *w1983 with prestressed bx bms rmvd rails & covered most of spandrels*

175
CR 1000S, E of CR 100E; Clifty Ck, N Bh
Filled-spandrel arch (2) c1925
Note: *original N rail rmvd*

176
Base Rd, at S Co line; Clifty Ck, N Bh
Filled-spandrel arch (2) c1920

(Rush #90)

(Rush #110)

Two Rush County arches (above) with similarly flat, heavy, and segmental rings which raise curiosity about how their designer reinforced them.

192
CR 1050S, S of Moscow; Flatrock R
Filled-spandrel arch (3) c1915
Note: *220 ft long; heavy, flat, segmental rings; rails & most of spandrels rmvd*

3014
SR #44, Rushville; Flat Rock R
Filled-spandrel arch (4) c1939
ISHC design
Note: *249 ft long; structure blt on curve; w1985 with cantilevered slab deck rmvd rails*

ST. JOSEPH [71]

NRC
GTW RR; Mishawaka Ave, Mishawaka
Filled-spandrel arch 1929
J A Heaman, F P Sisson & A N Laird, dsgrs
Foley Bros & Fulton & Peppard, bldrs
Note: *blt as part of relocation, double-tracking & elevation program; skewed; fairly flat ring; hollow or cellular abutments; paneled & bush-hammered spandrels & abutment pilasters; raised emblems on main rail posts; waterproofed surface; described in ENR (3 Oct 1929)*

12
Johnson Rd, SW of Crumstown; Geyer Dh
T-beam (2) c1925
Note: *rails sit on outer bm flange*

28
Quinn Rd, W of Walnut Rd; Pine Ck
T-beam (2) c1925
Note: *skewed; rails sit on outer bm flange*

58
Underwood Rd or W 1st St, Walkerton; Pine Ck
T-beam c1930
Note: *skewed; influenced by state design*

152
Kintz Rd, N of Indian Village; Judy Ck
Filled-spandrel arch c1920
J & O Kintz, dsgrs & bldrs
Note: *home-made; blt upon field stones; skewed; with fill largely absent, road travels over ring*

203 Mishawaka Ave Br NRC
Mishawaka Ave, Mishawaka; St Joseph R
Filled-spandrel arch (4) 1926
D B Luten, dsgr
Nat Concr, bldr
Note: *using Luten-design, Nat Concr underbid all others who relied on C W Cole's Co-approved plans with Melan system reinforcing; 389 ft long; cantilevered slab deck, balustraded rails & lightposts rmvd (1987); (II.4A)*

(St. Joseph #203)

Even under construction, Luten's Mishawaka Avenue Bridge shows its polished rings, bush-hammered spandrel walls, and paneled pier and abutment pilasters.

(St. Joseph - GTW RR)

Sometimes even railroads had to adjust their often narrowly utilitarian perspective when designing bridges for an urban area that had long built spans with City-Beautiful decor. The Grand Trunk Railroad crossed Mishawaka Avenue in Mishawaka with style and efficiency. The public could see its paneled spandrel walls, pier pilasters, and rails (above) as well as its decorated rail posts (below, right). On the inside, the railroad's design engineers appear to have used Thacher-style two-tier ring reinforcing without stirrups, and, to save a considerable amount of concrete, hollow (i.e., cellular or boxed) abutments (below, left).

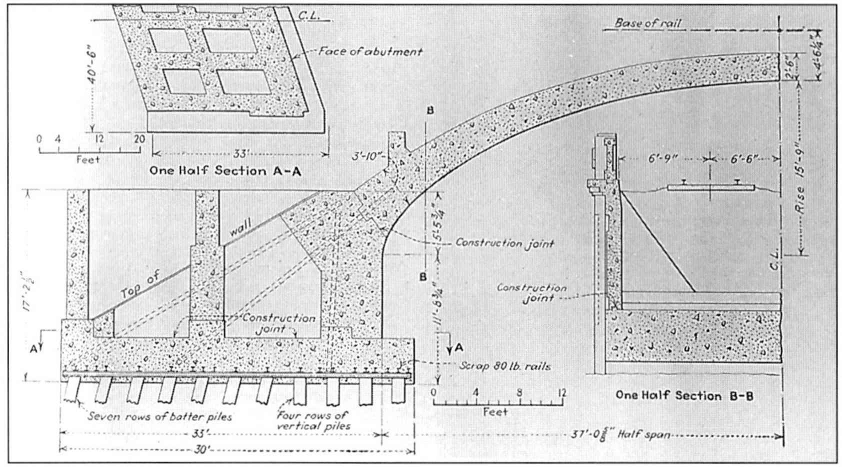

(St. Joseph - GTW RR; *ENR*, 1929)

(St. Joseph - GTW RR; *ENR*, 1929)

(St. Joseph - GTW RR)

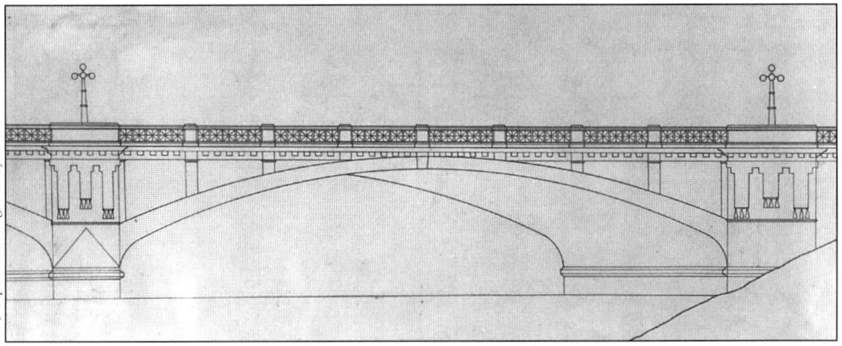

(St. Joseph #205; Lang, Feeney)

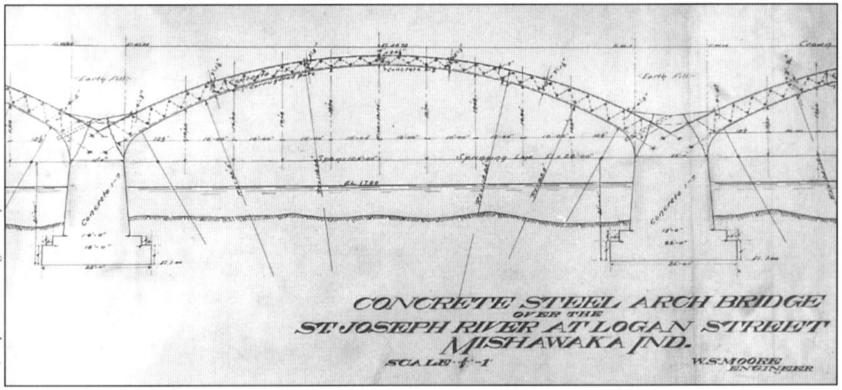

(St. Joseph #205; Lang, Feeney)

(St. Joseph #205)

A fine example of William S. Moore's "Concrete-Steel Arch Bridges" with von Emperger style reinforcing ribs for its rings and City-Beautiful decor (left). The bridge has been aesthetically-decapitated through rehabilitation (above).

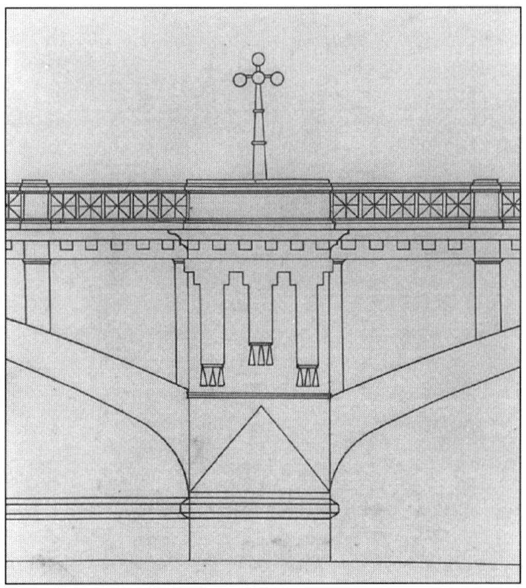

(St. Joseph #205; Lang, Feeney)

(St. Joseph #205)

205 Logan St/
Wm S Moore PE Mem Br NRC
Logan St, S Bend-Mishawaka;
St Joseph R
Filled-spandrel arch (4) 1913
W S Moore, dsgr
C L Copp, bldr
Note: Co offered bidders 2 sets of plans, a 5-span Luten design & a 4-span concrete-steel (Melan system) design of W S Moore; Moore's plans apparently brought the "lowest & best" bid; skewed; 467 ft long; 100-ft spans; w1976 with prestressed bx bms rmvd open-panel rails & light standards & obscured raised neoclassical emblems on pilasters

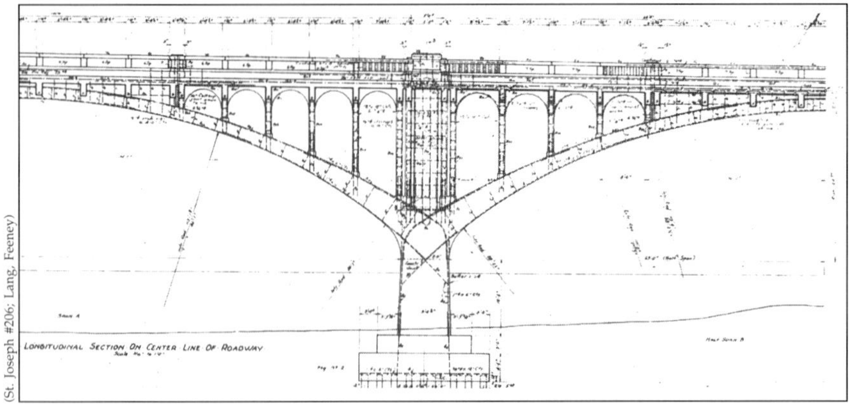

(St. Joseph #206; Lang, Feeney)

In the tradition of the time and place, Luten designed the Ernsperger or Ironwood Drive Bridge in City-Beautiful style (right), but the county has amended that through renovation (below). Luten used two-tier reinforcing with stirrups in the ribs and, as usual, extended the longitudinal rods elastically into the piers (above). The parsimonious designer did not waste concrete and steel on pier sections where there was no rib to bring them structural work.

(St. Joseph #206; Lang, Feeney)

(St. Joseph #206)

206 Ernsperger Ironwood Dr Br **NRC**
Ironwood Dr, S Bend; St Joseph R
Open-spandrel arch (3) 1928
Beam (8)
D B Luten, dsgr
Nat Concr, bldr
Note: *Co Engnr R J Lang prepared plans for 4-span filled-spandrel arch estimated to cost $231,650; Nat Concr successfully bid $159,500 to bld 3-span open-spandrel of Luten-design; 478 ft long; 122-130 ft spans; 2 ribs per span; spandrel piers; new cantilevered slab deck (1970) rmvd balustraded rails & lampposts with cast-bronze fixtures*

207 Kline St/ Twyckenham Dr Br **NRC**
Twyckenham Dr, S Bend; St Joseph R
Open-spandrel arch (4) 1929
W S Moore, dsgr
Rieth-Riley & E F Smith, bldrs
Note: *432 ft long; 89-115 ft spans; 4 ribs per span; walks carried on brackets; memorials to the soldiers killed in World War I rise above the pier & abutment columns; slab deck, walks, rails with open-arch panels & lights with cast-bronze fixtures rmvd (1983); (IX.2A)*

209 Jefferson St Br **NRC**
Jefferson Blvd, S Bend; St Joseph R
Filled-spandrel arch (4) 1906
A J Hammond, dsgr
Scribner & Heyworth, bldr
Wis Br & Iron, bldr
Note: *originally considered "one of the ornamental structures of the city"; "winding concr steps" lead into park; skewed; 490 ft long; 110-ft spans; Melan system reinforcing; neo-classical chalice raised on pier columns; deck, balustrated rails, refuge bays & light standards rmvd (1962); (I.3)*

210 LaSalle St Br **NRC**
La Salle St, S Bend; St Joseph R
Filled-spandrel arch (3) 1907
A J Hammond, dsgr
Concr-Steel Engrg, reinf dsgr
Marsh Br, bldr
Note: *279 ft long; patented Melan system reinforcing; rounded, capped cutwaters; piers & arch rings decorated like stone; frieze under rail coping; rail balusters gone, but posts remain; (I.3; IV.2)*

211 Angela Ave Br **NRC**
Angela Blvd, S Bend; St Joseph R
Filled-spandrel arch (3) 1940
W S Moore & F J Miller, dsgrs
H L Maddocks, bldr
Note: *Public Works Admin funding; 268 ft long; counterforted brackets carry sidewalks; rounded cutwaters shift into declining, streamlined pier pilasters; open-oval rail panels & raised-arch end posts; (IX.2A)*

3690 Leeper Br **NRC**
Michigan St, S Bend, or US #31; St Joseph R
Filled-spandrel arch (3) 1915
C W Cole, dsgr
Cleary-Kuert Contr, bldr
Note: *"one of the most artistic structures in the state"; 276 ft long; 116-ft center span; limestone-faced walls & columns; neoclassical decor; keystone inserts on rings; raised emblems on pier & abutment pilasters; open-arch rail panels; lampposts rmvd; (I.3)*

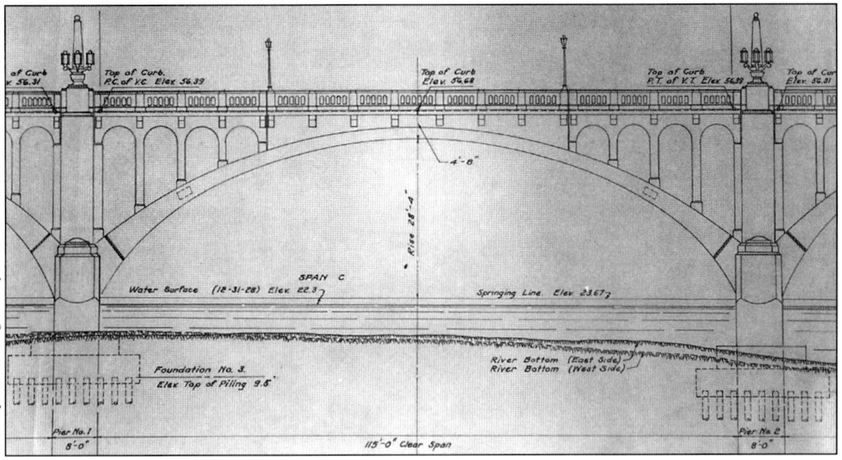

(St. Joseph #207; Lang, Feeney)

(St. Joseph #207)

For the Twyckenham Drive Bridge, Moore planned a substantial memorial to local soldiers who died in the First World War. He emphasized his commemorative obelisks with decorative lighting. While parts of the memorial have been professionally pillaged over time, its stately entrances remain essentially intact (right, top and bottom).

(St. Joseph #207)

(St. Joseph #207)

SCOTT [72]

29
CR 275S, W of US #31; Honey Run
Slab c1907

57
CR 200S, W of SR #203; Town Ck
[Continuous] slab (2) c1907
Note: *sits on stone abutments & pier of previous br; long slab spans (21 ft)*

100
Getty Rd or CR 875E, S of SR #56; Woods Fk
Girder/floor bms c1915

110
CR 150N, W of CR 200E; Flat Ck
Girder/floor bms c1907
Note: *skewed*

3435 **NRC**
SR #356, Vienna; Pigeon Roost Ck
Continuous framed slab (3) 1942
ISHC design
C A Kahl, bldr
Note: *skewed; slabs rigidly tied to piers; rails curve away from roadway at span end; (VIII.2B)*

SHELBY [73]

8 **NRC**
CR 600W, nr N Co line; Sugar Ck
Filled-spandrel arch (2) 1917
D B Luten, dsgr
Nat Concr, bldr
Note: *skewed; flat, linear rings; rail balusters rmvd*

17
CR 1000N, W of SR #9; Brandywine Ck
Filled-spandrel arch (2) c1915
Note: *triangular cutwater; inscribed spandrels somewhat arched to resonate with rings*

23 **NRC**
CR 400N, NE of Knighthood Village; Foreman Bh
Girder/floor bms [framed] c1920
Note: *girder reinforcing extends into abutments*

28
CR 350N, E of Knighthood Village; Little Blue R
Filled-spandrel arch (2) c1918

(Shelby #8)
The flat but flowing linear rings integrated with a slim pier mark Luten design.

36 **NRC**
CR 300S, W of CR 725E; Conns Ck
Filled-spandrel arch c1915
Note: *skewed; flat ring; neoclassical rail balusters*

49 Red Mills Br **NRC**
CR 400N, NW of Boggstown; Sugar Ck
Filled-spandrel arch (4) 1927
D B Luten, dsgr
Nat Concr, bldr
Note: *269 ft long; linear rings; deck extended on brackets; bush-hammered spandrels & pilaster panels; deck & balustraded rails rmvd*

99
CR 1200S, S Co line; Tough Ck
Filled-spandrel arch c1925
Note: *wide reinforcing straps run diagonally upward from ring's haunches to midspan through spandrels & rails*

100
S Co Line Rd, SW of Flat Rock; Ensley Dh
Filled-spandrel arch c1917
Note: *flat ring; deck rides on ring at crown*

The auto-induced destruction of one of the wrought-iron, bowstring-arch spans of the Red Mills Bridge opened the door to Luten-design concrete arches (below).

(Shelby #49)

118
CR 700S, E of SR #9; Lewis Ck, S Fk
Girder/floor bms [framed] c1917
Note: *girder reinforcement extends into abutments*

129 **NRC**
CR 50E; W of Wilson Cor; Lewis Ck, Mid Fk
Girder/floor bms 1927
[H B & W L, bldrs]
Note: *unusually tall girders (c10 ft) & long span (50 ft)*

183 **NRC**
CR 150W, NE of Flat Rock; Flat Rock R overflow
Slab (6) c1917
Note: *long slab br (124 ft); triangular cutwaters; no rails*

16
CR 650N, W of SR #9; Brandywine Ck
Slab (3) c1910
Note: *streambed pavement; rails rmvd*

SPENCER [74]

7
CR 420W or #21, N of CR 550S; Ohio R Dn
Girder/floor bms c1913
Note: *fill covers half of girders, leaving modest railing*

9
CR 450W or #21, S of CR 425S; Forked Pd Dh
Filled-spandrel arch c1907
Note: *rails rmvd*

14
CR 400W, S of SR #66; Willow Pd Dh
Filled-spandrel arch c1910
Note: *flat ring*

22
SR #70, E of SR #245; Waterman Bh
Slab (3) 1925
ISHC design
R P Olinger, bldr
Note: *skewed; w1960 with prestressed bx bms rmvd rails & enclosed slab*

26
SR #70, W of SR #66; Waterman Bh
Filled-spandrel arch 1925
ISHC design
R P Olinger, bldr
Note: *semicircular ring designed to carry fill above short spandrel walls; w1968 & lengthened with precast channel bms*

37
CR 80S or #96, W of Eureka; Clear Ck
Filled-spandrel arch c1907
Note: *flat ring; rails rmvd*

55
CR 400N, W of SR #45; Honey Ck
Filled-spandrel arch c1913
Note: *rails rmvd*

58
CR 200W or #33, W of Rockport; Lake Dn Dh
Filled-spandrel arch c1910
Note: *rails rmvd*

60
CR 100W, SW of Rockport; Garrett Ck
Filled-spandrel arch c1910
Note: *fairly flat ring*

76 **NRC**
CR 520W or #115, SW of Midway; Sweezer Dh
Slab/[girder] c1906
Note: *this & some other brs in the Co appear to be designed as hybrids of flat slabs & girder structures; very long as slab span (35 ft); heavy rails may double as girders & help with load-bearing; skewed*

79
CR 450W or #23, SW of Midway; Sweezer Dh, Bh
Slab/[girder] c1916
Note: *skewed; long as slab span (25 ft); heavy rails may double as girders & help with load-bearing*

80
CR 450W or #23, SW of Midway; Sweezer Dh
Slab/[girder] c1914
Note: *skewed; long as slab span (25 ft); heavy rails may double as girders & help with load-bearing*

82
CR 200E, NE of Chrisney; Chisney Dh, Bh
Girder/floor bms c1916

94
CR 700N or #42, SE of Midway; Sweezer Dh
Slab c1904
Note: *skewed; long slab span (23 ft)*

114
CR 1350N or #242, W of SR #545; Crooked Ck, E Fk
Slab c1907

132 **NRC**
CR 950N or #75, E of CR 1000E; Crooked Ck
Filled-spandrel arch c1913
Note: *flat ring; deck rides on ring at crown*

(Spencer #255)

On old nicely-crafted small country arch whose builder signed his work.

138
CR 1450N or #240, E of SR #245; Crooked Ck
T-beam c1920
Note: *rails sit on outer bm flange*

180
CR 455E or #219, Dale; Little Pigeon Ck, N Fk, Bh
Slab c1915

184
CR 2100N, NE of Mariah Hills; Hunley Ck
Girder/floor bms c1915

251
CR 1600N or #72, NW of Fulda; Crooked Ck, E Fk
Filled-spandrel arch c1905
Note: *fairly flat ring; token rails*

255 **NRC**
CR 100N or #306, W of Hatfield; Baker Ck or Willow Pond Dh
Filled-spandrel arch 1908
S P, bldr
Note: *small country arch; nrly semicircular ring; panels inscribed on rails*

258
CR 1275E or #287, N of Fulda; Swinging Ck
Slab c1905

259
CR 1100E or #83, N of Evanston; Crooked Ck, Mid Fk
Filled-spandrel arch c1905
Note: *skewed; flat ring*

268
CR 390S, S of Africa; Garrett Ck
Slab c1915

273
CR 250N or #108, E of CR 1000W; Baker Ck, Bh
Slab c1904
Note: *skewed; long slab span (22 ft)*

308
CR 700E, S of Kennedy; Crooked Ck, Bh
Filled-spandrel arch c1905
Note: *semicircular ring; E rail rmvd*

717
US #231, N of Chisney; Chrisney Dh
T-beam 1928
ISHC design
A G Ryan Contr, bldr
Note: *skewed; local Chisney bldr*

1353 **NRC**
US #231, Rockport; Lake Dn Ck
Filled-spandrel arch 1925
ISHC design
M & P Contr, bldr
Note: *nrly semicircular ring to carry fill above spandrels; w1932 rmvd spandrels & added rails*

3914
SR #62, W of US #231; Longs Bh, Bh
Slab (2) 1923
ISHC design
B B Hawkins, bldr
Note: *2nd oldest state-designed slab br; skewed; w1961 with prestressed bx bms rmvd rails & encased slab*

6281
SR #62, S of Gentryville; Buckhorn Ck, Bh
T-beam 1922
ISHC design
D Jenkins, bldr
Note: *old surviving state-design T-bm br; skewed; w1958 with precast channel bms rmvd rails*

STEUBEN [76]

62
Old US #27, E of Pleasant Lake; Pigeon Ck
T-beam 1927
ISHC design
Yost Bros, bldr

SULLIVAN [77]

6 **NRC**
S Co Line Rd or CR 1200S, W of CR 625E; Pollard Dh, Bh
Filled-spandrel arch c1918
Note: *2-arc ring approximates a semicircle; open-arch rail panels*

18 **NRC**
CR 1100S, W of CR 800E; Pollard Dh
T-beam 1921
W L Sisson, dsgr
Campbell, Snyder & Hart, bldr
Note: *40-ft span*

23
CR 900S, W of CR 500E; Maria Ck
Beam c1916

28
CR 500E, N of CR 975S; Maria Ck, Bh
T-beam c1927

(Sullivan #6)

(Sullivan #6)

Although the semicircular arc in his rails is more certain than in his ring, the craftsman succeeded in bringing grace to function, adding a touch of gentility in the countryside.

(Sullivan #177)

T-beams angled gently upward to midspan and sporting coped and paneled rails.

76
CR 450W, N of Rose Ch; Rogers Dh, Bh
T-beam c1930

77
CR 550S, W of Rose Ch; Rogers Dh, Bh
T-beam c1930
[D D H, bldr]
Note: *skewed*

86
CR 625W, NE of Merom Sta; Little Turtle Ck
Filled-spandrel arch c1930
Note: *nrly semicircular ring*

113
CR 250E, N of CR 625S; Mid Fk Ck
[Continuous] T-beam (2) c1916
Note: *skewed; rails sit on outer bm flange over span & angle away from roadway on abutments*

127 J I Adams Br **NRC**
CR 500S, E of CR 500E; Mid Fk Ck
[Continuous] slab (2) 1924
J E Collins, bldr
Note: *long slab span (22 ft)*

137 **NRC**
CR 400E, S of CR 500S; Mid Fk Ck
Girder/floor bms c1935
Note: *unusually wide girder structure; (II.3D)*

147 Four-Way Br
E Co Line Rd or CR 900E & 350S; Dh
T-beam c1930
Note: *intersects 2 roads & serves both; width of bm flanges varied to provide skew*

150
CR 250S, W of Victoria; Trout Lk Ck
T-beam c1927
Note: *shallower & thinner outer bms carry rails; rails angle gently upwards to midspan*

163
CR 700E, SE of Greenville; Big Bh Ck
Filled-spandrel arch c1917
Note: *considerable rise to ring*

166
CR 600E, S of Hymera; Sulphur Ck
Filled-spandrel arch 1916
[V R, H T, F T, bldrs]

177
Possum Hollow Rd or CR 825E, W of Gilmour; Big Bh or Possum Hollow Ck
T-beam c1930
Note: *long bm span (41 ft); shallower outer bms carry rails & both angle gently upward to midspan*

196
CR 50N, E of Sullivan; Morrison Ck
T-beam c1935
Note: *long bm span (40 ft); shallower rail-carrying outer bms*

238
CR 400W, N of CR 500N; Sugar Ck
Filled-spandrel arch c1925

243
CR 600N, W of SR #63; Turman Ck
T-beam c1925
W L Sisson, dsgr
Pedenshaus, bldr
Note: *long bm span (40 ft); bms angle slightly upward to midspan; shallower rail-carrying outer bms*

259
Old Kentucky Rd, NW of Fairbanks; Prairie Ck, Bh
T-beam c1917
Note: *skewed*

277
Thunderbird Pond Rd or CR 75W, N of SR #48; Turman Ck
Girder/floor bms (2) c1915

293 **NRC**
CR 950N, E of Baldridge; Busseron Ck, W Fk
Filled-spandrel arch 1916
W Sisson, dsgr
B Wood, bldr
Note: *flat ring*

295
CR 650N, Shelburn; Kettle Ck
T-beam c1926
Note: *long bm span (40 ft); shallow rail-carrying outer bms*

297
CR 900N, NW of Hymera; Busseron Ck
Filled-spandrel arch (2) c1917
Note: *triangular cutwater merging into pier pilaster*

303 Roy Drennon Br **NRC**
CR 1050N, E of CR 600E; Hooker Ck
T-beam (2) 1922
W L Sisson, dsgr
J W Dougherty & Sons, bldr
Note: *shallow rail-carrying outer bms*

305
CR 850E, S of CR 1100N; Boston Ck
Filled-spandrel arch c1927
Note: *flat ring*

320
CR 600N, SE of Hymera; Sulphur Ck
T-beam c1918
Note: *skewed; long bm span (41 ft); shallow rail-carrying outer bms*

321 Geo Bear Br
CR 750N, NE of Hymera; Sulphur Ck
Girder/floor bms 1923
J W Daugherty & Son, bldr

(Switzerland #59)

A Luten-design, spandrel-braced arch nestled into the countryside.

SWITZERLAND [78]

44 **NRC**
SR #56, Lamb; Green Valley Ck
Slab (2) 1922
ISHC design
J H Thornburgh, bldr
Note: *oldest extant state-designed slab br; skewed; w1958 with precast channel bms rmvd rails & enclosed slab, pier & abutments*

59 **NRC**
Turtle Ck Rd or CR #99, N of Florence; Turtle Ck
Spandrel-braced arch c1930
[D B Luten, dsgr]
Note: *slab deck cantilevered beyond solid ribs; (IX.1B)*

TIPPECANOE [79]

West Br **NRC**
Columbian Park, Lafayette
Filled-spandrel tied arch 1902
D B Luten, dsgr
Nat Br, bldr
Note: *advertised in ENR "as flattest elliptical arch br in the world," a claim contested (30 Oct 1902) & defended (13 Nov 1902); probably oldest surviving Luten structure; patented design; rail panels rmvd & spandrels covered; (II.3B; III.1Bbx)*

(Tippecanoe #173)

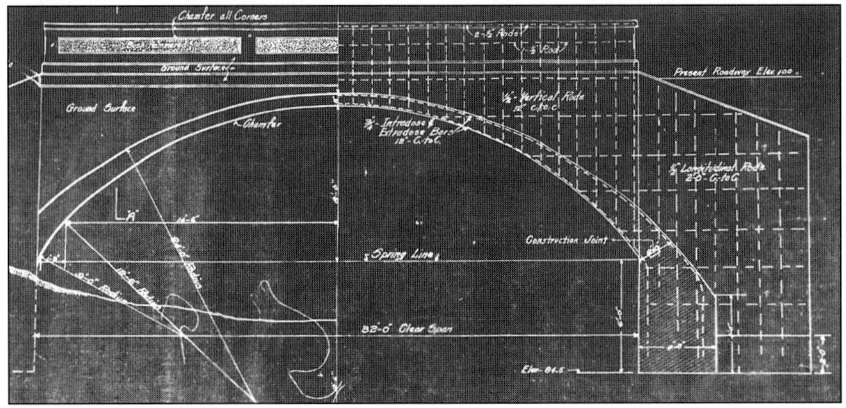

(Tippecanoe County Historical Society)

(Tippecanoe #151)

Walter DeLong's arch (above, top) lies partly encased in the box beams used to widen it. DeLong planned for two-tier ring reinforcing tied into that of the spandrel walls. He seems to have borrowed the pattern of coping and the bush-hammering of the panels from ISHC design.

19
CR 100EN of CR 1000S; Wea Ck
Filled-spandrel arch c1917
Note: w1984 with cantilevered slab deck rmvd rails

33
CR 200S, nr E Co line; Hog Run
Filled-spandrel arch c1920
Note: stone wingwall on SW from previous br; inscribed spandrels; w1982 with cantilevered slab deck rmvd rails which angled away from roadway over abutments

65
Lilly Rd, W of Lafayette; Wea Ck
Filled-spandrel arch (2) c1915
Note: site of Foresman-Hull mill; rather flat rings

82
Turner Rd, nr W Co line; Flint Ck
Filled-spandrel arch c1917
Note: stone wingwalls may be from former br; pipe rails

151 **NRC**
Pretty Prairie Rd, N of Americus; Tippecanoe R
Filled-spandrel arch (4) 1925
Note: 379 ft long; inscribed spandrels; deck & rails rmvd (1977)

173
CR 600N, W of SR #43; Burnetts Ck
Filled-spandrel arch 1928
W DeLong, dsgr
Note: w1973 with prestressed bx bms covered most of spandrels & rmvd rails

TIPTON [80]

6 **NRC**
S Co Line Rd or CR 600S, E of Ekin; Prairie Ck
T-beam (2) 1914
H Grifford, bldr

7
CR 725W, S of Goldsmith; Dixon Ck
Girder/floor bms c1930
Note: skewed

44
CR 75S, SE of Curtisville; Polywog Ck
Girder/floor bms c1916

93
S Co Line Rd or CR 600S, W of SR #37; Duck Ck
Filled-spandrel arch c1930

UNION [81]

19
Jobe Rd or CR 400W, N of Brownsville; Clay Ck
Filled-spandrel arch c1910
Note: heavy, flat ring

VANDERBURGH [82]

NRC
CCC&StL (Big 4) RR; Mohr Rd, E of St Joseph Ave
Filled-spandrel arch 1910
CCC&StL RR design
Note: semicircular ring designed to carry considerable fill

CCC&StL (Big 4) RR; Wimberg Rd, E of St Joseph Ave
Slab 1910
CCC&StL RR design
Note: deep slab (c2 ft) under fill

700
Golden Rule Rd, W of Seminary Rd; Duck Pond Dh, Bh
[Continuous] slab (2) c1910
Note: low rails

830 **NRC**
Heppler Rd, nr Martin; Barr Ck
T-beam (2) 1909
B Whitehead, bldr
Note: oldest known bm br; light outer, rail-carrying bms; pipe rails

1542 **NRC**
Old Henderson Rd, SE of Vaughn; Ohio R, Dn
Girder/floor bms (2) 1923
R W Meisinger, dsgr
M & P Contr, bldr
Note: abutment-to-pier outer-lower bms tie span together at stream level

1544
Old Henderson Rd, SE of Vaughn; Ohio R, Dn
Girder/floor bms (2) 1923
R W Meisinger, dsgr
M & P Contr, bldr
Note: abutment-to-pier outer-lower bms tie span together at stream level

(Tipton #6)

Hutio Gifford's T-beams.

(Vanderburgh #1544)

(Vanderburgh #1545)

Above – R. W. Meisinger cloned three girder structures to span seasonal drains to the Ohio River. To anchor securely the substructures of the two-span bridges, Meisinger tied the pier to the abutment through a streambed beam inserted longitudinally at the outer edges of each span.

1545 **NRC**
Old Henderson Rd, SE of Vaughn; Ohio R, Dn
Girder/floor bms (2) 1923
R W Meisinger, dsgr
M & P Contr, bldr
Note: *abutment-to-pier outer-lower bms tie span together at stream level*

1940
Schissler Rd, nr W Co line; Sanders Ck
Filled-spandrel arch c1908
Note: *skewed; flat ring*

1980
Seven Hills Rd, W of SR #57; Bluegrass Ck
Filled-spandrel arch c1908
Note: *skewed; arch tied; w1980 with precast channel bms rmvd rails*

1990
Slate Rd, SW of St Joseph; Big Ck, Fk
Filled-spandrel arch c1917
Note: *w1984 with precast channel bms rmvd rails*

2430
Wimberg Rd, E of St Joseph Ave; Locust Ck
Filled-spandrel arch 1906
B Whitehead, bldr
Note: *skewed; w1980 with precast channel bms covered spandrels & rmvd rails*

3443
SR #57, E of McCutchanville; Firlick Ck
T-beam c1941
ISHC design
Note: *skewed*

3445 **NRC**
SR #57 at Base Line Rd; Big Blue Grass Ck
Continuous framed slab (3) 1942
ISHC design
Hulzer Constr, bldr
Note: *slab reinforcing tied into piers; skewed; walks & open-arch rail panels; (VIII.2B)*

VERMILLION [83]

1
Old SR #63, N of Clinton; Norton Ck
Filled-spandrel arch 1934
ISHC design
H Tharp, bldr

15
Meridian Rd, SW of Universal; Gin Ck
Filled-spandrel arch c1925
Note: *flat ring, more linear than arched*

35
CR 225E, N of CR 1050S; Norton Ck
Filled-spandrel arch c1920
Note: *considerable rise to ring*

36
CR 200E, N of CR 1050S; Norton Ck
Girder c1916
Note: *skewed*

76 Patrick Ford Br **NRC**
CR 315N, nr W Co line; Little Vermilion R
Filled-spandrel tied arch (2) 1907
D B Luten, dsgr
Nat Concr, bldr
Note: *patented design; unsymmetrical rings*

524
6th St, Clinton; Feather Ck
Girder c1916

(Vermillion #76)

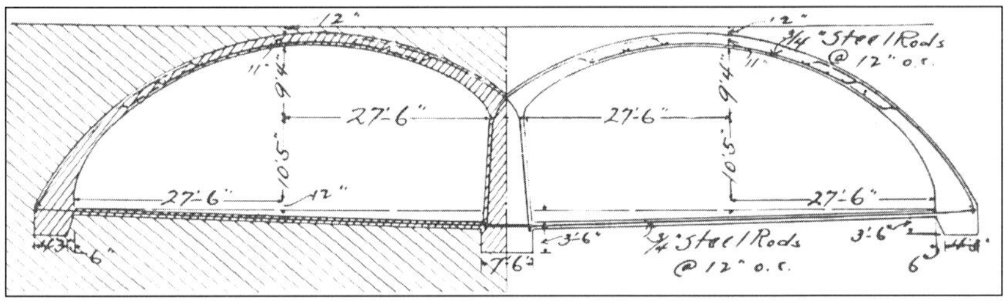

(Vermillion #76)

Luten's unsymmetrical tied arches when newly erected in 1907 at Patrick Ford.

VIGO [84]

59th Pl or CR #211, SE of Sandford; CCC&StL (Big 4) RR
Continuous T-beam (5) 1930
CCC&StL RR design
Note: *skewed; wide bm flanges; pipe rail*

CCC&StL (Big 4) RR; US #150, Marion Heights
Filled-spandrel arch c1929
CCC&StL RR design
Note: *skewed; semicircular ring designed to carry considerable fill*

6 Miller Br **NRC**
CR 50E or 147th Dr, SW of Black Hawk; Busseron Ck, W Fk
Girder/floor bms 1919
R E Gibbons, dsgr
Lloyd & Weeks, bldr
Note: *floor bms tied to girder top, placing loadbearing below roadway; girders & rails angled gently upward to midspan*

7 **NRC**
CR 89W or 159th Dr, E Co line; Watkins Ck
T-beam 1914
H C Anderson, dsgr
J J Nyhoff, bldr
Note: *neoclassical rail balusters*

12
109th St, SE of Black Hawk; Splunge Ck, Bh
Slab c1915
Note: *long slab span (25 ft); slab angled slightly upward to midspan; rail balusters in ditch*

17
CR 68E or 111th Dr, E of 103rd St; Owl Ck
Filled-spandrel arch c1910
Note: *low rails*

18
43rd St or CR 30S, N of SR #246; Busseron Ck, W Fk
Filled-spandrel arch c1900
Note: *skewed; fair rise to ring; pipe rail*

38
31st St or CR 18S, SE of Pimento; Turman Ck, Bh
Slab c1915

39
CR 13W or 111th Dr, W of US #41; Prairie Ck, Bh
Beam c1905

45
CR 17W or 123rd Dr, E of 18th Pl; Prairie Ck, Bh
Filled-spandrel arch c1915
Note: *open-rectangle rail panels*

54
CR 75W or 165th Dr, W of Farmersburg; Turman Ck, Bh
T-beam 1913
H C Anderson, dsgr
F Lloyd, bldr
Note: *shallow outer bm carries rail*

56
S Co Line Rd or 171th Dr, SE of Prairie Ck; Turman Ck, W Fk
T-beam c1910
Note: *shallow outer bm carries rail*

58
off S Co Line Rd or 171th Dr, SW of Prairie Ck; Prairie Ck overflow
T-beam c1905
Note: *shallow outer bm carries rail*

70
43rd Place or CR 35S, NE of Prairie Ck; Prairie Ck, Bh
Slab c1915

87 Wm F Altekruse Br **NRC**
Collins or Rigney Rd, NW of Allendale; Honey Ck
Filled-spandrel arch (2) 1917
H C Anderson, dsgr
Carpenter Const, bldr
Note: *skewed; rail balusters gone*

91 Henry Br
CR 6S or 81st Dr, at CR 34E, NE of Youngstown; Honey Ck, Bh
Girder/floor bms 1919
R E Gibbons, dsgr
Lloyd & Weeks, bldr
Note: *floor bms tied to girder top, placing load-bearing below roadway; bms & rails angle gently upward to midspan*

116 Hyde Br **NRC**
CR 92S or 103rd St, S of I-70; Honey Ck, Bh
Filled-spandrel arch (2) 1918
H C Anderson, dsgr
Allen & Losby, bldr
Note: *triangular cutwater*

159
Levee Rd, SE of Marion Heights; Little Sugar Ck, Bh
[Continuous] T-beam (3) 1930
Note: *skewed; wide bm flanges; heavier bms for center span; rails rmvd*

162 **NRC**
Shepardsville Rd or CR 26W, N of Tecumseh; S Salt Ck, Bh
T-beam 1915
H C Anderson, dsgr
Brelse M-g Const, bldr
Note: *neoclassical rail balusters; shallow outer bm under rail*

182
Gray Rd or 67th St, N of US #40; Lost Ck
T-beam 1915
H C Anderson, dsgr
H L Britton, bldr
Note: *some neoclassical rail balusters remain; shallow outer bm under rail*

187
E Co Line Rd or 139th St, SE of Coal Bluff; Sharon Bk
Girder c1915

188
81st St or Rio Grande Rd, S of SR #62; Otter Ck, N Bh
Filled-spandrel arch (2) c1921
Note: *w1980 with cantilevered slab deck rmvd rails & obscured spandrels & pilasters*

205
63rd Ave or CR 25W, E of US #41; Otter Ck
Filled-spandrel arch (2) c1919
Note: *raised spandrel panels; w1973 with cantilevered slab deck rmvd rails*

208
Downs Rd or CR 35N; Pit Run
Slab c1905
H C Anderson, dsgr
C O Landine, bldr

(Vigo #162)

The Vigo County Engineer did not want his T-beam span ignored.

(Vigo #321)

Luten won the opportunity to design a bridge named for the Vigo County Engineer.

242 **NRC**
Lafayette Ave, Spelterville; Otter Ck
Filled-spandrel arch (3) 1927
ISHC design
Walter Bros, bldr
Note: *brackets carry sidewalk to W; rare example of state use of neoclassically balustraded rails, although later rmvd for standard ISHC ones*

252 **NRC**
Old US #40, W of Larimer Hill; Clear Ck
Filled-spandrel arch (2) 1922
ISHC design
M & P Contr, bldr
Note: *among oldest unaltered state-design arches*

321 H C Anderson Br **NRC**
Lafayette Ave, Terre Haute; Lost Ck
Filled-spandrel arch 1917
D B Luten, dsgr
Nat Concr, bldr
Note: *neoclassical rail balusters*

1637
US #40, W of E Glenn; Lost Ck
Filled-spandrel arch 1924
ISHC design
Note: *semicircular ring designed for fill above spandrels; skewed; w1935 enclosed structure*

WABASH [85]

(See box on page 266)

41 Belden Br **NRC**
Schmalzried Rd or CR 100N, nr E Co line; Wabash R
Filled-spandrel arch (4) 1912
D B Luten, dsgr
Nat Concr, bldr
Note: *412 ft long; triangular cutwaters; rails rmvd; (II.4A)*

85
CR 600E, N of US #24; Rager Ck
Filled-spandrel arch c1915
D G, bldr
Note: *nrly semicircular ring*

119
CR 600E, S of Division Rd; Salamonie R, Bh
Filled-spandrel arch c1917
Note: *semicircular ring*

504 Mill St Br **NRC**
Mill St, Wabash; Charley Ck
Filled-spandrel arch 1903
D B Luten, dsgr
Nat Br, bldr
Note: *Luten named the br with an H rather than an M; skewed; illustrates patented spandrel wall help with load-bearing through cantilevering over abutments & with reinforcing for tension, allowing a reduction in ring size; rails rmvd*

505 Charley Ck Br **NRC**
Ferry St, Wabash; Charley Ck
Filled-spandrel arch (2) 1905
Trussed Concr Steel, dsgrs
J A Ross, bldr
Note: *240 ft long; rails rmvd*

(Wabash #504)

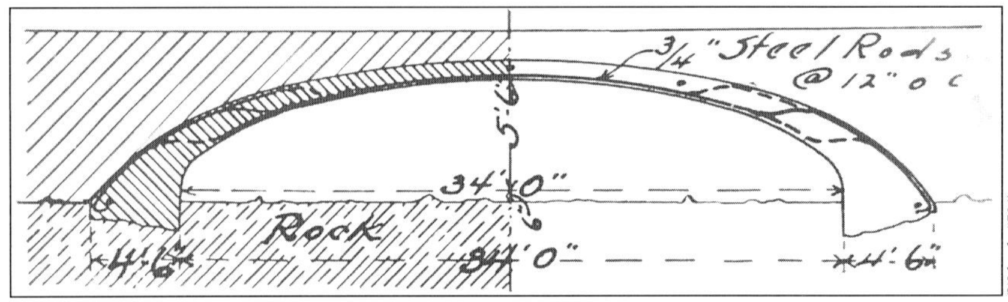

(Wabash #504)

Above – The Mill Street Bridge is one of Luten's oldest extant structures and reflects a number of features for which he would soon have patents. Wabash has grown enough since the horse-and-buggy days of this photograph to border the then-new concrete bridge and sidewalks with buildings.

(Wabash #41)

The Belden Bridge lacks some of the hallmarks of Luten style. The rings are not scored at their upper edge and polished, the spandrel walls not bush-hammered, and the piers not rounded. Nonetheless, basic structural characteristics remain constant: the rings are linear, the spandrels coped above and integrated with the rings below, and the piers kept elastically trim.

506 **NRC**
Huntington St, Wabash; Wabash R
Filled-spandrel arch (6) 1917
A W Grosvenor, dsgr
Janse, Boomer, Crain & Howe Constr, bldr
Note: *695 ft long; 109-ft spans; heavy, flat rings; w1982 with slightly extended slab deck rmvd rails with open-arch panels*

507
Miami St, Wabash; Charley Ck
Filled-spandrel arch c1915

THE CHARLEY CREEK BRIDGE: RARE KAHN-SYSTEM SPANS

Julius Kahn, like Daniel Luten, developed a reinforcing system for concrete structures which he advertised through his Trussed Concrete Steel Company as having been used in over a thousand structures, mostly buildings, by 1907. Kahn promoted his "trussed bar" as "ideal ... for beams, girders, and long-span construction." The Charley Creek Bridge in Wabash contains the only Kahn "trussed" bars known to have been used in Hoosier spans.[8]

Kahn's "trusses" consisted of diamond-shaped bars "with horizontal flanges or wings" rigidly attached in the shop "at diametrically opposite corners" to the "main" longitudinal rod. To address shear, Kahn had the "wings" bent to a forty-five degree angle to the longitudinal tension bar. Unlike Luten who used his "trusses" only with horizontal beams, Kahn applied his system to arch spans, at least in Wabash.

The rings of the Charley Creek Bridge follow "a parabolic arc" with two-tier reinforcing of Kahn trussed bars "so that their diagonals interlace across the depth of the arch ring" and "tie the concrete together as to give the effect of lattice girders" that will "not straighten out under stress." The spandrel walls also contained a grid of trussed bars bonded with the ring over the central pier. The designer believed his combined reinforcing turned the walls into "vertical cantilever slabs." While he would have considered Kahn's ring reinforcing as wasteful, Luten was at this same time developing ways to have the spandrel walls supplement the ring's load-bearing, some aspects of which Kahn appears to have been exploring, too.[9]

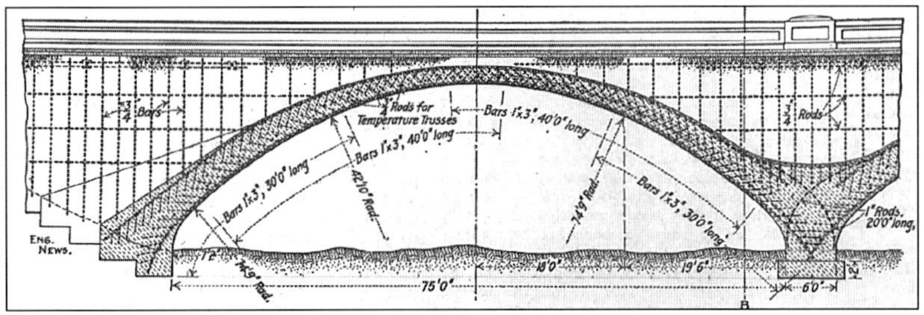

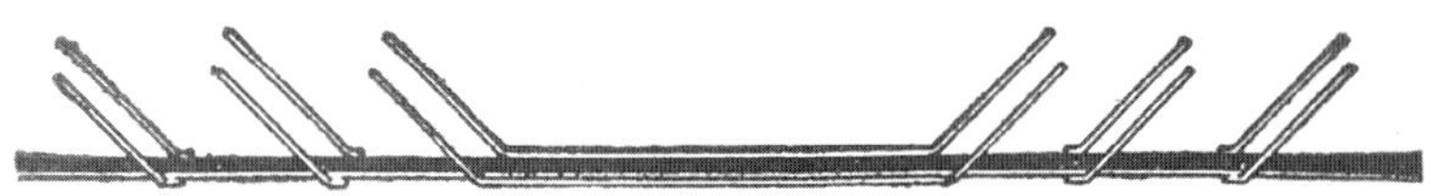

FIG. 34.—Kahn trussed bars.

The plan (above, top) indicates the placement of Kahn's "trussed bars" (above, bottom) in the two tiers of longitudinal reinforcement. The abutment did not need to be more substantial in order to address the very considerable amount of lateral thrust which the parabolic ring produced since the structure was seated at its end on bedrock.

WARREN [86]

49
CR 100N or #24, W of SR #63; Fall Ck
T-beam c1930
Note: seated on S stone abutment of previous br

53 **NRC**
CR 450N, E of US #41; Big Pine Ck
T-beam (4) c1917
J G Gregory, dsgr
Note: 145 ft long; shallow outer bm carries rail

55
CR 475N or #32, NE of Carbondale; Big Pine Ck, Bh
Slab c1930

66
CR 550E or #45, W of Independence; Kickapoo Ck
Beam (1/2) c1916
Note: very long original bm span (58 ft); added pier (1950) turns br into 2- span; skewed

(Warren #53)

(Warren #53)

In streamlining the outer sides of his T-beams and of his rails with long horizontal panels, the Warren County Engineer underscored the importance of this structure in the landscape.

1874
SR #263, N of CR 1025S; Opossum Ck
Filled-spandrel arch c1936
ISHC design
Note: skewed

WARRICK [87]

1583
SR #68, Lynnville; Mill Ck
T-beam c1933
ISHC design
Note: skewed

1819
SR #62, SW of Chandler; Caney Ck or Stollberg Dh
Filled-spandrel arch c1941
ISHC design
Note: skewed; fairly flat 3-centered ring; rather long filled-wall approach

3464
SR #62, W of Boonville; Cypress Ck
Continuous framed slab (3) 1942
ISHC design
R P Olinger, bldr
Note: slabs rigidly tied to pier columns; skewed; open-arch paneled rails; (VIII.2B)

WASHINGTON [88]

off SR #56, W of Salem; Goose Ck
Filled-spandrel arch c1917
Note: bypassed; used for private driveway

off SR #56, W of Salem; Mill Ck
Filled-spandrel arch (2) c1917
Note: abandoned; unsymmetrical rings; triangular cutwaters; (IV)

1
Hill Rd or CR #312, S of Fairview; Clifty Ck
Filled-spandrel arch c1915

9
Rush Ck Rd or CR #31, Rush Ck Valley; Rush Ck
Filled-spandrel arch (2) c1913
Note: skewed; fairly flat rings; w1989 prestressed bx bms on one side rmvd half of rails

46 **NRC**
Elk Ck Rd or CR #65, SW of Little York; Elk Ck
Filled-spandrel arch 1915
J Drager, bldr
Note: deck rides on ring at crown

59
CR #263, N of Canton; Blue R
Filled-spandrel arch c1915
Note: fairly flat ring

(Warrick #3464)

A graceful, ISHC-design, continuous slab bridge framed with trim bents.

(Washington #46)

Jon T. Drager could be proud of his craftsmanship.

(Washington #105)

An imposing deck carried over slender arches.

60 **NRC**
CR #261, N of Harristown; Blue R, Bh
[Continuous] slab (2) c1915
Note: *fairly deep slab (18 in)*

61
Bata Rd or CR #36, W of Harristown; Blue R, Fk
Filled-spandrel arch 1912
T F Wires, bldr
Note: *semicircular ring*

66
Saltillo Rd or CR #5, Claysville; Lost R
Filled-spandrel arch (2) c1912
Note: *unsymmetrical rings; skewed; w1978 to W rmvd spandrel & rails*

75
Martinsburg Rd or CR #51, W of Farabee; Blue R, Mid Fk
Filled-spandrel arch (2) c1910
Note: *unsymmetrical rings; triangular cutwaters*

78
Rubber Hill Rd or CR #197, NE of Salem; Blue R; W Fk
Filled-spandrel arch c1915

92 **NRC**
Bee Line Rd or CR #34, W of Salem; Mill Ck, Fk
Filled-spandrel arch 1911
H H O'Kaya, bldr
Note: *semicircular ring*

93
Bee Line Rd or CR #34, W of Salem; Mill Ck
Filled-spandrel arch 1911
J E Caubs, bldr
Note: *semicircular ring*

99
Vernon School Rd or CR #107, N of Livonia; Lost R, S Fk
Filled-spandrel arch c1916
Note: *rather flat ring*

105 **NRC**
Becks Mill Rd or CR #29, Becks Mill; Mill Ck
Filled-spandrel arch (2) c1920
J C Prow, dsgr
M & P Contr, bldr
Note: *skewed; deck extended considerably on brackets; neoclassically balustraded rails*

123
CR #49, W of New Pekin; Blue R, S Fk, Bh
[Continuous] slab (2) c1915
Note: *skewed*

133
Old Palmyra Rd or CR #10, NW of Martinsburg; Blue R
Filled-spandrel arch (2) c1910

136
Lockenour Rd, CR 700E or #24, E of Pekin; Blue R, S Fk
Filled-spandrel arch c1915
Note: *rather flat ring*

138
Bowers Knob Rd, CR 900E or #219, S of Blue R; Blue R, S Fk
Filled-spandrel arch c1916
Note: *deck rides on ring at crown*

161
Bee Line Rd or CR #34, Salem; Highland Ck
Filled-spandrel arch c1917
Note: *rather flat ring*

170
Bear Ck Rd or CR #6, W of Martinsburg; Bear Ck
Filled-spandrel arch c1912

172
Blue R Rd or CR #152, SW of Blue R; Blue River, Jeff Bh
Filled-spandrel arch c1920
Note: *w1973 to N rmvd spandrel; new slab deck rmvd rails*

1177
SR #135, S of Shorts Cor; Blue R
Filled-spandrel arch (2) c1931
ISHC design
Note: *skewed*

2001
Main St, New Pekin; Blue R
Filled-spandrel arch c1915

3069
SR #60, New Pekin; Blue R, S Fk
Filled-spandrel arch (3) c1937
ISHC design
Note: *skewed; slab deck extended on brackets*

WAYNE [89]

NRC
Milton St, Dublin; old Pa RR
Continuous girder/ floor bms (5) c1910
Pa RR design
Note: *same design used on this and the next 3 RR overpasses in Dublin; 120 ft long; gentle rise to deck maximizes clearance at center span, long (36 ft) span to accommodate 2 tracks; (VIII.1A)*

Johnson St, Dublin; old Pa RR
Continuous girder/ floor bms (5) c1910
Pa RR design
Note: *(see above)*

(Wayne #73)
A simple, small country arch with paneled lapels and pipe ties.

NRC
Davis St, Dublin; old Pa RR
Continuous girder/ floor bms (5) c1910
Pa RR design
Note: *(see above)*

Foundry Rd, Dublin; old Pa RR
Continuous girder/ floor bms (5) c1910
Pa RR design
Note: *(see above)*

NRC
old Pa RR; Foote St, Cambridge City
Filled-spandrel arch 1910
Pa RR design
Note: *heavy, segment-like ring decorated like stone; stairway suggests former entrance to rail platform; pipe rails; massive wingwalls to maintain RR embankment; with tracks & most fill gone, arch now marks a bygone transport center*

NRC
old Pa RR; Center St, Cambridge City
Filled-spandrel arch 1910
Pa RR design
Note: *largely a design duplicate of arch above, minus stairway but still with its fill*

old Pa RR; Washington Rd, nr Hiser
Filled-spandrel arch 1909
Pa RR design
Note: *largely a design duplicate of arches above, although this country copy lacks pipe railing*

10
Abington Twp Line Rd, S of Potter Shop Rd; Butlers Ck
Filled-spandrel arch c1915
Note: *skewed; most railing rmvd*

11
Potter Shop Rd, E of Abington Twp Line Rd; Butlers Ck
T-beam c1923
Note: *long bm span (40 ft)*

15
Watt Rd, E of Hiser Sta Rd; Nolands Fk
Filled-spandrel arch (3) c1915
Note: *257 ft long; pier & abutment pilasters*

36
Endsley Rd, E of US #27; White Bk
Filled-spandrel arch c1920
[R S, bldr]
Note: *flat ring*

39 Beelor Br **NRC**
Beelor Rd, SW of Richmond; Whitewater R, E Fk
Filled-spandrel arch (3) 1911
I E Smith, bldr
Note: *240 ft long; pier & abutment pilasters; rails rmvd (1987)*

52
Kempton Rd, N of Centerville; Nolands Fk
Filled-spandrel arch (2) c1910
Note: *skewed; pipe rails; base of slim pier now collared*

72
Moyer Rd, SW of Greens Fk; Greens Fk
Filled-spandrel arch (2) c1910
Note: *rails rmvd*

73
Moyer Rd, NE of Jacksonburg; Dry Bh
Filled-spandrel arch c1915
Note: *flat ring; paneled concr column & pipe rails*

76
Sample Rd, S of Pennville; Greens Fk
Filled-spandrel arch (2) c1910
Note: *pier pilaster; rails rmvd*

78
Hayes or Arboretum Rd, Richmond; Whitewater R, E Fk
Filled-spandrel arch c1915
Note: *flat ring; paneled spandrels*

121
Wallace Rd, SE of Whitewater; Whitewater R, Mid Fk
Filled-spandrel arch (2) c1909
Note: *skewed*

125
N Germantown Rd, N of E Germantown; Martindale Ck
Filled-spandrel arch (2) c1917
Note: *pipe rails; heavy, fairly flat rings; base of slim pier now collared*

135
Goose Heaven Rd, NW of Cambridge City; Symonds Ck
Filled-spandrel arch (2) c1915
Note: *skewed; flat rings; pier pilaster*

138 Lannerd Br **NRC**
Symonds Ck Rd, N of Dublin; Symonds Ck
Filled-spandrel arch 1911
T J & H F Burk, bldr
Note: *flat ring; metal rails*

141 **NRC**
Goose Heaven Rd, N of Cambridge City; W Whitewater R
T-beam (4) c1917
Note: *skewed; longest country bm br (176 ft); long bm spans (44 ft); wide bm flanges; rails sit on outer bm flange*

167
W E Oler Rd, SW of Williamsburg; Greens Fk
Filled-spandrel arch (2) c1910
Note: *flat rings; metal rails*

171 M T Fox Br
Fox Rd, E of Hagerstown; Martindale Ck
Filled-spandrel arch (2) 1912
P Hipskin, bldr
Note: *flat rings; pier pilaster; concr post & pipe rails*

172 D W Harris Br
Fox Rd, W of Sugar Grove Rd; Morgan Ck
Filled-spandrel arch 1911
T J & H F Burk, bldr
Note: *skewed; flat ring; rails rmvd*

173 **NRC**
Mineral Springs Rd, N of Greens Fk; Greens Fk
T-beam (3) c1913
Note: *skewed; long country bm br (165 ft); long bm spans (55 ft); rails sit on outer bm flange*

176 **NRC**
W E Oler Rd, E of Dingle Rd; Morgan Ck
Filled-spandrel arch 1912
J W Mueller, dsgr
Note: *flat ring; inscribed abutment pilasters; paneled concr post & pipe rails*

178
Clyde Oler Rd, E of Sugar Grove Rd; Morgan Ck
Filled-spandrel arch c1910
Note: *flat ring; rails rmvd*

185 **NRC**
Ted Davis Rd, N of Hoover Mill; Greens Fk
T-beam (3) c1917
Note: *skewed; long country bm br (165 ft); long bm spans (55 ft); rails sit on outer bm flange*

186 Simpson Br **NRC**
Swoveland Rd, E of Jacksonburg Rd; Martindale Ck
Filled-spandrel arch (2) 1911
T J & H F Burk, bldr
Note: *flat rings; rails rmvd*

191
Heiney Rd, S of Hagerstown; W Whitewater R
T-beam (3) c1915
Note: *long country bm br (165 ft); long bm spans (55 ft); wide bm flanges; rails sit on outer bm flange*

197
Main St, Hagerstown, Turnpike Rd, or old SR #38; Nettle Ck
Filled-spandrel arch c1910
Note: *ring decorated like stone*

225
Beard Rd or CR 1050N, NW of Economy; Martindale Ck
Filled-spandrel arch c1920
Note: *rails arced to resonate with ring; deck rides on ring at crown*

228 Mendenhall Br **NRC**
Mendenhall Rd, SW of Economy; Martindale Ck
Filled-spandrel arch (2) 1912
J W Mueller, dsgr
P Hipskin, bldr
Note: *skewed; rounded cutwater & pier pilaster*

245
Davis-Meyers Rd, NW of Fountain City; Morgan Ck
T-beam c1915
Note: *skewed; rails rmvd; artful abutment graffiti; (Epilogue.B)*

246
Round Barn N Rd, NW of Fountain City; Morgan Ck
Filled-spandrel arch c1911
Note: *flat ring; pipe rails rmvd*

338
E Main St, Cambridge City, or US #40; Whitewater R
Filled-spandrel arch (3) c1935
ISHC design
Note: *skewed; w1987 with cantilevered slab deck rmvd rails, but 1936 lampposts, donated by Cambridge City Centennial Committee, retained*

523
Green St, Cambridge City; Crietz Ck
[Continuous] slab (2) c1917
Note: *pipe rails*

701 South Side Br **NRC**
S G St, Richmond; Whitewater R
Open-spandrel arch (5) 1932
Filled-spandrel arch (6)
Note: *br sponsored by S Side Improvement Assoc; 706 ft long; 173-ft center spans; 2-rib open-spandrel rings; large & paneled cellular pier towers; w1990 with new cantilevered slab deck rmvd open-arch paneled rails with heart-like inserts; (Epilogue.A)*

709
N 17th St, Richmond; Whitewater R, E Fk
Filled-spandrel arch c1910
Note: *flat ring; inscribed spandrels; neo-classically balustrated rails*

4410 Main St Br **NRC**
US #40 or Main St, Richmond; Whitewater R, E Fk
Open-spandrel arch (3) 1920
Filled-spandrel arch (4)
J W Mueller, dsgr
I E Smith, bldr
Note: *among older open-spandrel arches; 599 ft long; 125-ft center spans; 2 parabolic ribs per open-spandrel ring; semicircular rings on filled-spandrel approach spans; large & paneled pier towers/pilasters; deck extended on floor bms to carry walks; open-oval rail panels; walled causeway approaches*

(Wayne #701)

The South Side Bridge (above and right) is one of two magnificent open-spandrel arch structures to complement Richmond's gorge. Before a partly-disfiguring facelift, this bridge displayed the most unique and modestly artful rail panels in the state (see Epilogue.A).

(Wayne #4410)

The open spandrels of the Main Street Bridge, Richmond, span the center of the gorge with grace and in style.

(Wayne #701)

WELLS [90]

22
CR 300E, SW of Reifsburg; Johns Dh
Slab c1923
Note: *skewed; long slab span (22 ft); low rails*

28
CR 700E, Domestic; Threemile Ck
Filled-spandrel arch c1917
[H C R, bldr]
Note: *flat ring*

45
CR 1000W, SE of Dillman; Owl Ck
Girder c1916
Note: *skewed*

84
CR 500S, W of SR #303; Rousch Dh
Slab c1925
Note: *long slab span (22 ft); low rails*

90
CR 300S, E of Co Home Rd; Sixmile Ck
Filled-spandrel arch c1920
Note: *skewed; flat ring; rails arced to resonate with ring*

99
Hoosier Hwy, S of Travisville; Neff Dh
Filled-spandrel arch c1930
Note: *inscribed spandrels; rails rmvd*

106 **NRC**
CR 100N, nr Rockford; Rock Ck
Filled-spandrel arch (2) c1919
Note: *unsymmetrical rings; pier pilaster*

114
CR 500N, E of SR #301; Eightmile Ck
Slab c1920
Note: *skewed; long slab span (22 ft); low rails mostly rmvd*

(Wells #230)

The Crosbie Bridge across the Wabash in Bluffton (above). The ISHC followed its conventional reinforcing pattern: two-tier reinforcing with stirrups in the rings extended elastically into the substructure (below). The state also exercised its preference for unsymmetrical outer spans as insurance for structural thrust. Since urban bridges built in the Thirties routinely contained sidewalks, the ISHC added brackets to its standard model and extended the walks beyond the rings and spandrel walls (bottom).

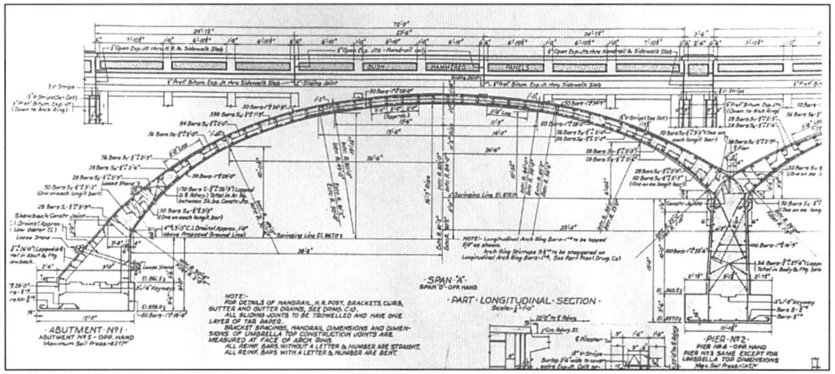

(Wells #230; INDOT)

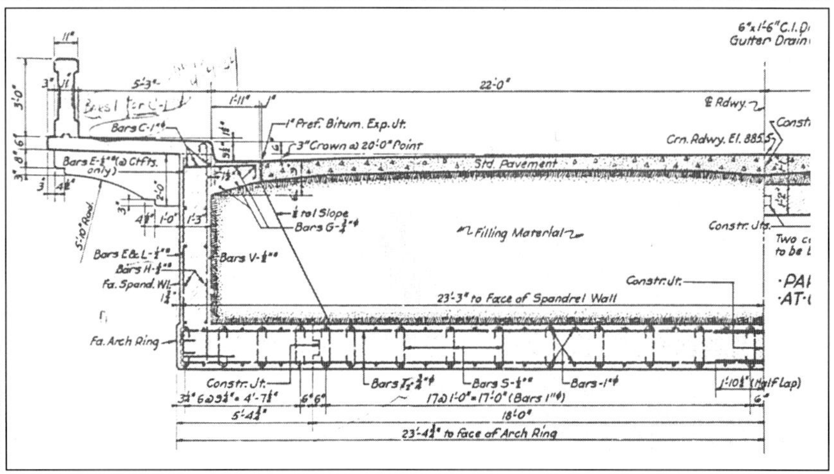

(Wells #230; INDOT)

130
CR 800N, W of old SR #303; Flat Ck
Girder c1916
Note: *skewed*

139 **NRC**
CR 100W, SE of Uniontown; Eightmile Ck
Filled-spandrel arch c1925
Note: *heavy, flat ring; inscribed spandrels*

151
CR 400W, S of CR 800N; Flat Ck
Girder c1917
Note: *skewed*

194
CR 800N, E of old SR #3; Flat Ck, Bh
Slab c1920
Note: *low rails*

230 Crosbie Br **NRC**
Main St, Bluffton, or SR #1; Wabash R
Filled-spandrel arch (4) 1934
ISHC design
J A Crosbie, bldr
Note: *skewed; 343 ft long; brackets carry sidewalks*

1421
SR #3, S of SR #5; Salamonie R
Filled-spandrel arch (3) c1932
ISHC design
Note: *skewed; 227 ft long; w1972 rmvd spandrels & rails*

WHITE [91]

270 **NRC**
Tower or Springboro Rd, Springboro; Spring Ck
Filled-spandrel arch c1935
Note: *only arch on Co highway system; coped railings with bush-hammered panels*

WHITLEY [92]

1584 **NRC**
SR #5, S Whitley; Eel R
Filled-spandrel arch (3) 1935
ISHC design
Note: *skewed; 213 ft long; deck & rails rpltd with stout copies (1990); lampposts retained; (VII.2A)*

(Whitley #1584)

(Whitley #1584)

(Whitley #1584)

The ISHC's South Whitley Bridge has undergone an aesthetically-thoughtful rehabilitation. Note, for example, the rails and the lights (original shown left; rehabilitation shown right). While heavier, less carefully coped, and with rail posts less pronounced, the new rails carry panels with the appearance of bush-hammering. The original light standards have been retained with somewhat different globes.

NOTES

[1]See Prologue. For the fullest retelling of these tales and legends, see Linda Degh, "The Negro in the Concrete," *Indiana Folklore* (1968) I: 61-67; Linda Degh, "The Haunted Bridges near Avon and Danville and Their Role in Legend Formation," *Indiana Folklore* (1969) II: 54-89; Ronald L. Baker, *Hoosier Folk Legends* (Bloomington, Ind., 1982), 63-66.

[2]Larry Hanson, "Hoosier Ghosts Who Haunt Halloween," *Indiana Rural News* (October 1975): 7; Steve Bell, "Ghost Stories," *Indianapolis Monthly* (October 1983): 46-47; "Local Tales Center on Eerie Bridge," *Indiana Preservationist* (September-October 1986): 6; Jamie Smith and Robin Bland "Avon's Famous Bridge Has Haunted Generations of Youth," *Hendricks County Flyer* (20 October 1991).

[3]Henry Grattan Tyrrell, *Artistic Bridge Design: A Systematic Treatise on the Design of Modern Bridges According to Aesthetic Principles* (Chicago, 1912), 105, 208-209.

[4]Mogens Ipsen, "Modernism in Bridge Design Reflected in Rigid-Frame Span," *ENR* (7 November 1935) 115: 647-48.

[5]Vincent H. Gaddis, *Story of Winona Lake* (Berne, Ind., 1949), 12-13, 31-32, 48-49.

[6]Thanks to Margaret Barondess of the Michigan Department of Transportation (Lansing, Michigan) for supplying photocopies of MDOT plans of several camelbacks. For photos and brief descriptions of a number of the Michigan spans, see Charles K. Hyde, *Historic Highway Bridges of Michigan* (Detroit, 1993), 135-41; Kathy Ataman, "Mottville's Camelback Bridge," *Michigan History* (May-June 1981) 65: 42-43.

[7]See I.5, II.3E.

[8]"Parabolic Reinforced Concrete Arch Bridge at Wabash, Ind.," "Foundations of Reinforced Concrete Arch Bridges," "Foundations for Concrete Arches and the Charley Creek Bridge," *ENR* (15 March, 5, 26 April 1906) 55: 290, 389-90, 473; "Kahn System" in Commercial Supplement, *ENR* (27 June 1907) 57: 47; Carl W. Condit, *American Building* (Chicago, 1968), 240-42; George A. Hool and Nathan C. Johnson, *Concrete Engineers' Handbook* (New York, 1918), 57-59.

[9]On Luten's spandrel design and his "truss," see II.2D, II.3C.

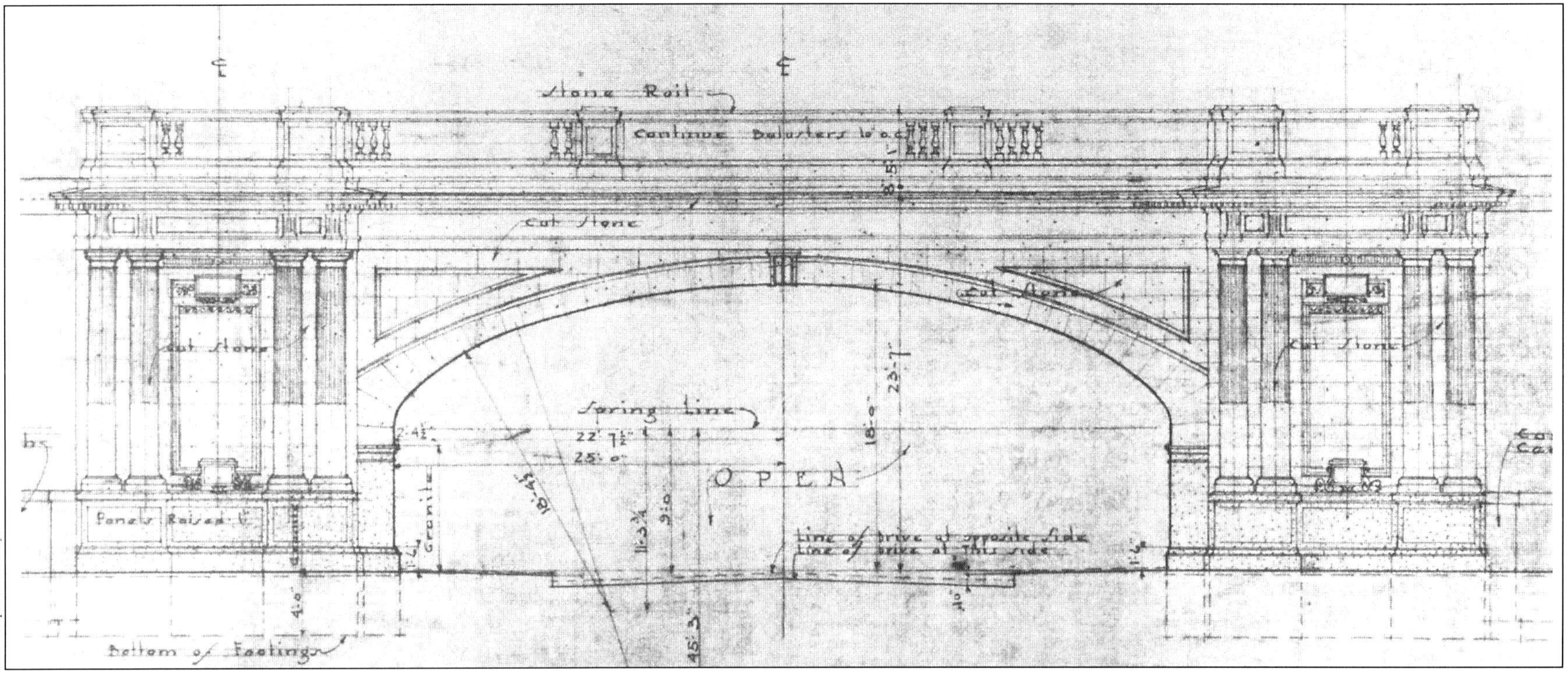

(#49 - Maple Rd Subway)

Maple Road Subway in Crown Hill Cemetery, Indianapolis.